Impacts of Molecular Structure on Nucleic Acid-Protein Interactions

Impacts of Molecular Structure on Nucleic Acid-Protein Interactions

Editors

Vaclav Brazda
Richard Bowater

MDPI • Basel • Beijing • Wuhan • Barcelona • Belgrade • Manchester • Tokyo • Cluj • Tianjin

Editors
Vaclav Brazda
Biophysical Chemistry and Molecular Oncology
Institute of Biophysics of the Czech Academy of Sciences
Brno
Czech Republic

Richard Bowater
School of Biological Sciences
University of East Anglia
Norwich
United Kingdom

Editorial Office
MDPI
St. Alban-Anlage 66
4052 Basel, Switzerland

This is a reprint of articles from the Special Issue published online in the open access journal *International Journal of Molecular Sciences* (ISSN 1422-0067) (available at: www.mdpi.com/journal/ijms/special_issues/Nucleic_Acid_Protein).

For citation purposes, cite each article independently as indicated on the article page online and as indicated below:

LastName, A.A.; LastName, B.B.; LastName, C.C. Article Title. *Journal Name* **Year**, *Volume Number*, Page Range.

ISBN 978-3-0365-7445-5 (Hbk)
ISBN 978-3-0365-7444-8 (PDF)

Contents

About the Editors

Vaclav Brazda

Václav Brázda is Professor of Molecular Biology and Genetics and Head of the Laboratory of Protein–DNA Interactions, Institute of Biophysics of the Czech Academy of Sciences in Brno, Czech Republic. He studied at the Faculty of Science of Masaryk University in Brno and completed a postdoctoral fellowship at the University Health Network in Toronto, Canada, in 2005-2006. He teaches at the Faculty of Chemistry of the Brno University of Technology. His research investigates nucleic acids, the relationship between the structure and function of DNA, and their interactions with proteins. He is a co-author of the broadly used bioinformatics server bioinformatics.ibp.cz.

Richard Bowater

Richard Bowater is Professor of Biochemistry and Molecular Biology Education in the School of Biological Sciences at the University of East Anglia where he teaches biochemistry and molecular biology to all levels of university students. Richard has authored many primary publications and reviews that focus on his research interests, and also delivered pedagogical commentaries about teaching biochemical concepts to diverse audiences. He became a Senior Fellow of the Higher Education Academy (SFHEA) in the UK in 2015, and is a member of the Biochemical Society, Microbiology Society, and the Royal Society of Biology.

Preface to "Impacts of Molecular Structure on Nucleic Acid-Protein Interactions"

This reprint presents a collection of research findings published in the Special Issue of the *International Journal of Molecular Sciences,* titled "Impacts of Molecular Structure on Nucleic Acid-Protein Interactions". These types of interactions are indispensable for many basic biological processes and the research findings reported here have important implications across a range of human diseases. We thank all authors for their contributions to the Special Issue and hope that you will find inspiration in their research findings.

Vaclav Brazda and Richard Bowater
Editors

International Journal of
Molecular Sciences

MDPI

Editorial

Impacts of Molecular Structure on Nucleic Acid–Protein Interactions

Richard P. Bowater [1,*] and Václav Brázda [2,*]

[1] School of Biological Sciences, University of East Anglia, Norwich Research Park, Norwich NR4 7TJ, UK
[2] Institute of Biophysics of the Czech Academy of Sciences, Královopolská 135, 612 00 Brno, Czech Republic
* Correspondence: r.bowater@uea.ac.uk (R.P.B.); vaclav@ibp.cz (V.B.)

Keywords: cruciform; DNA base sequence; DNA structure; G-quadruplex; nucleic acid–protein interactions; Z-DNA

Citation: Bowater, R.P.; Brázda, V. Impacts of Molecular Structure on Nucleic Acid–Protein Interactions. *Int. J. Mol. Sci.* **2023**, *24*, 407. https://doi.org/10.3390/ijms24010407

Received: 16 December 2022
Accepted: 21 December 2022
Published: 26 December 2022

Interactions between nucleic acids and proteins are some of the most important interactions in biology because they are the cornerstones for fundamental biological processes, such as replication, transcription, and recombination. Nucleic acids can adopt a wide range of structural conformations and this structural flexibility plays critical roles in their interactions with proteins [1]. This Special Issue of the International Journal of Molecular Sciences reports on diverse representatives of such interactions (Figure 1) across a wide range of biological systems.

RNA molecules can adopt great structural diversity due to the range of intramolecular interactions that can be formed within the single-stranded molecules [2]. DNA molecules are typically presented as double-stranded, right-handed B-form helices as their canonical structures, and this maximizes the thermodynamic stability of the molecule [3]. However, a significant body of research emphasizes that alternative, non-canonical DNA structures can exist, including double-stranded, left-handed Z-DNA, but also multi-strand structures such as G-quadruplexes (G4s), intercalated-motifs (i-motifs), triplexes, and cruciform structures. These non-B-DNA structures are usually characterized by the occurrence of single-stranded regions (loops) and/or sites of disrupted base pair stacking (junctions between continuous B-form DNA and the alternative structure) [3].

Variations in the structures of nucleic acids offer different binding sites for proteins, including interactions that focus on a range of sequence- and structure-specific nucleic acid targets. The structures of nucleic acids influence different aspects of biological activity, including physiological and pathological functions [4,5], themes that are addressed in this Special Issue. The collection of articles involve biophysical, biochemical, molecular biological and bioinformatics approaches and cover different biological systems, but there are some common themes among them: several refer to computational biology or bioinformatics approaches, or highlight additional information in DNA sequences [6–9]; several articles refer to different types of non-B DNA structures [10–12], with specific interests in quadruplexes [13–16]; several articles address different approaches and outcomes from proteins binding to DNA structures [11–13,17].

The wealth of DNA sequence information provided by genome-sequencing projects has brought new insights into the primary sequences of genomes and also about possible sequence-dependent local secondary structures [3,18]. Advances provided by such genome sequences are exemplified by the Human Genome Project, with complete telomere-to-telomere sequences being finalised in 2022 [19,20]. As highlighted, this Special Issue includes several articles that report on computational biology or bioinformatics studies of DNA sequences [6–9], identifying unexpected and additional information within them (Figure 1A). In a mini review, Bartas et al. summarize current knowledge about the amino acid composition of various nucleic-acid-binding proteins, highlighting differences across proteins that bind in a sequence-specific manner compared to those that recognize local non-B-DNA structures and those that recognize both types of properties of nucleic acids [6]. Bioinformatic studies of repetitive DNA sequences in *Drosophila melanogaster* polytene chromosomes show that chromatin structure plays a crucial role in the regulation of gene activity [7]. Recent advances reported by Choi et al. demonstrate that nucleic acids may provide useful tools for building complex logic circuits [8]. Finally, for this grouping of articles, Víglaský explores the organization of genetic information in nucleic acids using a novel orthogonal representation, which proves to be useful in predicting the likelihood of particular regions of nucleic acids to form non-canonical motifs [9].

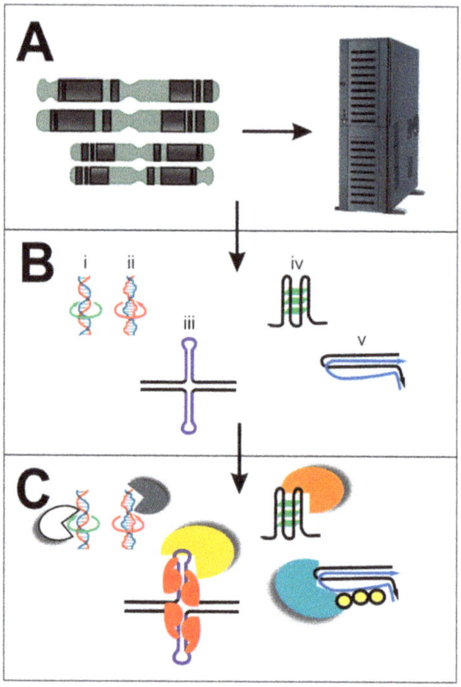

Figure 1. Nucleic acids provide a wide array of sequences and structures that are useful for biological processes. (**A**) Genome sequences store large amounts of information that can be accessed for biological processes and structure prediction by various computational algorithms or for building complex logic circuits. (**B**) Nucleic acids can adopt a range of structures, including those indicated. From left to right: (i) double-stranded, right-handed B-DNA; (ii) double-stranded, left-handed Z-DNA; (iii) two intramolecular hairpins can come together to form a cruciform; (iv) G-quadruplex, formed from four strands that can be parts of one molecule (as shown) or from different molecules; (v) a triplex can be formed when three strands come together, which can be parts of one molecule (as shown) or from different molecules. (**C**) The variety of structures of nucleic acids offer opportunities to be recognised by other molecules, such as proteins. The range of structures shown here may be recognised by different proteins, as indicated by the different colours.

From the earliest days of genome sequence analysis, it was recognized that natural DNA molecules contain a wide array of repeating sequences [3]. These types of sequences are particularly prone to adoption of non-canonical DNA structures, such as G4s, triplexes, and cruciforms (Figure 1B), which are all explored in this Special Issue [10–16]. Zhao and Usdin review the range of structures that can form in specific trinucleotide repeats, highlighting how their expansion in length is important in the pathology of fragile X-related disorders in humans [10]. Left-handed Z-helices can form in both DNAs and RNAs with appropriate sequences, and searching databases containing protein structures identified novel proteins predicted to bind them [11]. A different type of repetitive DNA sequence, inverted repeats, can adopt cruciform structures and many proteins have now been validated to bind to them [12]. A series of articles provide insights about G4s. Bezzi et al. suggest that putative G4s found in the SARS-CoV-2 RNA genome and the cellular proteins likely to interact with them may constitute interesting targets for antiviral drugs [13]. Putative G4s in viruses are explored further in a study that reveals a positive correlation between their frequencies in double-stranded DNA viruses and their hosts from archaea, bacteria, and eukaryotes, indicating that their close coevolution leads to reciprocal mimicking of genome organization [14]. The potential of compounds to target G4s was explored for Rhodamine 6G, which was shown to have high selectivity for G4s with parallel topology [15]. A bioinformatic study combined with circular dichroism measurements identified a stable G4 that is evolutionarily conserved amongst plants sensu lato (in Archaeplastida), and this may form an additional layer of regulatory networks [16].

The wide array of structures that can be adopted by nucleic acids offer different opportunities for proteins (and other molecules) to bind to, leading to different types of outcomes [4,5]. Some proteins recognise sequence-specific targets, but an increasing number are being shown to interact with non-canonical structural aspects of nucleic acids (Figure 1C). We have already referred to some articles in this Special Issue that describe such interactions [11–13]. Another study took advantage of available datasets and discovered new correlations between specific amino acid deviations in p53 proteins, showing a direct association between specific amino acid residues in the protein and changes in p53 functionality, and further highlighting the importance of p53 protein in processes that influence lifespan and aging [17].

To summarize, this Special Issue of the International Journal of Molecular Sciences reports on representatives of interactions between nucleic acids and proteins, with an emphasis on understanding how the structure of the nucleic acid influences such interactions. It is important to characterize these molecular complexes because many are essential requirements for the viability of cellular life due to their involvement in fundamental aspects of nucleic acid metabolism. It is now clear that the structural flexibility of the nucleic acids plays critical roles in their interactions with proteins, with important implications across a range of human diseases, including cancer and some infectious diseases. A deeper understanding of these molecular interactions will require the use of complementary methods and techniques [1]. As is described in this Special Issue, biophysical, biochemical, molecular biological and bioinformatics approaches will deliver useful advances across a wide range of biological systems.

Author Contributions: Writing—original draft preparation, review and editing, R.P.B. and V.B. All authors have read and agreed to the published version of the manuscript.

Funding: Václav Brázda is supported by the Czech Science Foundation, grant number 22-21903S.

Institutional Review Board Statement: Not applicable.

Informed Consent Statement: Not applicable.

Data Availability Statement: No new data were created or analyzed in this study. Data sharing is not applicable to this article.

Conflicts of Interest: Both authors declare no conflict of interest.

References

1. Cozzolino, F.; Iacobucci, I.; Monaco, V.; Monti, M. Protein–DNA/RNA Interactions: An Overview of Investigation Methods in the Omics Era. *J. Proteome Res.* **2021**, *20*, 3018–3030. [CrossRef] [PubMed]
2. Corley, M.; Burns, M.C.; Yeo, G.W. How RNA-Binding Proteins Interact with RNA: Molecules and Mechanisms. *Mol. Cell* **2020**, *78*, 9–29. [CrossRef] [PubMed]
3. Brazda, V.; Fojta, M.; Bowater, R.P. Structures and stability of simple DNA repeats from bacteria. *Biochem. J.* **2020**, *477*, 325–339. [CrossRef] [PubMed]

4. Wang, G.; Vasquez, K.M. Dynamic alternative DNA structures in biology and disease. *Nat. Rev. Genet.* **2022**, 1–24. [CrossRef] [PubMed]
5. Bansal, A.; Kaushik, S.; Kukreti, S. Non-canonical DNA structures: Diversity and disease association. *Front. Genet.* **2022**, *13*, 959258. [CrossRef] [PubMed]
6. Bartas, M.; Červeň, J.; Guziurová, S.; Slychko, K.; Pečinka, P. Amino Acid Composition in Various Types of Nucleic Acid-Binding Proteins. *Int. J. Mol. Sci.* **2021**, *22*, 922. [CrossRef] [PubMed]
7. Zhuravlev, A.V.; Zakharov, G.A.; Anufrieva, E.V.; Medvedeva, A.V.; Nikitina, E.A.; Savvateeva-Popova, E.V. Chromatin Structure and "DNA Sequence View": The Role of Satellite DNA in Ectopic Pairing of the Drosophila X Polytene Chromosome. *Int. J. Mol. Sci.* **2021**, *22*, 8713. [CrossRef] [PubMed]
8. Choi, S.; Lee, G.; Kim, J. Cellular Computational Logic Using Toehold Switches. *Int. J. Mol. Sci.* **2022**, *23*, 4265. [CrossRef] [PubMed]
9. Víglaský, V. Hidden Information Revealed Using the Orthogonal System of Nucleic Acids. *Int. J. Mol. Sci.* **2022**, *23*, 1804. [CrossRef] [PubMed]
10. Zhao, X.; Usdin, K. (Dys) function Follows Form: Nucleic Acid Structure, Repeat Expansion, and Disease Pathology in FMR1 Disorders. *Int. J. Mol. Sci.* **2021**, *22*, 9167. [CrossRef] [PubMed]
11. Bartas, M.; Slychko, K.; Brázda, V.; Červeň, J.; Beaudoin, C.A.; Blundell, T.L.; Pečinka, P. Searching for New Z-DNA/Z-RNA Binding Proteins Based on Structural Similarity to Experimentally Validated Zα Domain. *Int. J. Mol. Sci.* **2022**, *23*, 768. [PubMed]
12. Bowater, R.P.; Bohálová, N.; Brázda, V. Interaction of Proteins with Inverted Repeats and Cruciform Structures in Nucleic Acids. *Int. J. Mol. Sci.* **2022**, *23*, 6171. [CrossRef] [PubMed]
13. Bezzi, G.; Piga, E.J.; Binolfi, A.; Armas, P. CNBP Binds and Unfolds In Vitro G-Quadruplexes Formed in the SARS-CoV-2 Positive and Negative Genome Strands. *Int. J. Mol. Sci.* **2021**, *22*, 2614. [CrossRef] [PubMed]
14. Bohálová, N.; Cantara, A.; Bartas, M.; Kaura, P.; Šťastný, J.; Pečinka, P.; Brázda, V. Tracing dsDNA Virus–Host Coevolution through Correlation of Their G-Quadruplex-Forming Sequences. *Int. J. Mol. Sci.* **2021**, *22*, 3433. [CrossRef] [PubMed]
15. Trizna, L.; Janovec, L.; Halaganová, A.; Víglaský, V. Rhodamine 6G-Ligand Influencing G-Quadruplex Stability and Topology. *Int. J. Mol. Sci.* **2021**, *22*, 7639. [CrossRef] [PubMed]
16. Volná, A.; Bartas, M.; Karlický, V.; Nezval, J.; Kundrátová, K.; Pečinka, P.; Červeň, J. G-Quadruplex in Gene Encoding Large Subunit of Plant RNA Polymerase II: A Billion-Year-Old Story. *Int. J. Mol. Sci.* **2021**, *22*, 7381. [CrossRef] [PubMed]
17. Bartas, M.; Brázda, V.; Volná, A.; Červeň, J.; Pečinka, P.; Zawacka-Pankau, J.E. The Changes in the p53 Protein across the Animal Kingdom Point to Its Involvement in Longevity. *Int. J. Mol. Sci.* **2021**, *22*, 8512. [CrossRef] [PubMed]
18. Brázda, V.; Bartas, M.; Bowater, R.P. Evolution of Diverse Strategies for Promoter Regulation. *Trends Genet.* **2021**, *37*, 730–744. [CrossRef] [PubMed]
19. Church, D.M. A next-generation human genome sequence. *Science* **2022**, *376*, 34–35. [CrossRef] [PubMed]
20. Nurk, S.; Koren, S.; Rhie, A.; Rautiainen, M.; Bzikadze, A.V.; Mikheenko, A.; Phillippy, A.M. The complete sequence of a human genome. *Science* **2022**, *376*, 44–53. [CrossRef] [PubMed]

International Journal of *Molecular Sciences*

MDPI

Review

Amino Acid Composition in Various Types of Nucleic Acid-Binding Proteins

Martin Bartas, Jiří Červeň, Simona Guziurová, Kristyna Slychko and Petr Pečinka *

Department of Biology and Ecology, Institute of Environmental Technologies, Faculty of Science, University of Ostrava, 710 00 Ostrava, Czech Republic; martin.bartas@osu.cz (M.B.); jiri.cerven@osu.cz (J.Č.); P19061@student.osu.cz (S.G.); p19071@student.osu.cz (K.S.)

* Correspondence: petr.pecinka@osu.cz

Abstract: Nucleic acid-binding proteins are traditionally divided into two categories: With the ability to bind DNA or RNA. In the light of new knowledge, such categorizing should be overcome because a large proportion of proteins can bind both DNA and RNA. Another even more important features of nucleic acid-binding proteins are so-called sequence or structure specificities. Proteins able to bind nucleic acids in a sequence-specific manner usually contain one or more of the well-defined structural motifs (zinc-fingers, leucine zipper, helix-turn-helix, or helix-loop-helix). In contrast, many proteins do not recognize nucleic acid sequence but rather local DNA or RNA structures (G-quadruplexes, i-motifs, triplexes, cruciforms, left-handed DNA/RNA form, and others). Finally, there are also proteins recognizing both sequence and local structural properties of nucleic acids (e.g., famous tumor suppressor p. 53). In this mini-review, we aim to summarize current knowledge about the amino acid composition of various types of nucleic acid-binding proteins with a special focus on significant enrichment and/or depletion in each category.

Keywords: DNA; RNA; protein binding; G-quadruplex; triplex; i-motif; Z-DNA; Z-RNA; cruciform; amino acid composition

Citation: Bartas, M.; Červeň, J.; Guziurová, S.; Slychko, K.; Pečinka, P. Amino Acid Composition in Various Types of Nucleic Acid-Binding Proteins. *Int. J. Mol. Sci.* **2021**, *22*, 922. https://doi.org/10.3390/ijms22020922

Received: 29 December 2020
Accepted: 16 January 2021
Published: 18 January 2021

1. Introduction

Interactions between proteins and nucleic acids (DNA and RNA) are central to all aspects of maintaining and accessing genetic information. Nucleic acid-binding proteins are mostly composed of at least one DNA or RNA-binding domain where the interfacing with amino acids takes place in a specific or nonspecific manner [1]. Identification of nucleic acid-binding proteins is one of the most important tasks in molecular biology. Currently, nucleic acid-binding proteins can be identified and further characterized by several experimental techniques, including pull-down assays [2,3], yeast one-hybrid system [4,5], electrophoretic mobility shift assays [6,7], chromatin immunoprecipitation [8,9], and by other specialized techniques [10,11]. However, it is time-consuming and expensive to identify nucleic acid-binding proteins by experimental approaches [12]. With the easy availability of a large amount of protein sequence data, there is a rapid development of computational approaches and prediction tools that can rapidly and reliably identify nucleic acid-binding proteins [13,14]. Several such tools model nucleic acid-binding abilities based on protein amino acid composition [15,16]. There is a growing interest in so-called noncanonical nucleic acid structures and proteins that preferentially bind them [17–24]. Noncanonical nucleic acid structures are DNA and RNA structures different from their basic form, i.e., double-stranded right-handed DNA or single-stranded RNA, and are often formed by simple nucleotide repeats [25–28]. Physiologically, they are represented mainly by G-quadruplexes [29], i-motifs [30], triplexes [31], R-loops [32], slipped hairpins [33], DNA cruciforms [34], RNA hairpins [35], and Z-DNA [36]. These DNA/RNA structures have important biological functions [37–42] and contribute to many human diseases [43–46]. It became more and more evident, that proteins preferentially interacting with these structures share distinct amino acid features/fingerprints [47,48]. This mini-review aims to focus on the amino acid composition of various types of DNA and RNA-binding proteins and to compare the amino acid composition of proteins that prefer binding to different noncanonical forms of nucleic acids.

2. Amino Acid Composition of Nucleic Acid-Binding Proteins

According to the Gene Ontology (GO) knowledgebase, there are 5037 nucleic acid-binding proteins (filtering GO:0003676 term by "protein") with experimental evidence in *Homo sapiens* [49–51]. Of this number, 2572 are annotated as RNA-binding and 2439 as DNA-binding proteins (some proteins have both functions). 1768 human proteins are known to bind DNA in a sequence-specific manner. It would be interesting to quantify the overall amount of proteins binding nucleic acids in a structure-specific manner. Unfortunately, there is no such category yet. We strongly suggest revisions in this manner. Inspiration can be found in the following review papers/databases focused on specific properties of proteins binding to G-quadruplexes [19,52–54], cruciforms [55], and Z-DNA/Z-RNA [56].

2.1. History

Amino acid composition of some nucleic acid-binding proteins was intensively studied at the beginning of the 70s, when Koichi Iwai et al. determined that "calf-thymus histones comprise five main types which differ in amino acid composition and electrophoretic mobility: A glycine-rich, arginine-rich histone (also known as f2al or IV); a glutamic-acid-rich, arginine-rich histone (fe or III); a leucine-rich, intermediate type histone (f2a2 or IIb1); a serine-rich, slightly lysine-rich histone (f2b or IIb2); and an alanine-rich, very lysine-rich histone (f1 or I)" [57], by using specialized chromatographic technique followed by polyacrylamide gel electrophoresis. In 1975, from the comparison of 68 representative proteins and frequencies of 61 codons of the genetic code, it was found that the average amounts of lysine, aspartic acid, glutamic acid, and alanine are above the levels anticipated from the genetic code, and arginine, serine, leucine, cysteine, proline, and histidine are below such levels [58]. There are a couple of examples from the 90s and 2000s when amino acid substitution in nucleic acid-binding protein abolished its function, e.g., an arginine to lysine substitution in the bZIP (Basic Leucine Zipper) domain of an opaque-2 mutant in maize abolished specific DNA-binding [59], missense mutations (Met175Arg and Ser191Asn) abolishing DNA-binding of the osteoblast-specific transcription factor OSF2/CBFA1 in human patients with cleidocranial dysplasia [60], or impaired RNA-binding of fragile X mental retardation protein upon missense mutation Ile-304→Asn in one of its KH domain [61]. Recent advantages in sequencing and bioinformatic methods allow us to directly compare the amino acid composition of thousands of (not only) human nucleic acid-binding proteins [62,63]. One of the most popular programs for this purpose is, e.g., composition profiler [64], which is a web-based tool for semi-automatic discovery of enrichment or depletion of amino acids, either individually or grouped by their physicochemical or structural properties [64]. Scientists often find themselves in the situation when they only have a sequence of new "hypothetical" protein, derived mainly from transcriptome sequencing, and want to deduce its function [65]. In case that no meaningful alignment to protein with known function is available, there is still a way to get some useful information using only primary amino acid sequence and its composition. In 2003, Cai and Lin used a protein's amino acid composition and support vector machine (SVM) prediction to decide if protein belongs to one of three classes—rRNA-, RNA-, or DNA-binding [66]. Currently, there are also user-friendly web-based prediction tools called DNAbinder and PseDNA-Pro, which can predict if the submitted protein sequence has DNA-binding ability [12,67].

2.2. *Methods to Inspect the Amino Acid Composition of Proteins*

Several approaches are used to inspect the amino acid composition of nucleic acid-binding proteins. Basically, we can divide the methods into in vitro and in silico. In vitro approaches are necessary to obtain a sequence of the protein of interest. Although the development of large-scale genomic sequencing has greatly simplified the procedure of determining the primary structures of proteins, the genomic sequences of many organisms are still unknown, and also modifications such as post-translational events (citrullination, deamidation, polyglutamylation, . . .) may prevent proper determination of the protein sequence [68]. Then, the complete characterization of the primary protein structure often requires a mass spectrometry method with minimal assistance from genomic data, i.e., de novo protein sequencing [68,69]. In silico approaches are based mostly on previous knowledge about primary protein sequence. There is currently a plentitude of bioinformatics tools designed for that purpose, see, e.g., [64,70–73].

2.3. *Amino Acid Composition of Nucleic Acid-Binding Proteins*

Nucleic acid-binding proteins are traditionally divided into two categories. The first category comprises proteins with the ability to bind DNA, and the second category comprises proteins that bind to RNA. This division is quite outdated, mainly because, from the historical perspective, proteins that bind RNA were typically considered as functionally distinct from proteins that bind DNA and studied independently. Interestingly, current gene ontology analyses reveal that DNA-binding is potentially a major function of the mRNA-binding proteins [74]. Nonetheless, several studies inspecting amino acid composition of DNA and/or RNA-binding proteins were published [75,76] and find that particular amino acid residues are generally enriched or depleted within these protein categories (see Table 1).

Another, even more important division of nucleic acid-binding proteins is based on a so-called sequence or structure-specific type of binding. Proteins able to bind nucleic acids in a sequence-specific manner usually contain one or more of the well-defined structural motifs. One of such motifs, zinc-finger, binds DNA (or RNA) through specific interaction with nucleotides and sugar-phosphate backbone. Tandem repeating of slightly different zinc-finger motifs in protein then allows to recognizing its consensus nucleic acid-binding sequence specifically. Cysteine and histidine amino acid residues are crucially important to coordinate Zn^{2+} binding in the largest and best-characterized subgroup of zinc-finger binding proteins named the Cys_2His_2 fold subgroup [77,78]. Other well-defined sequence-specific motifs—leucine zipper, helix-turn-helix, or helix-loop-helix—are listed in Table 1, together with their common signatures of amino acid residues.

Table 1. Types of nucleic acid-binding proteins. This table summarizes the main categories of nucleic acid-binding proteins. There are two points of view. At first, we can simply divide these proteins into DNA and RNA-binding ones (and a relatively small category of proteins that are able to bind both DNA and RNA). Secondly (and more importantly), we can distinguish proteins that specifically bind known sequence motifs (sequence-specific DNA/RNA-binding) and proteins, which specifically bind local DNA/RNA structures. Besides, keep in mind that this table is very simplified, and categories are divided to be reader-friendly. In fact, many of the DNA/RNA-binding proteins combine sequence and structure-specific binding mechanisms.

	Important Notes	References
DNA-binding	Arginine, tryptophan, tyrosine, histidine, phenylalanine, and lysine residues enrichment. Glutamate, aspartate, and proline depletion in the protein-DNA interface.	[76,79]
RNA-binding	Arginine, methionine, histidine, and lysine residues enrichment. Glutamate, aspartate residues depletion in protein-RNA interface.	[75,76]
DNA and RNA-binding	Proteins that are able to bind both DNA and RNA.	[74,80]

Table 1. *Cont.*

	Important Notes	References
Sequence-specific		
Zinc finger proteins	Cysteine and histidine amino acid residues are crucially important to coordinate Zn^{2+} binding in the Cys_2His_2 subgroup of zinc-finger proteins	[77,78]
Helix-turn-helix (HTH)	Conserved "shs" and "phs" patterns, where 's' is a small residue, most frequently glycine in the first position, 'h' is a hydrophobic residue, and 'p' is a charged residue, most frequently glutamate. "shs" pattern lies in the turn between helix-2 and helix-3 of the core HTH structure, and "phs" is present in helix-2.	[81]
Basic Helix-loop-helix (bHLH)	Mostly arginine, lysine or histidine amino acid residues are present within conserved positions of this motif	[82,83]
Leucine zipper proteins	Leucine amino acid residues are crucial for leucine zipper motifs	[84,85]
Structure specific		
G-quadruplex binding proteins	Global enrichment for glycine, arginine, aspartic acid, asparagine, valine, and depletion for cysteine, histidine, leucine, proline, glutamine, and tryptophan residues	[47,86–88]
Cruciform binding proteins	Global enrichment for lysine and serine, and depletion for alanine, glycine, glutamine, arginine, tyrosine, and tryptophan residues	[48,55]
Triplex binding proteins	Global enrichment for asparagine, aspartic acid, isoleucine, tyrosine, and depletion for cysteine, histidine, and proline residues	[89]
Z-DNA/RNA-binding proteins	Global enrichment for isoleucine, aspartic acid, lysine, and depletion for cysteine residues	[89]

In contrast, many proteins do not recognize nucleic acid sequence but rather local DNA or RNA structures (G-quadruplexes, i-motifs, triplexes, cruciforms, left-handed DNA/RNA form, and others) [19,30,41,55,90]. Finally, there are also proteins recognizing both sequence and local structural properties of nucleic acids (e.g., famous tumor suppressor p53 [91], Myc-associated zinc finger protein (MAZ) [92,93], and many RNA-binding proteins [94])—these proteins usually contain sequence-specific binding domain(s) together with domain(s)/region(s) with preference to noncanonical nucleic acid structures [95–97]. In 2016, Wang et al. analyzed the abundance of intrinsic disorder in the DNA- and RNA-binding proteins in over 1000 species from Eukaryota, Bacteria, and Archaea domains of life [98]. They have revealed a very interesting phenomenon that DNA-binding proteins had significantly increased disorder content and were significantly enriched in disordered domains in Eukaryotes but not in Archaea and Bacteria. The RNA-binding proteins were significantly enriched in the disordered domains in Bacteria, Archaea, and Eukaryota, while the overall abundance of disorder in these proteins was significantly increased in Bacteria, Archaea, animals, and fungi [98]. Disordered domains or regions are also extensively present in chromatin-binding proteins [99,100]. Interestingly, some disordered proteins or regions show very high structural specificity to the different types of noncanonical nucleic acids. For instance, human protein SRSF1 (Serine/arginine-rich splicing factor 1) contains several intrinsically disordered regions [101], which are compositionally enriched in glycine (14.11% of overall amino acid residues) and arginine (17.39% of overall amino acid residues) content. It was previously shown that SRSF1 has a high affinity to RNA G-quadruplex structure [102]. Subsequent analyses have shown that the dataset of 77 G-quadruplex binding proteins is significantly globally enriched in arginine, glycine, aspartic acid, asparagine, and valine, and depleted in cysteine and other amino acid residues [47] (Figure 1). Finally, the common amino acid motif in the form of RGRGRGRGGGSGGSGGRGRG was derived, and most of the currently known G-quadruplex binding proteins contain at least some modification of it [47]. Using this motif, a new dataset of G-quadruplex binding proteins was predicted from the set of all human DNA/RNA-binding proteins [47], and some of them were independently experimentally validated (e.g., CIRBP, which is a cold-inducible RNA-binding protein in the study by Huang and colleagues [103]). A similar study focused on an amino acid composition of cruciform binding proteins was also published, and the significant enrichment for lysine and serine amino acid residues has been revealed [48]

(Figure 1). Unpublished results also indicate distinct amino acid profiles in Z-DNA/RNA and triplex binding proteins, both significantly enriched in aspartic acid and isoleucine and depleted in cysteine residues [89] (Figure 1). In future studies, it would be interesting to specifically analyze local amino acid composition (only in the nucleic acid interaction sites and their close neighborhood) in these proteins. Unfortunately for the vast majority of them, the knowledge about exact DNA/RNA binding site(s) is still missing.

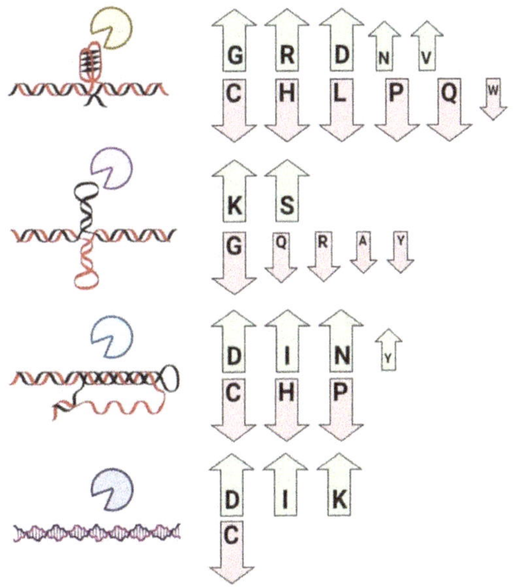

Figure 1. Significantly enriched and depleted amino acid residues (one letter aa code) in the dataset of G-quadruplex binding proteins (top), cruciform binding proteins, triplex binding proteins, and Z-DNA-binding proteins. Using Bonferroni correction, only values lower than 0.0025 were taken as significant (p <0.0025; p <0.0010; p <0.0001). The size of arrows indicates the significance of enrichment/depletion on scale (highest, moderate, lowest). Figure compiled using data from [47,48,89]. Created with BioRender.com.

As was shown above, proteins that preferentially recognize noncanonical nucleic acid structures often have a distinct amino acid composition with particular significant global enrichment and/or depletion of different amino acid residues. Noncanonical structures and proteins preferentially binding them often play a critical role in physiological molecular processes [32,104,105], but also in the progression of human diseases, such as various cancer types and neurodegenerative diseases, reviewed in [55,106,107]. Knowledge about the amino acid composition of various proteins binding noncanonical nucleic acids can be utilized as an additional clue/fingerprint in discovering novel noncanonical nucleic acid-binding protein candidates and therapeutically utilized [108–110].

The scheme below depicts sequence and structure-specific nucleic acid-binding phenomena in a nutshell (Figure 2).

Almost every year, multiple novel noncanonical nucleic acid-binding proteins are identified. This year was, for instance, found that Guanine Nucleotide-Binding Protein-Like 1 (GNL1) binds RNA G-quadruplex structures in genes associated with Parkinson's disease [111], or that Small Nuclear Ribonucleoprotein Polypeptide A (SNRPA) directly binds to the BAG-1 mRNA through the G-quadruplex which can modulate BAG-1 expression level [112] (anti-apoptotic BAG-1 protein is known to be overexpressed in colorectal cancers [113]). Prediction of proteins that preferentially bind noncanonical DNA/RNA structures, therefore, should be a logical first step towards rapid identification of novel therapeutic targets for future treatment of severe human diseases.

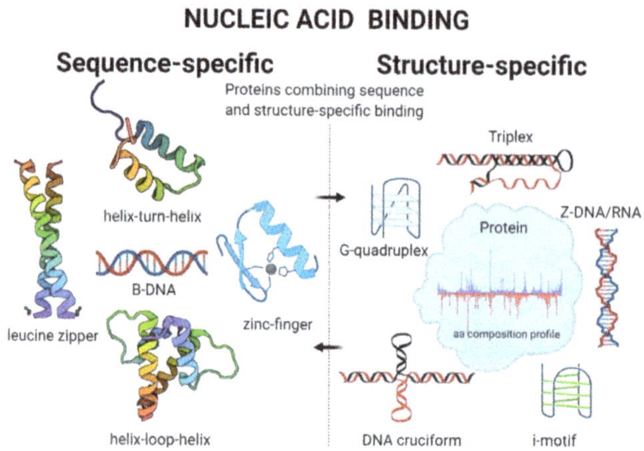

Figure 2. Types of nucleic acid-binding. The nucleic acid-binding mechanism can be basically divided into two main categories—sequence and structure-specific binding. (Left) Sequence-specific binding proteins recognize the variety of known DNA/RNA sequences via specific interaction with well-characterized protein motifs (zinc-fingers, helix-loop-helix, leucine zipper, helix-turn-helix, etc.). (Right) Structure-specific binding proteins recognize specific local structure(s) of nucleic acids, e.g., G-quadruplexes, i-motifs, cruciforms, triplexes, Z-DNA, and many others. In fact, it is a very common phenomenon that protein with the sequence-specific binding also prefers local DNA/RNA structure in its binding site or within the near neighborhood (e.g., p53), which is indicated by vertical black dashed line and arrows. Created with BioRender.com.

3. Closing Remarks

The global or local amino acid composition of nucleic acid-binding proteins is often overlooked and an unjustly underestimated parameter. Mainly statistically significant enrichment or depletion of particular amino acid residues may serve as a promising tool to predict novel proteins with a similar function, as it was confirmed e.g., for G-quadruplex binding proteins.

Author Contributions: Conceptualization, M.B. and P.P.; methodology, M.B.; resources, M.B., S.G., and K.S.; writing—original draft preparation, M.B., S.G, K.S., and J.Č.; writing—review and editing, J.Č and P.P.; visualization, M.B.; supervision, P.P. All authors have read and agreed to the published version of the manuscript.

Funding: This research was funded by University of Ostrava, SGS01/PřF/2020, and by National Agency for Agricultural Research (NAZV) of Czech Republic grant no. QK1810391 "Utilization of genomic and transcriptomic approaches to create genetic resources and breeding materials of poppy with specific traits".

Institutional Review Board Statement: Not applicable.

Informed Consent Statement: Not applicable.

Data Availability Statement: No new data were created or analyzed in this study. Data sharing is not applicable to this article.

Conflicts of Interest: The authors declare no conflict of interest. The funders had no role in the design of the study; in the collection, analyses, or interpretation of data; in the writing of the manuscript, or in the decision to publish the results.

Abbreviations

bZIP	Basic Leucine Zipper
CIRBP	Cold inducible RNA-binding protein
GNL1	Guanine Nucleotide-Binding Protein-Like 1
GO	Gene Ontology
bHLH	Helix-loop-helix
HTH	Helix-turn-helix
MAZ	Myc-associated zinc finger protein
SNRPA	Small Nuclear Ribonucleoprotein Polypeptide A
SRSF1	Serine/arginine-rich splicing factor 1
SVM	Support vector machine

References

1. Ghani, N.S.A.; Firdaus-Raih, M.; Ahmad, S. Computational Prediction of Nucleic acid-binding Residues From Sequence. In *Encyclopedia of Bioinformatics and Computational Biology*; Ranganathan, S., Gribskov, M., Nakai, K., Schönbach, C., Eds.; Academic Press: Oxford, UK, 2019; pp. 678–687. ISBN 978-0-12-811432-2.
2. Jutras, B.L.; Verma, A.; Stevenson, B. Identification of Novel DNA-Binding Proteins Using DNA-Affinity Chromatography/Pull Down. *Curr. Protoc. Microbiol.* **2012**, *24*, 1F.1.1–1F.1.13. [CrossRef]
3. Wang, I.X.; Grunseich, C.; Fox, J.; Burdick, J.; Zhu, Z.; Ravazian, N.; Hafner, M.; Cheung, V.G. Human Proteins That Interact with RNA/DNA Hybrids. *Genome Res.* **2018**, *28*, 1405–1414. [CrossRef] [PubMed]
4. Ouwerkerk, P.B.; Meijer, A.H. Yeast one-hybrid screens for detection of transcription factor DNA interactions. In *Plant Reverse Genetics*; Springer: Basel, Switzerland, 2011; pp. 211–227.
5. Gaudinier, A.; Tang, M.; Bågman, A.-M.; Brady, S.M. Identification of Protein–DNA Interactions Using Enhanced Yeast One-Hybrid Assays and a Semiautomated Approach. In *Plant Genomics: Methods and Protocols*; Busch, W., Ed.; Methods in Molecular Biology; Springer: New York, NY, 2017; pp. 187–215. ISBN 978-1-4939-7003-2.
6. Hellman, L.M.; Fried, M.G. Electrophoretic Mobility Shift Assay (EMSA) for Detecting Protein–Nucleic Acid Interactions. *Nat. Protoc.* **2007**, *2*, 1849. [CrossRef] [PubMed]
7. Seo, M.; Lei, L.; Egli, M. Label-Free Electrophoretic Mobility Shift Assay (EMSA) for Measuring Dissociation Constants of Protein-RNA Complexes. *Curr. Protoc. Nucleic Acid Chem.* **2019**, *76*, e70. [CrossRef] [PubMed]
8. Carey, M.F.; Peterson, C.L.; Smale, S.T. Chromatin Immunoprecipitation (Chip). *Cold Spring Harb. Protoc.* **2009**, *2009*, pdb-prot5279. [CrossRef] [PubMed]
9. de Barsy, M.; Herrgott, L.; Martin, V.; Pillonel, T.; Viollier, P.H.; Greub, G. Identification of New DNA-Associated Proteins from Waddlia Chondrophila. *Sci. Rep.* **2019**, *9*, 4885. [CrossRef]
10. Kunová, N.; Ondrovičová, G.; Bauer, J.A.; Bellová, J.; Ambro, L'.; Martináková, L.; Kotrasová, V.; Kutejová, E.; Pevala, V. The Role of Lon-Mediated Proteolysis in the Dynamics of Mitochondrial Nucleic Acid-Protein Complexes. *Sci. Rep.* **2017**, *7*, 631. [CrossRef]
11. Haronikova, L.; Coufal, J.; Kejnovska, I.; Jagelska, E.B.; Fojta, M.; Dvořáková, P.; Muller, P.; Vojtesek, B.; Brazda, V. IFI16 Preferentially Binds to DNA with Quadruplex Structure and Enhances DNA Quadruplex Formation. *PLoS ONE* **2016**, *11*, e0157156. [CrossRef]
12. Liu, B.; Wang, S.; Wang, X. DNA-binding Protein Identification by Combining Pseudo Amino Acid Composition and Profile-Based Protein Representation. *Sci. Rep.* **2015**, *5*, 15479. [CrossRef]
13. Fang, Y.; Guo, Y.; Feng, Y.; Li, M. Predicting DNA-Binding Proteins: Approached from Chou's Pseudo Amino Acid Composition and Other Specific Sequence Features. *Amino Acids* **2008**, *34*, 103–109. [CrossRef]
14. Wei, L.; Tang, J.; Zou, Q. Local-DPP: An Improved DNA-Binding Protein Prediction Method by Exploring Local Evolutionary Information. *Inf. Sci.* **2017**, *384*, 135–144. [CrossRef]
15. Liu, B.; Xu, J.; Lan, X.; Xu, R.; Zhou, J.; Wang, X.; Chou, K.-C. IDNA-Prot | dis: Identifying DNA-Binding Proteins by Incorporating Amino Acid Distance-Pairs and Reduced Alphabet Profile into the General Pseudo Amino Acid Composition. *PLoS ONE* **2014**, *9*, e106691. [CrossRef] [PubMed]
16. Choi, S.; Han, K. Prediction of RNA-Binding Amino Acids from Protein and RNA Sequences. *BMC Bioinform.* **2011**, *12*, S7. [CrossRef] [PubMed]
17. Brázda, V.; Coufal, J.; Liao, J.C.C.; Arrowsmith, C.H. Preferential Binding of IFI16 Protein to Cruciform Structure and Superhelical DNA. *Biochem. Biophys. Res. Commun.* **2012**, *422*, 716–720. [CrossRef] [PubMed]
18. Čechová, J.; Coufal, J.; Jagelská, E.B.; Fojta, M.; Brázda, V. P73, like Its P53 Homolog, Shows Preference for Inverted Repeats Forming Cruciforms. *PLoS ONE* **2018**, *13*, e0195835. [CrossRef]
19. Brázda, V.; Hároníková, L.; Liao, J.C.; Fojta, M. DNA and RNA Quadruplex-Binding Proteins. *Int. J. Mol. Sci.* **2014**, *15*, 17493–17517. [CrossRef]
20. Helma, R.; Bažantová, P.; Petr, M.; Adámik, M.; Renčiuk, D.; Tichý, V.; Pastuchová, A.; Soldánová, Z.; Pečinka, P.; Bowater, R.P. P53 Binds Preferentially to Non-B DNA Structures Formed by the Pyrimidine-Rich Strands of GaA· TTC Trinucleotide Repeats Associated with Friedreich's Ataxia. *Molecules* **2019**, *24*, 2078. [CrossRef]

21. Lyons, S.M.; Kharel, P.; Akiyama, Y.; Ojha, S.; Dave, D.; Tsvetkov, V.; Merrick, W.; Ivanov, P.; Anderson, P. EIF4G Has Intrinsic G-Quadruplex Binding Activity That Is Required for TiRNA Function. *Nucleic Acids Res.* **2020**, *48*, 6223–6233. [CrossRef]
22. Porubiaková, O.; Bohálová, N.; Inga, A.; Vadovičová, N.; Coufal, J.; Fojta, M.; Brázda, V. The Influence of Quadruplex Structure in Proximity to P53 Target Sequences on the Transactivation Potential of P53 Alpha Isoforms. *Int. J. Mol. Sci.* **2020**, *21*, 127. [CrossRef]
23. Oyoshi, T.; Masuzawa, T. Modulation of Histone Modifications and G-Quadruplex Structures by G-Quadruplex-Binding Proteins. *Biochem. Biophys. Res. Commun.* **2020**, *531*, 39–44. [CrossRef]
24. Bartas, M.; Brázda, V.; Bohálová, N.; Cantara, A.; Volná, A.; Stachurová, T.; Malachová, K.; Jagelská, E.B.; Porubiaková, O.; Červeň, J. In-Depth Bioinformatic Analyses of Nidovirales Including Human SARS-CoV-2, SARS-CoV, MERS-CoV Viruses Suggest Important Roles of Non-Canonical Nucleic Acid Structures in Their Lifecycles. *Front. Microbiol.* **2020**, *11*, 1583. [CrossRef] [PubMed]
25. Tateishi-Karimata, H.; Sugimoto, N. Chemical Biology of Non-Canonical Structures of Nucleic Acids for Therapeutic Applications. *Chem. Commun.* **2020**, *56*, 2379–2390. [CrossRef] [PubMed]
26. Cer, R.Z.; Donohue, D.E.; Mudunuri, U.S.; Temiz, N.A.; Loss, M.A.; Starner, N.J.; Halusa, G.N.; Volfovsky, N.; Yi, M.; Luke, B.T. Non-B DB v2. 0: A Database of Predicted Non-B DNA-Forming Motifs and Its Associated Tools. *Nucleic Acids Res.* **2012**, *41*, D94–D100. [CrossRef] [PubMed]
27. Brazda, V.; Fojta, M.; Bowater, R.P. Structures and Stability of Simple DNA Repeats from Bacteria. *Biochem. J.* **2020**, *477*, 325–339. [CrossRef]
28. Brázda, V.; Luo, Y.; Bartas, M.; Kaura, P.; Porubiaková, O.; Št'astný, J.; Pečinka, P.; Verga, D.; Da Cunha, V.; Takahashi, T.S. G-Quadruplexes in the Archaea Domain. *Biomolecules* **2020**, *10*, 1349. [CrossRef] [PubMed]
29. Rhodes, D.; Lipps, H.J. G-Quadruplexes and Their Regulatory Roles in Biology. *Nucleic Acids Res.* **2015**, *43*, 8627–8637. [CrossRef]
30. Zeraati, M.; Langley, D.B.; Schofield, P.; Moye, A.L.; Rouet, R.; Hughes, W.E.; Bryan, T.M.; Dinger, M.E.; Christ, D. I-Motif DNA Structures Are Formed in the Nuclei of Human Cells. *Nat. Chem.* **2018**, *10*, 631–637. [CrossRef]
31. Brázdová, M.; Tichý, V.; Helma, R.; Bažantová, P.; Polášková, A.; Krejčí, A.; Petr, M.; Navrátilová, L.; Tichá, O.; Nejedlý, K.; et al. P53 Specifically Binds Triplex DNA In Vitro and in Cells. *PLoS ONE* **2016**, *11*, e0167439. [CrossRef]
32. Chedin, F.; Benham, C.J. Emerging Roles for R-Loop Structures in the Management of Topological Stress. *J. Biol. Chem.* **2020**, *295*, 4684–4695. [CrossRef]
33. Xu, P.; Pan, F.; Roland, C.; Sagui, C.; Weninger, K. Dynamics of Strand Slippage in DNA Hairpins Formed by CAG Repeats: Roles of Sequence Parity and Trinucleotide Interrupts. *Nucleic Acids Res.* **2020**, *48*, 2232–2245. [CrossRef]
34. Fleming, A.M.; Zhu, J.; Jara-Espejo, M.; Burrows, C.J. Cruciform DNA Sequences in Gene Promoters Can Impact Transcription upon Oxidative Modification of 2′-Deoxyguanosine. *Biochemistry* **2020**, *59*, 2616–2626. [CrossRef] [PubMed]
35. Bevilacqua, P.C.; Ritchey, L.E.; Su, Z.; Assmann, S.M. Genome-Wide Analysis of RNA Secondary Structure. *Annu. Rev. Genet.* **2016**, *50*, 235–266. [CrossRef] [PubMed]
36. Shin, S.-I.; Ham, S.; Park, J.; Seo, S.H.; Lim, C.H.; Jeon, H.; Huh, J.; Roh, T.-Y. Z-DNA-Forming Sites Identified by ChIP-Seq Are Associated with Actively Transcribed Regions in the Human Genome. *DNA Res.* **2016**, *23*, 477–486. [CrossRef] [PubMed]
37. Spiegel, J.; Adhikari, S.; Balasubramanian, S. The Structure and Function of DNA G-Quadruplexes. *Trends Chem.* **2020**, *2*, 123–136. [CrossRef] [PubMed]
38. Varshney, D.; Spiegel, J.; Zyner, K.; Tannahill, D.; Balasubramanian, S. The Regulation and Functions of DNA and RNA G-Quadruplexes. *Nat. Rev. Mol. Cell Biol.* **2020**, *21*, 459–474. [CrossRef]
39. Kaushik, M.; Kaushik, S.; Roy, K.; Singh, A.; Mahendru, S.; Kumar, M.; Chaudhary, S.; Ahmed, S.; Kukreti, S. A Bouquet of DNA Structures: Emerging Diversity. *Biochem. Biophys. Rep.* **2016**, *5*, 388–395. [CrossRef]
40. Masai, H.; Tanaka, T. G-Quadruplex DNA and RNA: Their Roles in Regulation of DNA Replication and Other Biological Functions. *Biochem. Biophys. Res. Commun.* **2020**, *531*, 25–38. [CrossRef]
41. Herbert, A. Z-DNA and Z-RNA in Human Disease. *Commun. Biol.* **2019**, *2*, 1–10. [CrossRef]
42. Yuan, W.-F.; Wan, L.-Y.; Peng, H.; Zhong, Y.-M.; Cai, W.-L.; Zhang, Y.-Q.; Ai, W.-B.; Wu, J.-F. The Influencing Factors and Functions of DNA G-Quadruplexes. *Cell Biochem. Funct.* **2020**, *38*, 524–532. [CrossRef]
43. Bacolla, A.; Cooper, D.N.; Vasquez, K.M.; Tainer, J.A. Non-B DNA Structure and Mutations Causing Human Genetic Disease. In *eLS*; American Cancer Society: Atlanta, GA, USA, 2018; pp. 1–15. ISBN 978-0-470-01590-2.
44. Bacolla, A.; Tainer, J.A.; Vasquez, K.M.; Cooper, D.N. Translocation and Deletion Breakpoints in Cancer Genomes Are Associated with Potential Non-B DNA-Forming Sequences. *Nucleic Acids Res.* **2016**, *44*, 5673–5688. [CrossRef]
45. Cammas, A.; Millevoi, S. RNA G-Quadruplexes: Emerging Mechanisms in Disease. *Nucleic Acids Res.* **2017**, *45*, 1584–1595. [CrossRef] [PubMed]
46. Kharel, P.; Balaratnam, S.; Beals, N.; Basu, S. The Role of RNA G-Quadruplexes in Human Diseases and Therapeutic Strategies. *Wiley Interdiscip. Rev. RNA* **2020**, *11*, e1568. [CrossRef] [PubMed]
47. Brázda, V.; Cerveň, J.; Bartas, M.; Mikysková, N.; Coufal, J.; Pečinka, P. The Amino Acid Composition of Quadruplex Binding Proteins Reveals a Shared Motif and Predicts New Potential Quadruplex Interactors. *Molecules* **2018**, *23*. [CrossRef] [PubMed]
48. Bartas, M.; Bažantová, P.; Brázda, V.; Liao, J.; Červeň, J.; Pečinka, P. Identification of Distinct Amino Acid Composition of Human Cruciform Binding Proteins. *Mol. Biol.* **2019**, *53*, 97–106. [CrossRef]
49. Consortium, G.O. Expansion of the Gene Ontology Knowledgebase and Resources. *Nucleic Acids Res.* **2017**, *45*, D331–D338.

50. Consortium, G.O. Gene Ontology Consortium: Going Forward. *Nucleic Acids Res.* **2015**, *43*, D1049–D1056. [CrossRef]
51. Carbon, S.; Ireland, A.; Mungall, C.J.; Shu, S.; Marshall, B.; Lewis, S.; Hub, A.; Group, W.P.W. AmiGO: Online Access to Ontology and Annotation Data. *Bioinformatics* **2009**, *25*, 288–289. [CrossRef]
52. Mishra, S.K.; Tawani, A.; Mishra, A.; Kumar, A. G4IPDB: A Database for G-Quadruplex Structure Forming Nucleic Acid Interacting Proteins. *Sci. Rep.* **2016**, *6*, 38144. [CrossRef]
53. Moccia, F.; Platella, C.; Musumeci, D.; Batool, S.; Zumrut, H.; Bradshaw, J.; Mallikaratchy, P.; Montesarchio, D. The Role of G-Quadruplex Structures of LIGS-Generated Aptamers R1.2 and R1.3 in IgM Specific Recognition. *Int. J. Biol. Macromol.* **2019**, *133*, 839–849. [CrossRef]
54. Riccardi, C.; Napolitano, E.; Platella, C.; Musumeci, D.; Melone, M.A.B.; Montesarchio, D. Anti-VEGF DNA-Based Aptamers in Cancer Therapeutics and Diagnostics. *Med. Res. Rev.* **2021**, *41*, 464–506. [CrossRef]
55. Brázda, V.; Laister, R.C.; Jagelská, E.B.; Arrowsmith, C. Cruciform Structures Are a Common DNA Feature Important for Regulating Biological Processes. *BMC Mol. Biol.* **2011**, *12*, 33. [CrossRef] [PubMed]
56. Kim, C. How Z-DNA/RNA-binding Proteins Shape Homeostasis, Inflammation, and Immunity. *BMB Rep.* **2020**, *53*, 453–457. [CrossRef] [PubMed]
57. Iwai, K.; Ishikawa, K.; Hayashi, H. Amino-Acid Sequence of Slightly Lysine-Rich Histone. *Nature* **1970**, *226*, 1056–1058. [CrossRef] [PubMed]
58. Jukes, T.H.; Holmquist, R.; Moise, H. Amino Acid Composition of Proteins: Selection against the Genetic Code. *Science* **1975**, *189*, 50–51. [CrossRef] [PubMed]
59. Aukerman, M.J.; Schmidt, R.J.; Burr, B.; Burr, F.A. An Arginine to Lysine Substitution in the BZIP Domain of an Opaque-2 Mutant in Maize Abolishes Specific DNA-binding. *Genes Dev.* **1991**, *5*, 310–320. [CrossRef]
60. Lee, B.; Thirunavukkarasu, K.; Zhou, L.; Pastore, L.; Baldini, A.; Hecht, J.; Geoffrey, V.; Ducy, P.; Karsenty, G. Missense Mutations Abolishing DNA-binding of the Osteoblast-Specific Transcription Factor OSF2/CBFA1 in Cleidocranial Dysplasia. *Nat. Genet.* **1997**, *16*, 307–310. [CrossRef]
61. Siomi, H.; Choi, M.; Siomi, M.C.; Nussbaum, R.L.; Dreyfuss, G. Essential Role for KH Domains in RNA-binding: Impaired RNA-binding by a Mutation in the KH Domain of FMR1 That Causes Fragile X Syndrome. *Cell* **1994**, *77*, 33–39. [CrossRef]
62. Cheng, S.; Melkonian, M.; Smith, S.A.; Brockington, S.; Archibald, J.M.; Delaux, P.-M.; Li, F.-W.; Melkonian, B.; Mavrodiev, E.V.; Sun, W.; et al. 10KP: A Phylodiverse Genome Sequencing Plan. *GigaScience* **2018**, *7*. [CrossRef]
63. Kriventseva, E.V.; Kuznetsov, D.; Tegenfeldt, F.; Manni, M.; Dias, R.; Simão, F.A.; Zdobnov, E.M. OrthoDB V10: Sampling the Diversity of Animal, Plant, Fungal, Protist, Bacterial and Viral Genomes for Evolutionary and Functional Annotations of Orthologs. *Nucleic Acids Res.* **2019**, *47*, D807–D811. [CrossRef]
64. Vacic, V.; Uversky, V.N.; Dunker, A.K.; Lonardi, S. Composition Profiler: A Tool for Discovery and Visualization of Amino Acid Composition Differences. *BMC Bioinform.* **2007**, *8*, 211. [CrossRef]
65. Sivashankari, S.; Shanmughavel, P. Functional Annotation of Hypothetical Proteins – A Review. *Bioinformation* **2006**, *1*, 335–338. [CrossRef] [PubMed]
66. Cai, Y.; Lin, S.L. Support Vector Machines for Predicting RRNA-, RNA-, and DNA-Binding Proteins from Amino Acid Sequence. *Biochim. Et Biophys. Acta (Bba) - Proteins Proteom.* **2003**, *1648*, 127–133. [CrossRef]
67. Kumar, M.; Gromiha, M.M.; Raghava, G.P. Identification of DNA-Binding Proteins Using Support Vector Machines and Evolutionary Profiles. *BMC Bioinform.* **2007**, *8*, 463. [CrossRef] [PubMed]
68. Standing, K.G. Peptide and Protein de Novo Sequencing by Mass Spectrometry. *Curr. Opin. Struct. Biol.* **2003**, *13*, 595–601. [CrossRef] [PubMed]
69. Vitorino, R.; Guedes, S.; Trindade, F.; Correia, I.; Moura, G.; Carvalho, P.; Santos, M.A.S.; Amado, F. De Novo Sequencing of Proteins by Mass Spectrometry. *Expert Rev. Proteom.* **2020**, *17*, 595–607. [CrossRef] [PubMed]
70. Gasteiger, E.; Hoogland, C.; Gattiker, A.; Wilkins, M.R.; Appel, R.D.; Bairoch, A. Protein identification and analysis tools on the ExPASy server. In *The proteomics protocols handbook*; Springer: Basel, Switzerland, 2005; pp. 571–607.
71. Cao, D.-S.; Xu, Q.-S.; Liang, Y.-Z. Propy: A Tool to Generate Various Modes of Chou's PseAAC. *Bioinformatics* **2013**, *29*, 960–962. [CrossRef]
72. Vishnoi, S.; Garg, P.; Arora, P. Physicochemical N-Grams Tool: A Tool for Protein Physicochemical Descriptor Generation via Chou's 5-Step Rule. *Chem. Biol. Drug Des.* **2020**, *95*, 79–86. [CrossRef]
73. Zuo, Y.; Li, Y.; Chen, Y.; Li, G.; Yan, Z.; Yang, L. PseKRAAC: A Flexible Web Server for Generating Pseudo K-Tuple Reduced Amino Acids Composition. *Bioinformatics* **2017**, *33*, 122–124. [CrossRef]
74. Hudson, W.H.; Ortlund, E.A. The Structure, Function and Evolution of Proteins That Bind DNA and RNA. *Nat. Rev.. Mol. Cell Biol.* **2014**, *15*, 749–760. [CrossRef]
75. Terribilini, M.; Lee, J.-H.; Yan, C.; Jernigan, R.L.; Honavar, V.; Dobbs, D. Prediction of RNA-binding Sites in Proteins from Amino Acid Sequence. *RNA* **2006**, *12*, 1450–1462. [CrossRef]
76. Zhang, J.; Ma, Z.; Kurgan, L. Comprehensive Review and Empirical Analysis of Hallmarks of DNA-, RNA-and Protein-Binding Residues in Protein Chains. *Brief. Bioinform.* **2019**, *20*, 1250–1268. [CrossRef]
77. Michalek, J.L.; Besold, A.N.; Michel, S.L.J. Cysteine and Histidine Shuffling: Mixing and Matching Cysteine and Histidine Residues in Zinc Finger Proteins to Afford Different Folds and Function. *Dalton Trans.* **2011**, *40*, 12619–12632. [CrossRef]
78. Laity, J.H.; Lee, B.M.; Wright, P.E. Zinc Finger Proteins: New Insights into Structural and Functional Diversity. *Curr. Opin. Struct. Biol.* **2001**, *11*, 39–46. [CrossRef]

79. Yesudhas, D.; Batool, M.; Anwar, M.A.; Panneerselvam, S.; Choi, S. Proteins Recognizing DNA: Structural Uniqueness and Versatility of DNA-Binding Domains in Stem Cell Transcription Factors. *Genes* **2017**, *8*, 192. [CrossRef] [PubMed]
80. Ahmad, M.; Xu, D.; Wang, W. Type IA Topoisomerases Can Be "Magicians" for Both DNA and RNA in All Domains of Life. *RNA Biol.* **2017**, *14*, 854–864. [CrossRef] [PubMed]
81. Aravind, L.; Anantharaman, V.; Balaji, S.; Babu, M.M.; Iyer, L.M. The Many Faces of the Helix-Turn-Helix Domain: Transcription Regulation and Beyond. *FEMS Microbiol Rev* **2005**, *29*, 231–262. [CrossRef]
82. Atchley, W.R.; Fitch, W.M. A Natural Classification of the Basic Helix–Loop–Helix Class of Transcription Factors. *Proc. Natl. Acad. Sci. USA* **1997**, *94*, 5172–5176. [CrossRef]
83. Casey, B.H.; Kollipara, R.K.; Pozo, K.; Johnson, J.E. Intrinsic DNA-binding Properties Demonstrated for Lineage-Specifying Basic Helix-Loop-Helix Transcription Factors. Available online: http://genome.cshlp.org (accessed on 2 January 2021).
84. Hakoshima, T. Leucine Zippers. In *eLS*; American Cancer Society: Atlanta, GA, USA, 2014; ISBN 978-0-470-01590-2.
85. Miller, M. The Importance of Being Flexible: The Case of Basic Region Leucine Zipper Transcriptional Regulators. *Curr. Protein Pept. Sci.* **2009**, *10*, 244–269. [CrossRef]
86. Yagi, R.; Miyazaki, T.; Oyoshi, T. G-Quadruplex Binding Ability of TLS/FUS Depends on the β-Spiral Structure of the RGG Domain. *Nucleic Acids Res.* **2018**, *46*, 5894–5901. [CrossRef]
87. Ishiguro, A.; Kimura, N.; Noma, T.; Shimo-Kon, R.; Ishihama, A.; Kon, T. Molecular Dissection of ALS-Linked TDP-43 – Involvement of the Gly-Rich Domain in Interaction with G-Quadruplex MRNA. *FEBS Lett.* **2020**, *594*, 2254–2265. [CrossRef]
88. Takahama, K.; Oyoshi, T. Specific Binding of Modified RGG Domain in TLS/FUS to G-Quadruplex RNA: Tyrosines in RGG Domain Recognize 2′-OH of the Riboses of Loops in G-Quadruplex. *J. Am. Chem. Soc.* **2013**, *135*, 18016–18019. [CrossRef] [PubMed]
89. Bartas, M.; Červeň, J.; Pečinka, P. Identification of Distinct Amino Acid Composition of Z-DNA/RNA and Triplex-Binding Proteins. *Mol. Bio. 53*, 97–106.
90. Ribeiro de Almeida, C.; Dhir, S.; Dhir, A.; Moghaddam, A.E.; Sattentau, Q.; Meinhart, A.; Proudfoot, N.J. RNA Helicase DDX1 Converts RNA G-Quadruplex Structures into R-Loops to Promote IgH Class Switch Recombination. *Mol. Cell* **2018**, *70*, 650–662.e8. [CrossRef] [PubMed]
91. Cai, B.-H.; Chao, C.-F.; Huang, H.-C.; Lee, H.-Y.; Kannagi, R.; Chen, J.-Y. Roles of P53 Family Structure and Function in Non-Canonical Response Element Binding and Activation. *Int. J. Mol. Sci.* **2019**, *20*, 3681. [CrossRef] [PubMed]
92. Bossone, S.A.; Asselin, C.; Patel, A.J.; Marcu, K.B. MAZ, a Zinc Finger Protein, Binds to c-MYC and C2 Gene Sequences Regulating Transcriptional Initiation and Termination. *Proc. Natl. Acad. Sci. USA* **1992**, *89*, 7452–7456. [CrossRef] [PubMed]
93. Cogoi, S.; Zorzet, S.; Rapozzi, V.; Géci, I.; Pedersen, E.B.; Xodo, L.E. MAZ-Binding G4-Decoy with Locked Nucleic Acid and Twisted Intercalating Nucleic Acid Modifications Suppresses KRAS in Pancreatic Cancer Cells and Delays Tumor Growth in Mice. *Nucleic Acids Res.* **2013**, *41*, 4049–4064. [CrossRef] [PubMed]
94. Dominguez, D.; Freese, P.; Alexis, M.S.; Su, A.; Hochman, M.; Palden, T.; Bazile, C.; Lambert, N.J.; Van Nostrand, E.L.; Pratt, G.A.; et al. Sequence, Structure, and Context Preferences of Human RNA-binding Proteins. *Mol. Cell* **2018**, *70*, 854–867. [CrossRef]
95. Laptenko, O.; Tong, D.R.; Manfredi, J.; Prives, C. The Tail That Wags the Dog: How the Disordered C-Terminal Domain Controls the Transcriptional Activities of the P53 Tumor-Suppressor Protein. *Trends Biochem. Sci.* **2016**, *41*, 1022–1034. [CrossRef]
96. Petr, M.; Helma, R.; Polášková, A.; Krejčí, A.; Dvořáková, Z.; Kejnovská, I.; Navrátilová, L.; Adámik, M.; Vorlíčková, M.; Brázdová, M. Wild-Type P53 Binds to MYC Promoter G-Quadruplex. *Biosci. Rep.* **2016**, *36*. [CrossRef]
97. Inukai, S.; Kock, K.H.; Bulyk, M.L. Transcription Factor–DNA-binding: Beyond Binding Site Motifs. *Curr. Opin. Genet. Dev.* **2017**, *43*, 110–119. [CrossRef]
98. Wang, C.; Uversky, V.N.; Kurgan, L. Disordered Nucleiome: Abundance of Intrinsic Disorder in the DNA- and RNA-Binding Proteins in 1121 Species from Eukaryota, Bacteria and Archaea. *Proteomics* **2016**, *16*, 1486–1498. [CrossRef]
99. Watson, M.; Stott, K. Disordered Domains in Chromatin-Binding Proteins. *Essays Biochem.* **2019**, *63*, 147–156. [CrossRef] [PubMed]
100. Turner, A.L.; Watson, M.; Wilkins, O.G.; Cato, L.; Travers, A.; Thomas, J.O.; Stott, K. Highly Disordered Histone H1−DNA Model Complexes and Their Condensates. *Proc. Natl. Acad. Sci. USA* **2018**, *115*, 11964–11969. [CrossRef] [PubMed]
101. Serrano, P.; Aubol, B.E.; Keshwani, M.M.; Forli, S.; Ma, C.-T.; Dutta, S.K.; Geralt, M.; Wüthrich, K.; Adams, J.A. Directional Phosphorylation and Nuclear Transport of the Splicing Factor SRSF1 Is Regulated by an RNA Recognition Motif. *J. Mol. Biol.* **2016**, *428*, 2430–2445. [CrossRef] [PubMed]
102. Von Hacht, A.V.; Seifert, O.; Menger, M.; Schütze, T.; Arora, A.; Konthur, Z.; Neubauer, P.; Wagner, A.; Weise, C.; Kurreck, J. Identification and Characterization of RNA Guanine-Quadruplex Binding Proteins. *Nucleic Acids Res.* **2014**, *42*, 6630–6644. [CrossRef]
103. Huang, Z.-L.; Dai, J.; Luo, W.-H.; Wang, X.-G.; Tan, J.-H.; Chen, S.-B.; Huang, Z.-S. Identification of G-Quadruplex-Binding Protein from the Exploration of RGG Motif/G-Quadruplex Interactions. *J. Am. Chem. Soc.* **2018**, *140*, 17945–17955. [CrossRef]
104. Rigo, R.; Palumbo, M.; Sissi, C. G-Quadruplexes in Human Promoters: A Challenge for Therapeutic Applications. *Biochim. Et Biophys. Acta (Bba)-Gen. Subj.* **2017**, *1861*, 1399–1413. [CrossRef]
105. Poggi, L.; Richard, G.-F. Alternative DNA Structures In Vivo: Molecular Evidence and Remaining Questions. *Microbiol. Mol. Biol. Rev.* **2020**, *85*. [CrossRef]
106. Sissi, C.; Gatto, B.; Palumbo, M. The Evolving World of Protein-G-Quadruplex Recognition: A Medicinal Chemist's Perspective. *Biochimie* **2011**, *93*, 1219–1230. [CrossRef]

107. Brázda, V.; Coufal, J. Recognition of Local DNA Structures by P53 Protein. *Int. J. Mol. Sci.* **2017**, *18*, 375. [CrossRef]
108. Sun, Z.-Y.; Wang, X.-N.; Cheng, S.-Q.; Su, X.-X.; Ou, T.-M. Developing Novel G-Quadruplex Ligands: From Interaction with Nucleic Acids to Interfering with Nucleic Acid–Protein Interaction. *Molecules* **2019**, *24*, 396. [CrossRef] [PubMed]
109. Kharel, P.; Becker, G.; Tsvetkov, V.; Ivanov, P. Properties and Biological Impact of RNA G-Quadruplexes: From Order to Turmoil and Back. *Nucleic Acids Res.* **2020**, *48*, 12534–12555. [CrossRef] [PubMed]
110. Lee, T.; Pelletier, J. The Biology of DHX9 and Its Potential as a Therapeutic Target. *Oncotarget* **2016**, *7*, 42716–42739. [CrossRef] [PubMed]
111. Turcotte, M.-A.; Garant, J.-M.; Cossette-Roberge, H.; Perreault, J.-P. Guanine Nucleotide-Binding Protein-Like 1 (GNL1) Binds RNA G-Quadruplex Structures in Genes Associated with Parkinson's Disease. *RNA Biol.* **2020**, 1–15. [CrossRef] [PubMed]
112. Bolduc, F.; Turcotte, M.-A.; Perreault, J.-P. The Small Nuclear Ribonucleoprotein Polypeptide A (SNRPA) Binds to the G-Quadruplex of the BAG-1 5′UTR. *Biochimie* **2020**, *176*, 122–127. [CrossRef]
113. Clemo, N.K.; Collard, T.J.; Southern, S.L.; Edwards, K.D.; Moorghen, M.; Packham, G.; Hague, A.; Paraskeva, C.; Williams, A.C. BAG-1 Is up-Regulated in Colorectal Tumour Progression and Promotes Colorectal Tumour Cell Survival through Increased NF-KB Activity. *Carcinogenesis* **2008**, *29*, 849–857. [CrossRef]

International Journal of *Molecular Sciences*

MDPI

Article

Chromatin Structure and "DNA Sequence View": The Role of Satellite DNA in Ectopic Pairing of the *Drosophila* X Polytene Chromosome

Aleksandr V. Zhuravlev [1,*], Gennadii A. Zakharov [1,2], Ekaterina V. Anufrieva [3], Anna V. Medvedeva [1], Ekaterina A. Nikitina [1,3] and Elena V. Savvateeva-Popova [1]

1 Pavlov Institute of Physiology, Russian Academy of Sciences, 199034 Saint Petersburg, Russia; gennadiy.zakharov@gmail.com (G.A.Z.); avmed56@mail.ru (A.V.M.); 21074@mail.ru (E.A.N.); esavvateeva@mail.ru (E.V.S.-P.)

2 EPAM Systems Inc., Saint Petersburg 197110, Russia

3 Faculty of Biology, Herzen State Pedagogical University of Russia, 191186 Saint Petersburg, Russia; Kate.an21@yandex.ru

* Correspondence: beneor@mail.ru; Tel.: +7-(931)-330-3129

Abstract: Chromatin 3D structure plays a crucial role in regulation of gene activity. Previous studies have envisioned spatial contact formations between chromatin domains with different epigenetic properties, protein compositions and transcription activity. This leaves specific DNA sequences that affect chromosome interactions. The *Drosophila melanogaster* polytene chromosomes are involved in non-allelic ectopic pairing. The mutant strain agn^{ts3}, a *Drosophila* model for Williams–Beuren syndrome, has an increased frequency of ectopic contacts (FEC) compared to the wild-type strain *Canton-S* (*CS*). Ectopic pairing can be mediated by some specific DNA sequences. In this study, using our Homology Segment Analysis software, we estimated the correlation between FEC and frequency of short matching DNA fragments (FMF) for all sections of the X chromosome of *Drosophila CS* and agn^{ts3} strains. With fragment lengths of 50 nucleotides (nt), *CS* showed a specific FEC–FMF correlation for 20% of the sections involved in ectopic contacts. The correlation was unspecific in agn^{ts3}, which may indicate the alternative epigenetic mechanisms affecting FEC in the mutant strain. Most of the fragments that specifically contributed to FMF were related to 1.688 or 372-bp middle repeats. Thus, middle repetitive DNA may serve as an organizer of ectopic pairing.

Keywords: *Drosophila*; polytene chromosomes; *Canton-S*; *agnostic*; ectopic pairing; 1.688 repeats; 372-bp repeats

Citation: Zhuravlev, A.V.; Zakharov, G.A.; Anufrieva, E.V.; Medvedeva, A.V.; Nikitina, E.A.; Savvateeva-Popova, E.V. Chromatin Structure and "DNA Sequence View": The Role of Satellite DNA in Ectopic Pairing of the *Drosophila* X Polytene Chromosome. *Int. J. Mol. Sci.* **2021**, *22*, 8713. https://doi.org/10.3390/ijms22168713

Academic Editors: Richard Bowater and Vaclav Brazda

Received: 30 July 2021
Accepted: 11 August 2021
Published: 13 August 2021

1. Introduction

Spatial organization of the cell nucleus is an important factor defining the regulation of gene activity, as well as the processes of DNA replication, recombination and reparation. During interphase, chromosomes occupy separate territories in the nucleus, being radially arranged: gene-rich chromosome territories are localized toward the interior, while gene-poor territories are close to periphery [1,2]. In human cells, regions of increased gene expression (ridges) are clustered in spatially distinct gene-enriched domains characterized by irregular forms and low chromatin condensation. These domains are predominantly located toward the nuclear interior. On the contrary, antiridges are relatively gene poor, condensed, transcriptionally inactive and localize closer to the cell envelope. Mechanisms behind the formation and maintenance of such 3D structures are still unclear, possibly due to the interaction between some unknown DNA sequences with nuclear matrix, specific proteins and/or non-coding RNAs [3]. Studying such mechanisms is necessary for understanding the process of gene regulation at the system level. As gene juxtaposition in nuclei facilitates specific chromosomal translocations, 3D chromatin structures can also predict the genetic rearrangements leading to carcinogenesis [4].

Chromosome territories are not rigidly fixed spatial units and show a significant percentage of intermingling, mostly at their borders, that is influenced by gene transcription [5]. Some genes are able to change their location in nuclei, being brought together with the help of actin and myosin motor proteins [6]. Transcription mainly occurs within the nuclear areas enriched in RNA polymerase II (RNAPII), known as transcription factories [7]. The expression level for given genes depends on their proximity to such a factory. The constitutively active genes may nucleate the factory, whereas the others relocalize to it upon their induction [4,7].

In addition to diploid cells, some organisms, such as *Diptera* species, also have polytene cells where chromatids are not segregated after multiple duplications. The giant polytene chromosomes of *Drosophila melanogaster* 3rd instar larvae are characterized by specific banding patterns, which can be revealed by electron and light microscopy. The densely packed thick "black" bands with high DNA content are transcriptionally repressed chromatin areas with low gene density. The largest among them are the intercalary heterochromatin (IH) bands, being late-replicating, under-replicated genomic areas prone to chromosomal breaks, constrictions and non-allelic ectopic contacts formation. *Drosophila* polytene chromosomes harbor about 250 IH sites. The "grey" bands are partially decondensed and more transcriptionally active compared to the IH. Interbands are the most active and the least condensed genomic areas, characterized by the "open" chromatin structure. Bands are united into cytological sections, such as 1A, 1B, 1C ... and up to 20F for the X chromosome [8–10]. Each type of band contains specific proteins and is enriched for specific genetic elements. Interbands mostly contain the promoters of the constantly active housekeeping genes, being associated with open chromatin proteins such as CHRIZ/Chromator. The "grey" bands contain multiple active genes but lack CHRIZ, being enriched with RNApol II. The IH bands are composed of tissue-specific genes, being associated with SUUR, D1, lamin B and histone H1 proteins [11]. Notably, such structures are not unique for polytene nuclei, as the chromatin folding and protein composition are conserved in different fly tissues, being closely related to the morphology of the polytene chromosomes. At the same time, the ability to form distant contacts in polytene chromosomes is restricted by their lack of flexibility [12]. Replication timing is similar between polytene and diploid *Drosophila* cells. The late replicating black bands in various tissues correspond to silent chromatin types, with borders enriched for SUUR, lamin and H3K27me3 [12]. This makes polytene nuclei a convenient model to study 3D nuclear organization.

As previously shown by Horchstrasser et al. [13], polytene chromosomes occupy specific spatial domains similar to chromosome territories in diploid nuclei. Chromosomes extend across the nucleus in a Rabl orientation (i.e., their centromeres group near one pole of the nucleus and telomeres near the opposite pole). Chromosomes are coiled in a right-handed fashion, with their 2L and 2R arms being mostly next to each other, as well as the 3L and 3R arms. The loci enriched in IH and ectopic contacts are oriented toward the envelope. However, there are no stable intrachromosomal interactions beyond the distance of two cytological divisions. Thus, chromosomal configuration varies significantly, even though the contacts between the distant loci through the ectopic fibers were not addressed in that study.

Chromosome conformation capture technologies make it possible to investigate nuclear 3D organization with a resolution of the order of one to tens kb. The chromosome conformation capture (3C) method is used to estimate the average frequency of a contact between the two known chromosomal loci in a cell population. Chromosome conformation capture-on-chip (4C) technology allows spatial contacts of a selected genomic site to be assayed with all unknown distant sites. In the chromosome conformation capture carbon copy (5C) method, a massive analysis of contacts between specific loci across the entire genome is performed. The Hi-C method reveals spatial contacts at the level of the whole genome [14]. The complex net of such physical contacts uncovers the chromosome topology in detail, assaying interactions between genes and their regulatory elements.

Using Hi-C technology, the spatial structure of the *Drosophila* genome was studied for both diploid and polytene nuclei with a resolution of 15 kb [15]. Polytene bands nearly correspond to topologically associated domains (TADs) with a mean size of 195 kb. They are conserved for polytene and diploid nuclei. TADs are persistent throughout fly development, and are formed by the axial condensation of the chromatin fiber. The putative role of a TAD is DNA compaction rather than regulation of gene activity. Stable interactions between the different TADs were not observed, with is consistent with the variable long-range chromosomal conformation in the nucleus [13]. In the other study on *Drosophila* embryonic nuclei, the long-range interactions between Polycomb-repressed domains were found. The hierarchically organized domains were associated with active and repressive epigenetic modifications of chromatin [16]. Ectopic pairing was not considered in Hi-C studies, as its frequency was low and such long-range contacts were beyond the limit of the resolution. Thus, light and electron microscopy remain the most appropriate methods by which to study the ectopic pairing.

Ectopic contact is morphologically observed as an intimate association of IH bands or as an unstructured fiber connecting two IH bands. As the paired chromosome sections seem to be covalently linked due to the chromatids recombination/reparation, ectopic pairing is not disrupted upon squashing with acetic fixation [12,17]. The method of squashed preparations was used to estimate frequency of ectopic contacts (FEC) in several *Drosophila* strains. FEC can be expressed as the total number of ectopic contacts between a given section pair.

agn^{ts3} is a *D. melanogaster* mutant with a dysfunction of LIM kinase 1 (LIMK1), the main regulator of actin polymerization in nervous cells. This mutant strain shows multiple cognitive impairments, being the model object for Williams–Beuren syndrome [18]. a/t-rich *agnostic* locus (X:11AB) is predisposed to mutations. Its length varies, probably due to spontaneous unequal recombination [19]. Impairment of LIMK1 and actin dynamics affects the spatial organization of chromatin [20]. For *agn*^{ts3}, as well as the wild-type strains *Canton-S (CS)*, *Berlin* and *Oregon-R*, multiple polymorphisms were found in the *limk1* gene and flanking sequences. In *agn*^{ts3}, there is also mobile S-element insertion downstream *limk1* [21,22]. The *agn*^{ts3} FECs are significantly higher compared to the wild-type strains [22]. Though profiles of ectopic pairing have shown a significant inter-strain variation, the pairing often occurs at the same loci. Thus, DNA sequence itself may define which loci can form a contact, whereas the epigenetic factors and/or activity of specific genes, such as *limk1*, affect FEC values by changing chromatin properties and tendency to pair.

There are several models of ectopic contact formation with the IH bands: (1) Pairing of "sticky ends" of the short repeated sequences within the areas of the DNA breaks that occur due to under-replication. (2) Pairing between the extended homologous DNA sequences. Generally, ectopic pairing occurs between the areas of chromosomes that do not show a significant homology, though it has been observed for some bands. (3) DNA branch migration upon replication mistakes, presumably due to the restricted homology between the associating sequences. (4) Pairing mediated by specific heterochromatin-associated proteins [9]. At least in the first three cases, presence of identical DNA sequences is crucial for pairing according to the principle of complementarity. In other words, high FECs should correlate with high frequencies of matching short DNA fragments (FMF) for the contacting regions.

The method of squashed preparations is imprecise, as it permits the localization of areas of contacts with a resolution of tens to hundreds of thousands bp. Hence, it gives no information about the specific DNA sequences involved in ectopic pairing. The bioinformatics approach helps to handle this problem. To estimate correlations between Drosophila FEC and FMF, we designed software called Homology Segment Analysis [23]. The current version of the software performs the following:

1. DNA sequences of the chromosome sections (A) are taken one by one, searching for short single-stranded fragments (k-mers) of a given length that are the same as fragments of the other sections (B). For each pair of sections (A-B), FMF is calculated as the total number of matching fragments for both DNA chains. To increase matching specificity, short DNA repeats (microsatellites) can be excluded at this stage.
2. For each section A, the rho value of the Spearman correlation between FEC and FMF is computed. Both specific and unspecific correlations are considered (FEC and FMF values correspond to the same or different section pairs, respectively).
3. For all sections A, the average rho values (R) and the proportion of statistically significant FEC-FMF correlations (P) are calculated at different fragment lengths, for different Drosophila strains, and statistically analyzed.
4. For each A-B pair, the list of the matching fragments is generated and ordered according to their numbers of occurrence. This lets us reveal short DNA sequences that specifically impact FEC-FMF correlation.

Steps 1–3 can be performed for sections within chromosome parts of different sizes.

Using a previous version of the software, we showed a positive correlation between *Berlin*/*agn*^{ts3} FEC and FMF for identical short (30–50 nt) DNA fragments. In that research, we specifically focused on the X:11AB region and its contacts with the other sections of the X chromosome. Most of the fragments found to putatively make impact into ectopic pairing were similar to the middle repetitive DNA 372-bp sequence and the 1.688 g/mL satellite DNA family [24]. The distribution and properties of 372-bp indicate its possible role in *Drosophila* dosage compensation and primary sex determination [25]. The 1.688 satellite DNA is abundant in *Drosophila* genome (2%), being localized both in heterochromatin and in euchromatin domains (1860 and 168 copies, respectively), mainly on the X chromosome. This satellite family includes 360/359-bp, 353-bp and 257-bp subfamilies [26]. The 1.688 satellite DNA is known to produce small RNAs that participate in the localization of male-specific lethal complex (MSL) on the X chromosome, increasing male survival [27]. The above point to a striking connection between the ectopic pairing and non-coding RNA-dependent processes of dosage compensation. 359-bp also produces a long non-coding RNA that interacts with centromeres of all major chromosomes, participating in their mitotic segregation [28].

In this study, we calculated FEC-FMF correlations for all pairs of the X chromosome sections (1A–20F) in *CS* and *agn*^{ts3} strains at fragment lengths of 10–60 nt. Each section begins with an IH band, hence all of them are theoretically able to participate in ectopic pairing. The effect of the proximity of chromosome sections on FEC-FMF correlation value was estimated by analyzing the average correlation values within the chromosome parts of different sizes. For fragments specifically making the contribution to FMF, the biological nature was determined using NCBI Blast software. Most of them showed a high percentage of identity with 1.688 and 372-bp repeats, as well as related genes. These repeated sequences were either concentrated in genomic regions predisposed to ectopic pairing or governing pairing themselves, via DNA-DNA complementary binding or indirectly with the help of some unknown protein or RNA factors.

2. Results

2.1. FEC-FMF Correlations for the Whole X Chromosome

The average values of statistically significant FEC-FMF Spearman rho correlations (R) were calculated for all the pairs of sections of the X chromosome (Figure 1). R specific (R_{SP}) varied within 0.18–0.3, corresponding to a rather weak positive FEC-FMF correlation [29]. R_{SP} was slightly higher for *agn*^{ts3} compared to *CS*, especially at a fragment length (L) of 45–50 nt (L45-50), possibly due to a larger FEC in *agn*^{ts3}. Repeats exclusion did not significantly affect R_{SP}, except for its small decrease at L25 in *CS*. For both *CS* and *agn*^{ts3} strains, there were no significant R_{SP} differences from those calculated at L50.

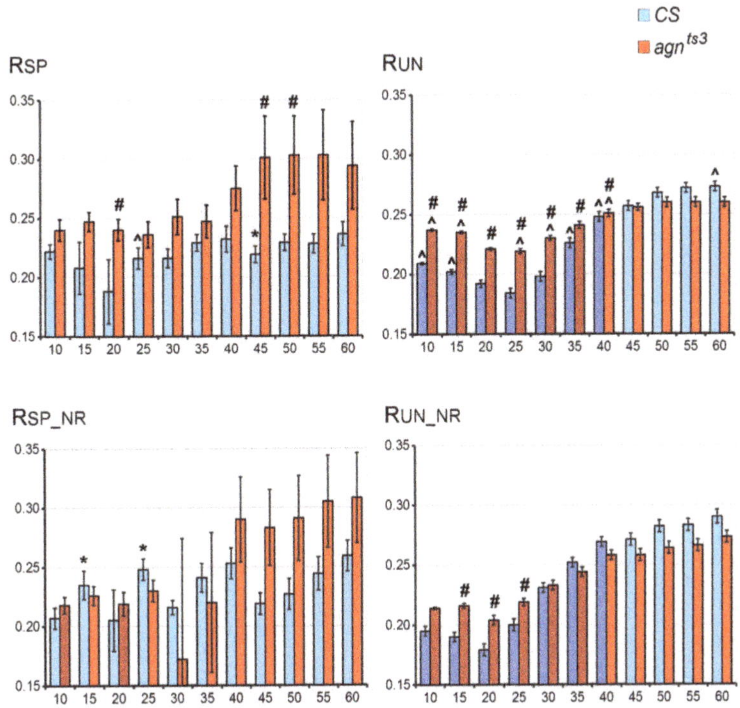

Figure 1. R values at different fragment lengths. *X* axis: L (nt). *Y* axis: R (conventional units). Difference: # from *CS*, ^ from the case with excluded repeats, * from the case with unspecific correlation, shading – difference from R calculated at L50 (two-sided Mann–Whitney U-test; $p < 0.05$). Standard error of mean is shown. Here and below: SP—specific, SP_NR—specific with repeats exclusion, UN—unspecific, UN_NR—unspecific with repeats exclusion. Sampling number: for specific correlations, n = 6–20 (*CS*), 5–28 (agn^{ts3}); for unspecific correlations, n = 568–2235 (*CS*), 667–2987 (agn^{ts3}). Total number of R estimations: for specific correlations, n = 90 (*CS*), 95 (agn^{ts3}); for unspecific correlations, n = 10,620 (*CS*), 11,210 (agn^{ts3}).

To check the correlation specificity, R unspecific (R_{UN}) was calculated as the average value of statistically significant rho correlations for all the different section pairs A and B. Such a shuffle of sections permits estimation of the "false" correlation between the inappropriate FEC and FMF values. R_{UN} was nearly the same as R_{SP}; thus, if rho correlation appears to be statistically significant by chance, its average value does not differ from that of the "true" correlation. R_{UN} was higher for agn^{ts3} compared to *CS* at L10–40; however, there were no interstrain difference at a larger L. For both strains, R_{UN} grew with the length of fragments, reaching maximum values at L50–55. The exclusion of short repeats reduced R_{UN} values for small Ls (10–20). Thus, microsatellites seem to have a significant impact on unspecific correlations. At a larger L (45–60), repeats exclusion did not significantly affect R_{UN}. For both strains, it was mostly impossible to distinguish between R_{SP} and R_{UN}. Thus, the average rho value rather weakly reflected the probability of ectopic pairing.

The picture was different for the proportions (P) of statistically significant correlations (their share of all correlations). P specific (P_{SP}) showed a non-linear variation along with L growth (Figure 2). For agn^{ts3}, the first P_{SP} maximum (about 0.3) was observable at L15 (i.e., 30% of all sections forming the ectopic contacts had a significant FEC-FMF correlation); then, P_{SP} dropped to 0.12 at L30, returning to about 0.15 at L35. Finally, P_{SP} dropped to 0.05 at L50, being nearly equal to the probability of finding an FEC-FMF correlation by chance ($p < 0.05$). This corresponds to the virtually complete absence of section-to-section matching for fragments longer than 60 nt (FMF was zero for the most sections). For *CS*, the whole picture was the same, but P_{SP} changes with L were not significant. Most importantly, we observed a striking interstrain P_{SP} difference at L45–50: P_{SP} value remained rather high for *CS* but not for agn^{ts3}. The fragment length of 50 nt has been previously shown as optimal to detect FEC-FMF correlations for the other *Drosophila* wild-type strain, *Berlin* [24]. Thus, at least in some strains, high FMF (L50) can serve as a predictor of ectopic pairing with a probability of about 20%.

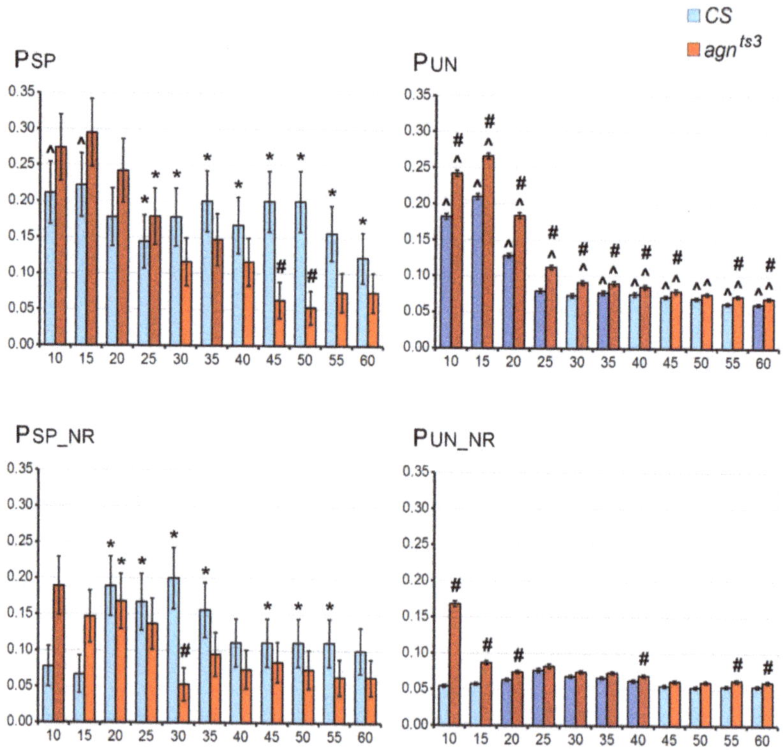

Figure 2. P values at different fragment lengths. *X* axis: L (nt). *Y* axis: P (conventional units). Difference: # from *CS*, ^ from the case with excluded repeats, * from the case with unspecific correlation, shading-difference from P calculated at L50 (Chi-square test; $p < 0.05$). Standard error of sample proportion is shown.

P unspecific (P_{UN}) values showed a decrease along with the L increase, down to 0.05–0.07, which was close the theoretically expected *p* value of 0.05. For *CS*, P_{SP} is significantly higher than P_{UN} at L25–60. For agn^{ts3}, there were no significant differences between P_{SP} and P_{UN}, except at L25. Hence, agn^{ts3} strain clearly demonstrated less FEC-FMF correlation specificity compared to the wild-type strain. Repeats exclusion significantly decreased P_{UN}. This also proves that short simple repeats abundant in *Drosophila* genome make a significant impact into unspecific FMF-FEC correlations. The exclusion of microsatellites decreased the interstrain P_{SP} difference, though P_{SP} remained significantly higher for *CS* at L30. Putatively, short repeats are parts of some longer DNA fragments that specifically have an impact on FEC-FMF correlation, therefore repeat filtration can simultaneously remove these fragments from samplings used to calculate FMF.

The observable P-L dependence seems to be the summary effect of the following trends: (1) Both P_{SP} and P_{UN} are high for short fragments that are randomly distributed within genome. (2) Both P_{SP} and P_{UN} are low for long fragments, but P_{UN} drops more rapidly along with L increase. Exclusion of short repeats lowers the probability of FEC-FMF correlation, so we do not recommend it be performed for large Ls. High FMF calculated at L50 seems to indicate the higher probability of specific ectopic pairing for *CS* X chromosome sections. However, the share of statistically significant correlations remains rather low (about 20%). Thus, ectopic pairing is mostly governed by some other factors not taken into account in the FMF calculated according to the above scheme.

2.2. FEC-FMF Correlation for Chromosomal Regions of the Different Length

Ectopic contacts are mainly formed between sections that are not too far from each other (i.e., are separated by not more than 10–20 sections) [30]. To check the influence of the intersection distance (D) on R and P values, we performed FEC-FMF calculations for sections within chromosome zones of specific length D (for a more detailed description, see Section 4.2, Stage 9).

The average R values for different D are shown in Figure 3. In *CS*, R_{SP}–R_{UN} difference (R_{DIFF}) was about 0.05–0.1 for small fragments (L10) and small D (10–25). However, in most cases it was insignificant because of the high R variance. For larger D, R_{DIFF} was small but statistically significant, save for the highest D values obtained for very small samplings. The highest R_{DIFF} could be observed at L30, with its maximum at D40–45. At L50, there was no significant *CS* R_{DIFF}, save for a few D values. The picture was similar for *agn*^{ts3}, but its R_{DIFF} was smaller at L30 and even became negative at L50 (D < 50). The interstrain difference was small in most cases, but at L30 *CS* had apparently lower R_{UN} compared to the mutant strain. Generally, the highest R_{SP} and R_{DIFF} could be observed for rather small Ds (up to 40), confirming the hypothesis that ectopic contacts mainly occur between spatially close bands.

The exclusion of microsatellites affected the *CS* R_{DIFF}, which approached zero at L30 and increased at L50 with most D values (Supplementary Materials, Figure S1). R_{DIFF} (L50) remained maximal for small D. Hence, the highest FEC-FMF positive correlation can be observed within the short areas of the X chromosome. In *agn*^{ts3}, the area of D with positive R_{DIFF} shrinked at L10 and increased at L50, probably due to the increase in correlation specificity after repeats filtration. However, the range of D with the negative R_{DIFF} increased at L30. Hence, *agn*^{ts3} shows pronounced negative FEC-FMF correlations for relatively small Ds.

For the *CS* P values, we can see an obvious increase in P_{SP}–P_{UN} difference (P_{DIFF}) along with increases in D and L values (Figure 4). At L50, occurrence of specific correlations became significantly higher for almost the whole range of D. The highest values of both P_{SP} and P_{UN} could be observed in *agn*^{ts3} at L10, but P_{DIFF} was relatively small. There was also an increase in *agn*^{ts3} P_{DIFF} along with D growth at L30, but it completely vanished at L50 and even became negative for the mean D values. At L50, the interstrain P_{SP} difference was maximum and P_{UN} was equal for both strains, being close to the theoretically predicted value of 0.05. This confirms results obtained for the X chromosome without division onto D zones. Thus, the most specific FEC-FMF correlation for the whole X chromosome can be observed in *CS* at L50.

The exclusion of microsatellites did not generally change the picture for *CS* P_{DIFF}, the value and significance of which increased along with the growth of D and L (Figure S2). P_{UN} was about 0.05, similar to that for the whole X chromosome (see Figure 2). P_{SP} (L50) was somewhat lower compared to the case without repeats exclusion, probably because of simultaneous filtration of long fragments specifically making impact into FMF. For *agn*^{ts3}, P_{DIFF} was mostly negative at L30. This may indicate some inverse relationship between FEC and FMF for that strain. At L50, the mutant P_{DIFF} was negative for D < 30 and positive for D > 60, though in both cases its values were small. Thus, in contrast to *CS*, *agn*^{ts3} does not show a stable increase in P_{DIFF} along with D and L growth.

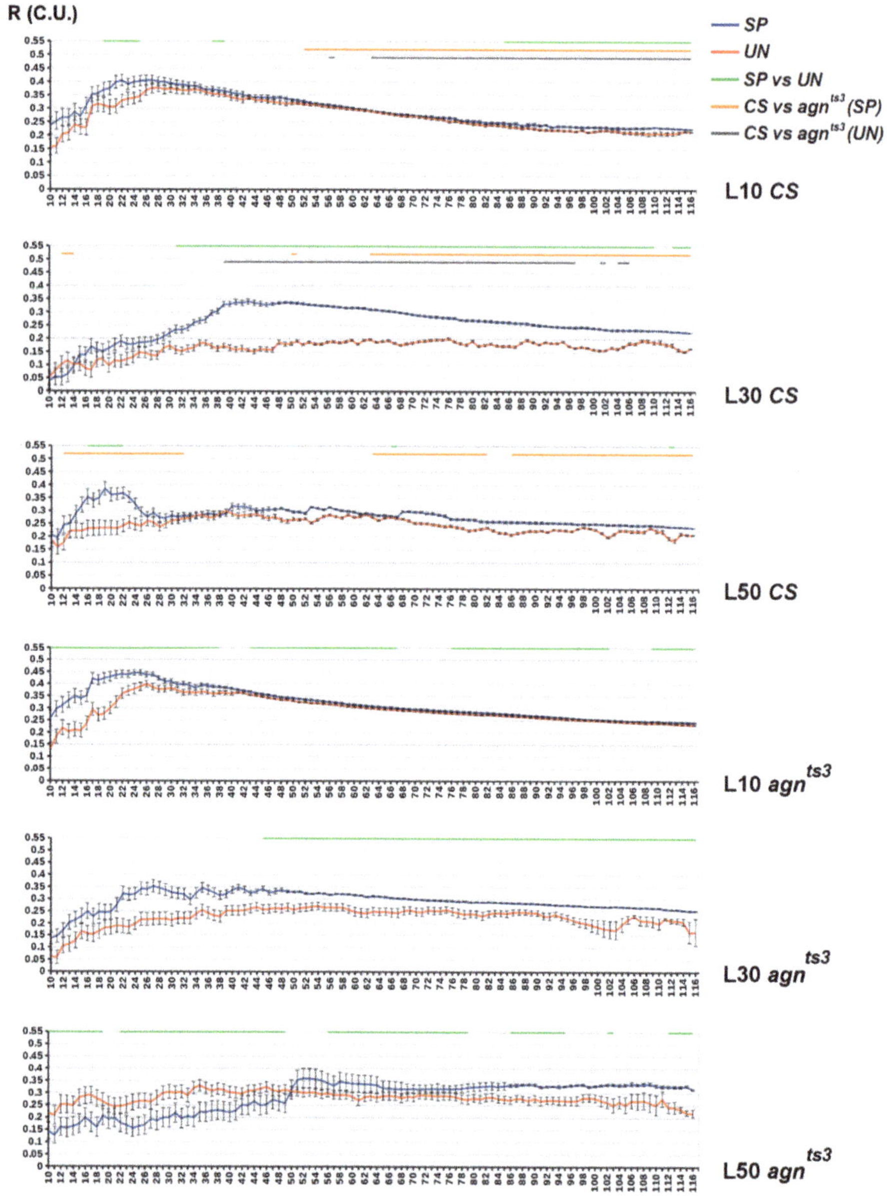

Figure 3. The average R values for the chromosomal zones of different intersection distances (D). X axis: D—zone length (in sections). Y axis: R (conventional units). Standard error of mean is shown. The areas of D with statistical differences are shown by straight lines above the diagram: green—SP vs. UN, orange—SP (*CS*) vs. SP (*agn*^{ts3}), grey—UN (*CS*) vs. UN (*agn*^{ts3}) (two-sided Mann–Whitney U-test; $p < 0.05$, n = 120–D).

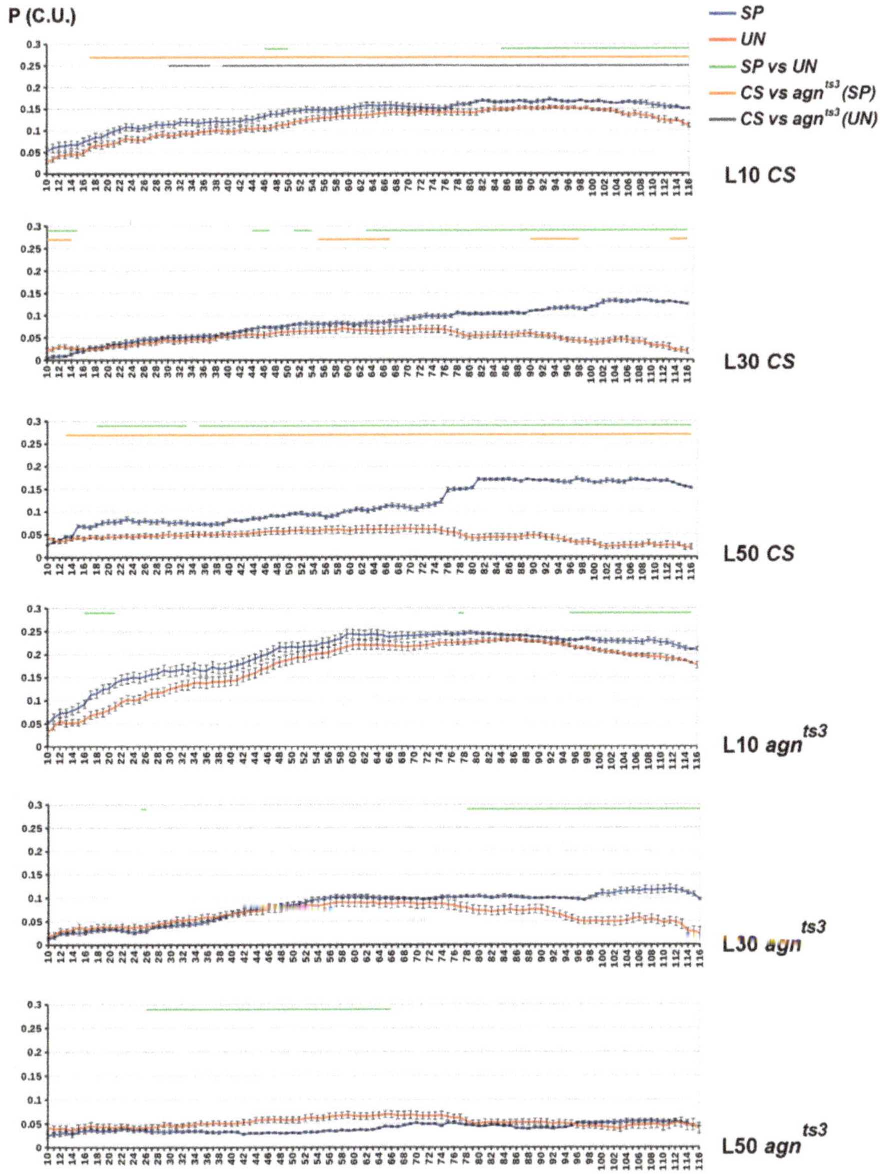

Figure 4. The average P values for the chromosomal zones of different intersection distances (D). *X* axis: D—zone length (in sections). *Y* axis: P (conventional units). Standard error of mean is shown. The areas of D with statistical differences are shown by straight lines above the diagram: green—SP vs. UN, orange—SP (*CS*) vs. SP (*agn*^{ts3}), grey—UN (*CS*) vs. UN (*agn*^{ts3}) (two-sided Mann–Whitney U-test; $p < 0.05$, $n = 120–D$).

2.3. Sections Prone to Ectopic Pairing

Some chromosomal areas are known to be prone to ectopic pairing. The total FEC number (FEC_{TOT}) is a number of ectopic contacts between a given section and all the other sections of the X chromosome. There was a strong positive FEC_{TOT} correlation for *CS* and *agn*^{ts3}: rho = 0.757 ($p < 0.001$, $n = 119$). After replacing the exact FEC with the indices of presence (1) or absence (0) of contacts, the interstrain correlation decreased but remained significant: rho = 0.540 ($p < 0.001$). Thus, in both strains, ectopic contacts are mainly formed by the same sections, and sections with a high FEC are usually localized at the same position (see also [30]).

According to [31], there are five chromatin features (F) increasing the probability of the *Drosophila* X chromosome ectopic pairing, such as *Dm225* and *Dm234b* genes hybridization sites, ectopic conjugation, weak points, late replication and giant palindromes. For each F, we assigned the value of 1 or 0 to sections depending on whether they had or did not have a specific F. Then we calculated the F index (Ind) as the sum of values for each section (F_{SUM}) divided by 10. As expected, there was a positive correlation between FEC_{TOT} and F_{SUM}: rho (*CS*) = 0.357 ($p < 0.001$); rho (agn^{ts3}) = 0.252 ($p < 0.01$). Replacing the exact values with indices of presence (1) or absence (0) made correlation insignificant.

At the same time, we did not observe any correlation between FEC_{TOT} and the total FMF number for each section (FMF_{TOT}) calculated at L30 or L50. FEC-FMF correlation was specifically observed only for a set of the X chromosome sections (Figure 5). Three of them (10B, 11D and 18B) coincided for both *Drosophila* strains. Only the minor part of sections (26.1% for *CS*, and 18.2% for agn^{ts3}) had chromatin features predisposing them to ectopic pairing. Considering a rather moderate FEC_{TOT}-F_{SUM} correlation, we state that the indicated specific chromatin properties can facilitate ectopic pairing, but are not necessary for that. For most sections, ectopic pairing seems to be governed by some other mechanisms possibly related to specific DNA sequences within the interacting areas.

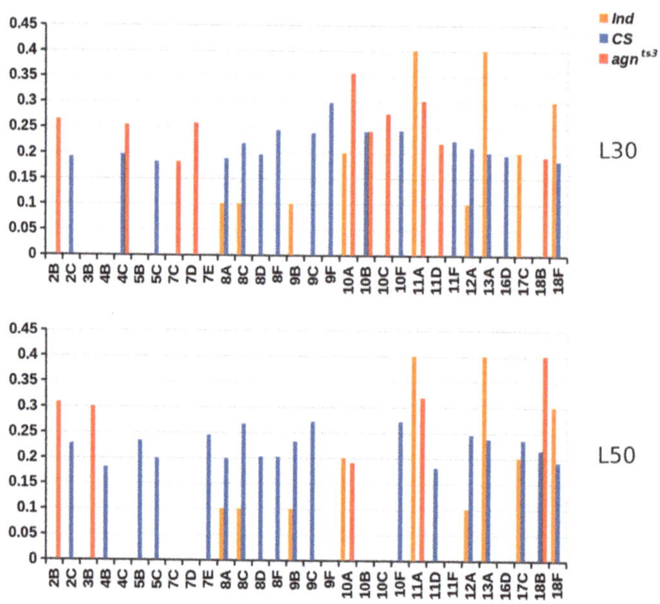

Figure 5. X chromosome sections showing statistically significant FEC-FMF correlations. Ind—the index of chromatin features (F). Y value: R_{CORR} (for *CS* and agn^{ts3}) and Ind values (C.U.).

2.4. *The Biological Nature of Sequences Making Impact into Ectopic Pairing*

To reveal the molecular nature of the fragments most contributing to FMF, we analyzed their sequences using NCBI Blast. The first fifty L30 and L50 fragments with the maximum number of occurrences (NOs) were assayed, both for all sections (set I) and for a set of sections showing *CS*- or agn^{ts3}-specific FEC-FMF correlations (sets II and III, respectively). To simplify the analysis, we did not consider fragments with NOs ≤ 10 and considered only one of the fragments with the equal NO values.

BLAST analysis made it possible to divide all matching fragments into six classes: (1) Microsatellites, such as a_n, t_n, $(at)_n$, $(aat)_n$, $(gata)_n$, $(agata)_n$, $(tcccag)_n$ and so on, the maximum motif length being of six. (2) Fragments showing a high percentage identity (90–100%) with 1.688 and 372-bp repeats, as well as with some genes such as *c11.1*, *kl22 Drak* intron and related. (3) Fragments showing a high percentage identity with sequences of *c11.1* and *kl22 Drak* but not with 1.688 or 372-bp repeats. (4) Fragments of transposon HB1. (5) Fragments of long non-coding RNA genes and retrotransposon roo-900. (6) Other; mostly the sequences with an unknown molecular nature, or those for which BLAST returned no result. Up to 31% of fragments (with NOs < 50) remained non-annotated. The results are present in Figure 6. The full list of fragments, along with their NO, molecular nature and BLAST sequence identity, is shown in Table S1.

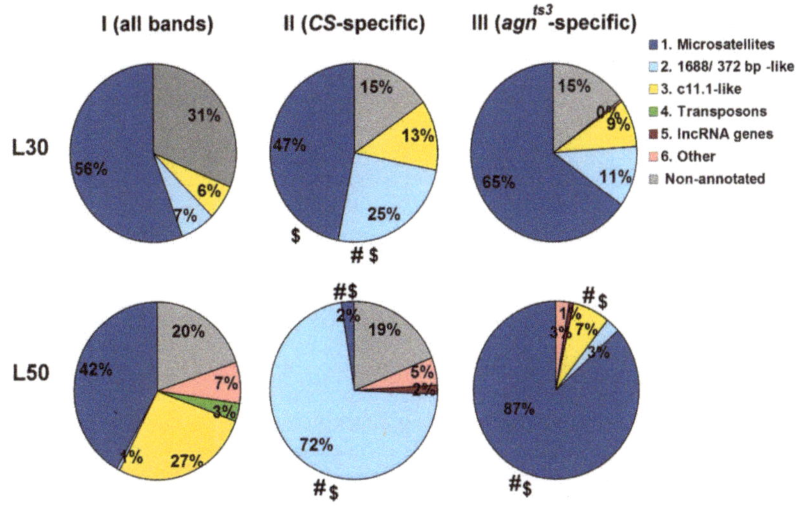

Figure 6. Short DNA fragments that contribute to FMF. Difference: # from I, non-annotated sequences included when calculating the proportions; $ from I, non-annotated sequences excluded (Chi-square test of two proportions; $p < 0.05$, $n > 200$).

In set I, microsatellites constituted the major class of fragments. The above is not surprising, as the program searches fragments one by one, with a step of one nucleotide, generating a large number of identical short sequences from an extended area of repeats. At the same time, the simplest repeats, such as $(at)_n$ and $(ta)_n$, prevailed at L30, but not at L50. The microsatellite $(tcccag)_n$, appeared to have a maximum NO at L50 and a third-rank NO at L30, indicating its abundance in *Drosophila* genome. The NO was also high for fragments belonging to class 3, while class 2 was nearly absent at L50. Transposon parts constituted at least 3% of L50 fragments.

In set II, containing fragments of sections with *CS*-specific FEC-FMF correlations, NOs for microsatellites significantly decreased, especially at L50. At the same time, NOs for 1.688/372-bp-related sequences increased many times (up to 72% at L50). P_{SP} was at its maximum for *CS* at L50 (see Figures 2 and 4). This clearly indicates that microsatellites mostly contribute to unspecific FMF, and 1.688/372-bp-related sequences mostly contribute to FEC-FMF specific correlation. The latter is in agreement with our previous data obtained for *Berlin* [24]. $(at)_n$ was the only microsatellite that seemed to have an impact on *CS*-specific correlation. The fact that $(at)_n$ and complementary sequences constituted 66% of microsatellites and 31% of all *CS*-specific fragments at L30 explains why the NO was high for *CS*-specific sets of selected microsatellite fragments. Sequences of class 3 also contributed to FEC-FMF correlation at L30.

On the contrary, for agn^{ts3}-specific fragments (set III) we observed an increase in microsatellite NOs, which reached their maximum $(at)_n$ at L30 and $(gata)_n$ at L50. For classes 2 and 3, the NOs insignificantly grew at L30, although for class 3 the NO dropped at L50. In agn^{ts3}, P_{SP} was only slightly higher than P_{UN} at L30, with no difference at L50 (see Figures 2 and 4). The positions of sections with specific FEC-FMF correlations were different for *CS* and agn^{ts3} (Figure 5). Taken together, our data show that 1.688/372-bp-related sequences are associated with ectopic pairing in *CS*, with either a minor or no association for agn^{ts3}.

3. Discussion

Ectopic pairing is a long-range interaction that occurs with a relatively low frequency between the IH bands of the *Drosophila* polytene chromosomes. Though ectopic contacts can be easily observed using light microscopy, their molecular nature and functional role in the nuclear 3D organization remain obscure. The agn^{ts3} strain was shown to have a significantly higher FEC compared to the wild-type fly strains. At the same time, some IH bands are more prone to form ectopic contacts compared to the others. In *Drosophila* reciprocal hybrids of *Berlin* and agn^{ts3}, specific bands of the X chromosome demonstrate either matroclinic or patroclinic inheritance of high FEC values [24]. Thus, ectopic pairing obviously has both genetic and epigenetic basis.

As molecular processes of ectopic pairing are not studied in detail, we cannot say a priori how long a zone of local DNA pairing should be to initiate a contact formation. Neither can we say whether g/c rich sequences will interact preferably over a/t rich due to the higher binding energy or if the tendency will be reversed, as a/t regions are often nucleosome free, easy to melt, frequent in genome, and therefore may be more prone to misalign. In our study, we have made a simple assumption that all cases of local DNA match will increase the probability of the ectopic contact formation. The optimal length of the areas should be the tread off the specificity of matching and the probability to find a relatively long specific area in genome. We have revealed the optimal fragment length to be about 50 nt. The ability to consider cases of incomplete matching would let us work with longer fragments, but the current version of Homology Segment Analysis software does not permit us to work with local mismatches. FMF values do not depend on the relative orientation of fragments (e.g., they are the same for the sequential parts of long $(at)_n$ repeats and several short non-overlapping $(at)_n$ repeats). The fragments of all chromosome sections were considered, though only the IH bands participate in ectopic pairing. We believe that the above has a rather small effect on FMF, as the IH bands are densely packed, containing much more DNA than interbands. FEC was zero for most of the section pairs, and the range of FEC variation was rather narrow: in most cases, the FEC equaled 1 or 2 in its Spearman rank correlation. In addition to DNA sequence, epigenetic factors greatly influence FEC values.

As a result, we obtained a rather moderate average FEC-FMF Spearman correlation rho value (R of 0.3), as well as the proportion of section pairs for which the correlation was significant (P of 0.2). Nevertheless, for *CS*, P appeared to be higher than that of unspecific correlations, both for the whole X chromosome and for almost the entire range of the lengths of its parts. Moreover, we found that not all DNA sequences, but mainly those related to 1.688 or 372-bp repeats, contribute to FEC-FMF correlations. This cannot be explained by the predominance of such sequences in the *Drosophila* X chromosome, as, according to our data, the most common matching repeats were $(at)_n$ at L30 and $(tcccag)_n$ at L50 (Table S1). The other microsatellites also contributed to FMF, but only $(at)_n$ seemed to make an impact on the *CS*-specific FEC-FMF correlation. In agn^{ts3}, FEC-FMF correlations are generally unspecific, despite the higher FEC for this strain. Similarly, the X—X:11AB FEC-FMF correlation was lower in agn^{ts3} compared to *Berlin* [24]. At L30, specific correlations still could be observed for about 10% of bands participating in agn^{ts3} ectopic pairing. Microsatellites mostly contributed to agn^{ts3}-specific sets of matching fragments.

Satellites are multi-copy tandem DNA repeats classified according to the length of their monomers: microsatellites (1–10 bp), minisatellites (10–100 bp) and satellites (>100 bp). *Drosophila* satellites are involved in multiple cell processes, such as dosage compensation, heterochromatin establishment, gene activity regulation, maintenance of genomic architecture, chromosome segregation and development [32]. Satellite DNA is one of the most abundant and fast evolving components of genome, being the major part of the constitutive heterochromatin in eukaryotes. There are at least 14 families of highly repeated *D. melanogaster* satellite DNA [33]. The most abundant satellite repeats of the X chromosome are aatat, aagag, 359/372/260-bp, and IGS [34].

The 372-bp satellite is a conservative a/t rich *Drosophila* repeat concentrated on the euchromatin of the X chromosome, located mainly between cytogenetic regions 4 and 15, with about 300–400 copies per haploid genome. Its distribution and sequence features suggest its participation in the primary fly sex determination and dosage compensation [25]. This satellite DNA is homologous to the 1.688 g/mL class of satellites predominantly localized to the centromere heterochromatin of the X chromosome [34,35]. The 1.688-3F satellite expresses a siRNA-generating hairpin dsRNA that increases males survival, regulating the male-specific lethal complex (MSL) positioning on the X chromosome. It is possible that siRNA affects dosage compensation by modifying chromatin at the 1.688 repeat [27]. It should be noted that the *agnostic* locus in agn^{ts3} does not show dosage compensation [36,37]. Along with a lack of FEC-FMF correlation, this may indicate some deregulation in 1.688 or 327-bp activity in agn^{ts3}. The 1.688 sequence shows a significant intraspecific nucleotide divergence: 10% for heterochromatin and 27% for euchromatin. Hence, the chromatin structure seems to influence the rate of 1.688 evolution [26]. In our study, the identity between the matching fragments with strain-specific FEC-FRF correlations and 1.688 satellites from 3C and 10Ep bands [35] was about 100% at L30 and slightly below 100% at L50.

Polytene chromosomes normally have low flexibility, extending across the nucleus in a Rabl orientation, with no stable interaction between loci 1-2 cytological divisions apart [13]. The dysfunction of the two interband-associated proteins, Chromator and JIL-1 kinase, leads to dramatic impairment of polytene chromosome morphology (i.e., band misalignment, curling and numerous ectopic contact formations) [38]. As the pattern of ectopic contacts is determined early in development, it should rather reflect the morphology of diploid embryonic nuclei with long-range interactions (e.g., between Polycomb-repressed domains) [16]. Thus, the functional role of ectopic pairing in chromatin spatial organization is questionable. At the same time, it seems to represent some aspects of nuclear 3D structures typical of the early stages of fly development.

The molecular processes that govern ectopic pairing are still poorly understood. In the *Drosophila* interphase nucleus, blocks of heterochromatin tend to associate with each other. The association does not require similar sequences such as $(aagag)_n$ satellite to be located within the contacting areas. Presumably, their associations are mediated by proteins that recognize the general features of heterochromatin, such as specific histone modifications, repetitiveness, late replication and low activity [39]. Nevertheless, the restricted homology at the IH areas containing DNA breaks may be important for ectopic contact formations. Probably, it results from the joining and ligation of truncated DNA ends between the IH bands. The level of suppressor of under-replication *(SuUR)* gene expression positively correlates with FEC. *SuUR* overexpression enhances the IH under-replication and ectopic pairing only before the third instar larval stage [17].

In accordance with the above, FEC as a phenotypic trait is determined at the embryonic stage: the high temperature (37 °C) applied at that stage increases FEC in *CS* without an effect in agn^{ts3}. This seems to be connected to strain-specific properties of heterochromatin that begins to form at this stage [22]. High FEC values in agn^{ts3} may reflect the increase in strain-specific recombination/reparation activity, as well as the activity of some chromatin proteins, such as the Polycomb group, HP1, Chromator and JIL-1. It is interesting to note that both JIL and agn^{ts3} LIMK1 genes bring the insertion of a mobile S-element that may theoretically cause some interaction between these genes [21,40]. *agn*^{ts3} is also characterized by significant changes in miRNA expression profile compared to the wild-type strains *CS* and *Berlin* [21,24]. The mutant-specific heterochromatin properties may affect the wide profile expressions of genes, including non-coding RNA genes.

On the contrary, miRNAs may affect the expression of heterochromatin proteins and the tendency of the IH bands to form ectopic contacts. The targets of some of miRNAs participating in the development of human neurodegenerative disorders are Swi/Snf-like chromatin remodeling complex, REST factor that recruits histone deacetylases (HDAC), and SIRT1, a NAD$^+$-dependent HDAC involved in heterochromatin formation. miR-124 and miR-34c negatively regulate HDAC1/2 and SIRT1, respectively [41]. *Drosophila* miR-124 and miR-34 are decreased in *agn*^{ts3} compared to *CS* [21]. Hence, the high percentage of heterochromatin in *agn*^{ts3} may be due to the increase in HDAC activity.

agn^{ts3} is shown to impair the activity of LIMK1, the main regulator of actin remodeling [18]. The filamentous and globular actin differently affect chromatin conformation, as well the activity of HDAC [42]. Actin is widely involved in the regulation of genetic apparatus, including transcription machinery [20]. Active forgetting also depends on an LIMK1-dependent signaling cascade [43]. This reveals a possible connection between the molecular processes of heterochromatin formation in the early embryogenesis of fruit flies and the *Drosophila* ability to learn, memorize and forget.

Heterochromatin proteins such as the Polycomb group and HP1 may influence ectopic pairing by bringing the IH bands closer together in space or linking them to the envelope [17]. HP1 promotes the formation of chromosome loops, facilitating the coalescence of dispersed middle repeats such as micropia retrotransposon and non-coding RNA gene $\alpha\gamma$ [44]. H3K9me3 modification of 1.688 satellite creates a binding site for HP1 [45,46]. Hence, FEC-FMF correlation can be theoretically explained by HP1-dependent juxtaposition of the 1.688-containing under-replicated IP bands, followed by ligation of the double-stranded DNA ends. A similar role of 1.688-3C in ectopic pairing of the spermatocyte X chromosome was proposed in [34,47]. Previously, 1.688 was proposed to influence chromatin architecture by interacting with proteins of the nuclear matrix, such as Topoisomerase II and satellite binding protein. The 1.688 satellite may also participate in long-range interactions regulated by siRNA-dependent chromatin modifications [48]. Though 372-bp repeats are localized to euchromatin, they may also play a role in bringing together chromosome sections that form ectopic contacts.

In summary, our computational data confirm the hypothesis that *Drosophila* satellite DNA such as 1.688 and related sequences, can participate in long-range interactions between the IH bands. The lower FEC-FMF correlation for *agn*^{ts3} relative to *CS* may indicate less specificity of ectopic pairing in the mutant strain, similar to Chromator/JIL-1 mutants. Some proteins or non-coding RNAs, possibly produced by 1.688-like repeats, can mediate the interaction. Alternatively, they can affect heterochromatin properties and/or DNA replication within IH bands, making them more prone to pairing. Further studies are necessary to prove this experimentally.

4. Materials and Methods

4.1. FEC Matrices

FEC matrices for the X chromosome sections of *CS* and *agn*^{ts3} strains were taken from [30]. Files containing FECs for all X chromosome sections of both strains were supplied with Homology Segment Analysis software. Each matrix was built on the data obtained by orcein staining of squashed preparations of *Drosophila* 3rd instar larvae. For each strain, 30 larvae were taken, and about 20 nuclei were examined for each larva. FECs were calculated as the total number of ectopic contacts between the given section pair. The example of an ectopic contact is shown in Figure S3.

4.2. Bioinformatics Analysis

FEC-FMF correlations were estimated using Homology Segment Analysis software (Zhuravlev A.V., Zakharov G.A.; Pavlov Institute of Physiology, Saint Petersburg, Russia). The program is written on Python3. The calculations in the paper were performed using the latest version of the program, freely available at [23]. For both strains, we used the *D. melanogaster* X chromosome sequence genome assembly Release 6 (dm6) [49]. As there are no full genomic sequences for these strains, and S-elements are relatively short and can be found at various positions in different *Drosophila* strains, we do not consider it justified to include *limk1* S-element in agn^{ts3} genome sequence to compute its FMF. All interstrain differences are expressed here as FEC differences. The borders of the X sections were chosen according to Flybase data (www.flybase.org; accessed on 20 February 2021): 1A–20F, except 20B missing some nucleotides, totally 119 sections.

To install and run all the scripts, see the Readme file in the main directory.

The computational algorithm is as follows:

1. Sequential selection (with a step of 1 nt) of short DNA fragments of a given length (L) from one specific section of the X chromosome.
2. Search of the section fragments, as well as the complimentary fragments, within all the other sections of the X chromosome, using Aho-Corasick algorithm.
3. For each X chromosome section (B) except A: calculation of normalized frequency of all fragments matching for A and B (FMF(A-B)). The average FMF for all chunks of 10 kb length is equal to 1. The list of the localized fragments is saved for each B.
4. Stages 1–3. are repeated for all X chromosome sections.

For all B except A: FMF(A) is a set of FMF(A-B); FEC(A) is a set of FEC(A-B).

FEC(A-B) is a value obtained from FEC matrix for a given *Drosophila* strain, being the number of contacts between sections A and B.

5. Specific correlation computation:
    a. For each section A: calculation of FEC(A)-FMF(A) Spearman rank correlation coefficient (rho; $p < 0.05$, $n = 119$).
    b. Calculation of the average rho value (R) and proportion of statistically significant FEC-FMF cases (P). P is calculated as follows: $P = n\ (p < 0.05)/\text{total } n$ of estimations for which Spearman correlation data were obtained.
6. Unspecific correlation computation: For all different sections A and B: calculation of FEC(A)-FMF(B) Spearman rank correlation rho and p. R and P values are calculated as in 5b.
7. Stages 1–6. are repeated at different fragment lengths (L) (10–60 nt, with a step of 5 nt).
8. Excluding DNA microsatellites: Stages 1–7 are repeated with fragment samplings excluding fragments that contain DNA repeats. By default, a repeat contains identical elements in a row: four nucleotides or three dinucleotides or two trinucleotides.
9. Estimation of section proximity effect. By default, the distance (D) between the boundary sections varies from 10 to 116, and only fragments of a specific L (10, 30, 50 nt) are used to compute FEC-FMF correlations. The procedure is performed as follows:
    a. For each D: a sequential selection of the X chromosome zones (Z) of D length, with a step of 1 section (e.g., for D = 30, 90 different Z are selected, starting from 1 (Z1, or 1A–5F) and up to 90 (Z90, or 15F–20F)). The section notations A–F are equivalent to 1–6, so Z90 of D30 is also denoted as 156–206.
    b. For each Z(D): specific and unspecific R and P calculation, as described in Stages 5–8, taking into account only the sections within Z. Currently, analysis of unspecific correlations is time consuming, taking up about 3 h for each L. So some cases (e.g., with specific L values or excluded repeats) may be omitted to speed up the processes. For each D, R(D) and P(D) values constitute samplings for further statistical analysis. The sampling size n is equal to the number of Z(D): $n = 120–D$.

The scheme of the Stages 1–9. is also given in Figure S4.

4.3. Statistical Analysis

All analyses were performed using scripts included in the Homology Segment Analysis software package.

a. For results obtained at Stages 5–8.: R are compared using a two-sided Mann–Whitney U-test, P are compared using a Chi-square test for two sample proportion comparisons. The parameters of analysis are automatically varied: strain (CS/agn^{ts3}); type of correlation (specific/unspecific); repeats exclusion ("no"/"yes"); chromosome regions (with/without division into sections); type of analysis (comparison of data obtained for different Ls using the same parameters/comparison of data obtained for the same L using different parameters).
b. For results obtained at Stage 9.: R and P and compared using a two-sided Mann–Whitney U-test. Samplings obtained for different Ds and Ls are analyzed independently. The parameters of analysis are automatically varied: strain (CS/agn^{ts3}); type of correlation (specific/unspecific); repeats exclusion ("no"/"yes").

4.4. BLAST Analysis of Fragments Contributing to FMF

a. For the given L values (here, L30 or L50), the full list of fragments of all sections making contributions to FMF are generated and arranged according to the number of fragment occurrences (NO) > 10, starting from the maximum NO. If the NO is equal for different fragments, only the first fragment is chosen.
b. The same procedure is performed for a set of sections showing statistically significant FEC-FMF correlations for the given strain and L value (see Stage 5.).
c. The biological nature of the first 50 fragments in each list is revealed using NCBI Blast (http://blast.ncbi.nlm.nih.gov; accessed on 20 February 2021): BLASTN, database—Nucleotide collection, species—*Drosophila melanogaster*, max target sequences—100, other parameters—by default.

5. Conclusions

Using our Homologous Segment Analysis software, we have shown a specific positive correlation between FEC and FMF for about 20% of the *CS* X chromosome sections involved in ectopic pairing. Most of the 50 nt fragments specifically contributing to FMF appeared to be related to 372-bp or 1.688 middle repeats. Thus, our bioinformatics approach lets us to handle the problem caused by the low resolution of the method of squashed preparations, which does not give information about the specific sequences involved in ectopic pairing. Using the experimental data on chromatin properties obtained by Hi-C and other 3C-related methods with significantly higher resolutions can substantially increase correlation values and validity. Moreover, Homology Segment Analysis can be easily applied to search correlations between FMF and every feature associated with pairs of genomic regions both in *Drosophila* and in other species. For example, it can be used to find DNA motifs involved in contact formations, as well as the binding of proteins or RNA that mediate such interactions and, thereby, define nuclear 3D organization.

Supplementary Materials: The following are available online at https://www.mdpi.com/article/10.3390/ijms22168713/s1.

Author Contributions: Conceptualization, A.V.Z.; Methodology, A.V.Z.; Software, A.V.Z. and G.A.Z.; Validation, A.V.Z.; Investigation, A.V.Z. and E.V.A.; Resources, E.V.S.-P.; Data Curation, A.V.Z. and E.V.A.; Writing—Original Draft Preparation, A.V.Z., E.V.A., A.V.M., G.A.Z. and E.V.S.-P.; Writing—Review & Editing, A.V.Z., E.A.N., A.V.M., G.A.Z. and E.V.S.-P.; Visualization, A.V.Z. and G.A.Z.; Supervision, E.V.S.-P. and E.A.N.; Project Administration, E.V.S.-P.; Funding Acquisition, E.V.S.-P. All authors have read and agreed to the published version of the manuscript.

Funding: This research was funded by the State Program of the Russian Federation 47 GP "Scientific and Technological Development of the Russian Federation" (2019-2030) (63.1) and the Russian Foundation for Basic Research (N 20-015-00300 A).

Institutional Review Board Statement: Not applicable.

Informed Consent Statement: Not applicable.

Data Availability Statement: The FEC data used in this paper are freely available at: https://drive.google.com/drive/folders/1i_SOIR3cxFN1951akkAMGcOgXwd71NG5?usp=sharing (accessed on 12 August 2021).

Conflicts of Interest: The authors declare no conflict of interest.

References

1. Croft, J.A.; Bridger, J.M.; Boyle, S.; Perry, P.; Teague, P.; Bickmore, W.A. Differences in the Localization and Morphology of Chromosomes in the Human Nucleus. *J. Cell Biol.* **1999**, *145*, 1119–1131. [CrossRef]
2. Ranade, D.; Koul, S.; Thompson, J.; Prasad, K.B.; Sengupta, K. Chromosomal Aneuploidies Induced upon Lamin B2 Depletion Are Mislocalized in the Interphase Nucleus. *Chromosoma* **2017**, *126*, 223–244. [CrossRef] [PubMed]
3. Goetze, S.; Mateos-Langerak, J.; Gierman, H.J.; de Leeuw, W.; Giromus, O.; Indemans, M.H.G.; Koster, J.; Ondrej, V.; Versteeg, R.; van Driel, R. The Three-Dimensional Structure of Human Interphase Chromosomes Is Related to the Transcriptome Map. *Mol. Cell. Biol.* **2007**, *27*, 4475–4487. [CrossRef] [PubMed]
4. Osborne, C.S.; Chakalova, L.; Mitchell, J.A.; Horton, A.; Wood, A.L.; Bolland, D.J.; Corcoran, A.E.; Fraser, P. Myc Dynamically and Preferentially Relocates to a Transcription Factory Occupied by Igh. *PLoS Biol.* **2007**, *5*, e192. [CrossRef] [PubMed]
5. Branco, M.R.; Pombo, A. Intermingling of Chromosome Territories in Interphase Suggests Role in Translocations and Transcription-Dependent Associations. *PLoS Biol.* **2006**, *4*, e138. [CrossRef]
6. Kumaran, R.I.; Thakar, R.; Spector, D.L. Chromatin Dynamics and Gene Positioning. *Cell* **2008**, *132*, 929–934. [CrossRef] [PubMed]
7. Osborne, C.S.; Chakalova, L.; Brown, K.E.; Carter, D.; Horton, A.; Debrand, E.; Goyenechea, B.; Mitchell, J.A.; Lopes, S.; Reik, W.; et al. Active Genes Dynamically Colocalize to Shared Sites of Ongoing Transcription. *Nat. Genet.* **2004**, *36*, 1065–1071. [CrossRef]
8. Zhimulev, I.F.; Belyaeva, E.S.; Vatolina, T.Y.; Demakov, S.A. Banding Patterns in Drosophila Melanogaster Polytene Chromosomes Correlate with DNA-Binding Protein Occupancy. *Bioessays* **2012**, *34*, 498–508. [CrossRef] [PubMed]
9. Zhimulev, I.F. Polytene Chromosomes, Heterochromatin, and Position Effect Variegation. *Adv. Genet.* **1998**, *37*, 1–566. [CrossRef] [PubMed]
10. Kolesnikova, T.D. Banding Pattern of Polytene Chromosomes as a Representation of Universal Principles of Chromatin Organization into Topological Domains. *Biochemistry* **2018**, *83*, 338–349. [CrossRef]
11. Zhimulev, I.F.; Zykova, T.Y.; Goncharov, F.P.; Khoroshko, V.A.; Demakova, O.V.; Semeshin, V.F.; Pokholkova, G.V.; Boldyreva, L.V.; Demidova, D.S.; Babenko, V.N.; et al. Genetic Organization of Interphase Chromosome Bands and Interbands in Drosophila Melanogaster. *PLoS ONE* **2014**, *9*, e101631. [CrossRef]
12. Kolesnikova, T.D.; Goncharov, F.P.; Zhimulev, I.F. Similarity in Replication Timing between Polytene and Diploid Cells Is Associated with the Organization of the Drosophila Genome. *PLoS ONE* **2018**, *13*, e0195207. [CrossRef]
13. Hochstrasser, M.; Mathog, D.; Gruenbaum, Y.; Saumweber, H.; Sedat, J.W. Spatial Organization of Chromosomes in the Salivary Gland Nuclei of Drosophila Melanogaster. *J. Cell Biol.* **1986**, *102*, 112–123. [CrossRef] [PubMed]
14. De Wit, E.; de Laat, W. A Decade of 3C Technologies: Insights into Nuclear Organization. *Genes Dev.* **2012**, *26*, 11–24. [CrossRef] [PubMed]
15. Eagen, K.P.; Hartl, T.A.; Kornberg, R.D. Stable Chromosome Condensation Revealed by Chromosome Conformation Capture. *Cell* **2015**, *163*, 934–946. [CrossRef] [PubMed]
16. Sexton, T.; Yaffe, E.; Kenigsberg, E.; Bantignies, F.; Leblanc, B.; Hoichman, M.; Parrinello, H.; Tanay, A.; Cavalli, G. Three-Dimensional Folding and Functional Organization Principles of the Drosophila Genome. *Cell* **2012**, *148*, 458–472. [CrossRef]
17. Belyaeva, E.S.; Demakov, S.A.; Pokholkova, G.V.; Alekseyenko, A.A.; Kolesnikova, T.D.; Zhimulev, I.F. DNA Underreplication in Intercalary Heterochromatin Regions in Polytene Chromosomes of Drosophila Melanogaster Correlates with the Formation of Partial Chromosomal Aberrations and Ectopic Pairing. *Chromosoma* **2006**, *115*, 355–366. [CrossRef] [PubMed]
18. Nikitina, E.; Medvedeva, A.; Zakharov, G.; Savvateeva-Popova, E. Williams Syndrome as a Model for Elucidation of the Pathway Genes—the Brain—Cognitive Functions: Genetics and Epigenetics. *Acta Nat.* **2014**, *6*, 9–22. [CrossRef]
19. Nikitina, E.; Medvedeva, A.; Zakharov, G.; Savvateeva-Popova, E. The Drosophila Agnostic Locus: Involvement in the Formation of Cognitive Defects in Williams Syndrome. *Acta Nat.* **2014**, *6*, 53–61. [CrossRef]
20. Manetti, F. LIM Kinases Are Attractive Targets with Many Macromolecular Partners and Only a Few Small Molecule Regulators. *Med. Res. Rev.* **2012**, *32*, 968–998. [CrossRef]

21. Savvateeva-Popova, E.V.; Zhuravlev, A.V.; Brázda, V.; Zakharov, G.A.; Kaminskaya, A.N.; Medvedeva, A.V.; Nikitina, E.A.; Tokmatcheva, E.V.; Dolgaya, J.F.; Kulikova, D.A.; et al. Drosophila Model for the Analysis of Genesis of LIM-Kinase 1-Dependent Williams-Beuren Syndrome Cognitive Phenotypes: INDELs, Transposable Elements of the Tc1/Mariner Superfamily and MicroRNAs. *Front. Genet.* **2017**, *8*, 123. [CrossRef]
22. Medvedeva, A.; Molotkov, D.; Nikitina, E.; Popov, A.; Karagodin, D.; Baricheva, E.; Savvateeva-Popova, E. Systemic Regulation of Genetic and Cytogenetic Processes by a Signal Cascade of Actin Remodeling: Locus Agnostic in Drosophila. *Russ. J. Genet.* **2008**, *44*, 669–681. [CrossRef]
23. Zhuravlev, A.; Zakharov, G. Homology Segment Analysis. Available online: https://bitbucket.org/beneor/homology-segment-analysis/src/master/ (accessed on 30 July 2021).
24. Medvedeva, A.; Tokmatcheva, E.; Kaminskaya, A.; Vasileva, S.; Nikitina, E.; Zhuravlev, A.; Zakharov, G.; Zatsepina, O.; Savvateeva-Popova, E. Parent-of-Origin Effects on Nuclear Chromatin Organization and Behavior in Drosophila Model for Williams-Beuren Syndrome. Vavilovskii Zhurnal Genet. *Selektsii* **2021**, in press.
25. Waring, G.L.; Pollack, J.C. Cloning and Characterization of a Dispersed, Multicopy, X Chromosome Sequence in Drosophila Melanogaster. *Proc. Natl. Acad. Sci. USA* **1987**, *84*, 2843–2847. [CrossRef] [PubMed]
26. De Lima, L.G.; Hanlon, S.L.; Gerton, J.L. Origins and Evolutionary Patterns of the 1.688 Satellite DNA Family in Drosophila Phylogeny. *G3 Genes Genomes Genet.* **2020**, *10*, 4129–4146. [CrossRef]
27. Menon, D.U.; Coarfa, C.; Xiao, W.; Gunaratne, P.H.; Meller, V.H. SiRNAs from an X-Linked Satellite Repeat Promote X-Chromosome Recognition in Drosophila Melanogaster. *Proc. Natl. Acad. Sci. USA* **2014**, *111*, 16460–16465. [CrossRef]
28. Rošić;, S.; Köhler, F.; Erhardt, S. Repetitive Centromeric Satellite RNA Is Essential for Kinetochore Formation and Cell Division. *J. Cell Biol.* **2014**, *207*, 335–349. [CrossRef]
29. Akoglu, H. User's guide to correlation coefficients. *Turk. J. Emerg. Med.* **2018**, *18*, 91–93. [CrossRef]
30. Savvateeva-Popova, E.; Peresleni, A.; Scharagina, L.; Medvedeva, A.; Korochkina, S.; Grigorieva, I.; Dyuzhikova, N.; Popov, A.; Baricheva, E.; Karagodin, D.; et al. Architecture of the X Chromosome, Expression of LIM Kinase 1, and Recombination in the Agnostic Mutants of Drosophila: A Model for Human Williams Syndrome. *Russ. J. Genet.* **2004**, *40*, 605–624. [CrossRef]
31. Zhimulev, I.; Semeshin, V.; Kulichkov, V.; Belyaeva, E. Intercalary Heterochromatin in Drosophila. *Chromosoma* **1982**, *87*, 197–228. [CrossRef]
32. Shatskikh, A.S.; Kotov, A.A.; Adashev, V.E.; Bazylev, S.S.; Olenina, L.V. Functional Significance of Satellite DNAs: Insights From Drosophila. *Front. Cell Dev. Biol.* **2020**, *8*, 312. [CrossRef] [PubMed]
33. Lohe, A.R.; Brutlag, D.L. Multiplicity of Satellite DNA Sequences in Drosophila Melanogaster. *Proc. Natl. Acad. Sci. USA* **1986**, *83*, 696–700. [CrossRef] [PubMed]
34. Jagannathan, M.; Warsinger-Pepe, N.; Watase, G.J.; Yamashita, Y.M. Comparative Analysis of Satellite DNA in the Drosophila Melanogaster Species Complex. *G3 Genes Genomes Genet.* **2017**, *7*, 693–704. [CrossRef] [PubMed]
35. DiBartolomeis, S.M.; Tartof, K.D.; Jackson, F.R. A Superfamily of Drosophila Satellite Related (SR) DNA Repeats Restricted to the X Chromosome Euchromatin. *Nucleic Acids Res.* **1992**, *20*, 1113–1116. [CrossRef]
36. Savvateeva-Popova, E.; Peresleny, A.; Scharagina, L.; Tokmacheva, E.; Medvedeva, A.; Kamyshev, N.; Popov, A.; Ozersky, P.; Baricheva, E.; Karagodin, D.; et al. Complex Study of Drosophila Mutants in the Agnostic Locus: A Model for Coupling Chromosomal Architecture and Cognitive Functions. *J. Evol. Biochem. Physiol.* **2002**, *38*, 706–733. [CrossRef]
37. Savvateeva, E. Genetic Control of Second Messenger Systems and Their Role in Learning. Usp. *Sovr. Genet.* **1991**, *17*, 33–99.
38. Rath, U.; Ding, Y.; Deng, H.; Qi, H.; Bao, X.; Zhang, W.; Girton, J.; Johansen, J.; Johansen, K. The Chromodomain Protein, Chromator, Interacts with JIL-1 Kinase and Regulates the Structure of Drosophila Polytene Chromosomes. *J. Cell. Sci.* **2006**, *119*, 2332–2341. [CrossRef]
39. Sage, B.T.; Csink, A.K. Heterochromatic Self-Association, a Determinant of Nuclear Organization, Does Not Require Sequence Homology in Drosophila. *Genetics* **2003**, *165*, 1183–1193. [CrossRef]
40. Nikitina, E.A.; Medvedeva, A.V.; Gerasimenko, M.S.; Pronikov, V.S.; Surma, S.V.; Shchegolev, B.F.; Savvateeva-Popova, E.V. A Weakened Geomagnetic Field: Effects on Genomic Transcriptiln Activity, Learning, and Memory in Drosophila Melanogaster. *Neurosci. Behav. Phys.* **2018**, *48*, 796–803. [CrossRef]
41. Bourassa, M.W.; Ratan, R.R. The Interplay between MicroRNAs and Histone Deacetylases in Neurological Diseases. *Neurochem. Int.* **2014**, *77*, 33–39. [CrossRef]
42. Klages-Mundt, N.L.; Kumar, A.; Zhang, Y.; Kapoor, P.; Shen, X. The Nature of Actin-Family Proteins in Chromatin-Modifying Complexes. *Front. Genet.* **2018**, *9*, 398. [CrossRef]
43. Davis, R.L.; Zhong, Y. The Biology of Forgetting-A Perspective. *Neuron* **2017**, *95*, 490–503. [CrossRef]
44. Seum, C.; Delattre, M.; Spierer, A.; Spierer, P. Ectopic HP1 Promotes Chromosome Loops and Variegated Silencing in Drosophila. *EMBO J.* **2001**, *20*, 812–818. [CrossRef] [PubMed]
45. Lachner, M.; O'Carroll, D.; Rea, S.; Mechtler, K.; Jenuwein, T. Methylation of Histone H3 Lysine 9 Creates a Binding Site for HP1 Proteins. *Nature* **2001**, *410*, 116–120. [CrossRef] [PubMed]
46. Usakin, L.; Abad, J.; Vagin, V.V.; de Pablos, B.; Villasante, A.; Gvozdev, V.A. Transcription of the 1.688 Satellite DNA Family Is under the Control of RNA Interference Machinery in Drosophila Melanogaster Ovaries. *Genetics* **2007**, *176*, 1343–1349. [CrossRef]

47. Tartof, K.D.; Hobbs, C.; Jones, M. A Structural Basis for Variegating Position Effects. *Cell* **1984**, *37*, 869–878. [CrossRef]
48. Menon, D.U.; Meller, V.H. Identification of the Drosophila X Chromosome: The Long and Short of It. *RNA Biol.* **2015**, *12*, 1088–1093. [CrossRef] [PubMed]
49. Zerbino, D.R.; Achuthan, P.; Akanni, W.; Amode, M.R.; Barrell, D.; Bhai, J.; Billis, K.; Cummins, C.; Gall, A.; Girón, C.G.; et al. Ensembl 2018. *Nucleic Acids Res.* **2018**, *46*, D754–D761. [CrossRef] [PubMed]

International Journal of
Molecular Sciences

MDPI

Article

Cellular Computational Logic Using Toehold Switches

Seungdo Choi, Geonhu Lee and Jongmin Kim *

Department of Life Sciences, Pohang University of Science and Technology, 77 Cheongam-ro, Pohang 37673, Gyeongbuk, Korea; choisd@postech.ac.kr (S.C.); kunhu0213@postech.ac.kr (G.L.)
* Correspondence: jongmin.kim@postech.ac.kr; Tel.: +82-54-279-2322

Abstract: The development of computational logic that carries programmable and predictable features is one of the key requirements for next-generation synthetic biological devices. Despite considerable progress, the construction of synthetic biological arithmetic logic units presents numerous challenges. In this paper, utilizing the unique advantages of RNA molecules in building complex logic circuits in the cellular environment, we demonstrate the RNA-only bitwise logical operation of XOR gates and basic arithmetic operations, including a half adder, a half subtractor, and a Feynman gate, in *Escherichia coli*. Specifically, de-novo-designed riboregulators, known as toehold switches, were concatenated to enhance the functionality of an OR gate, and a previously utilized antisense RNA strategy was further optimized to construct orthogonal NIMPLY gates. These optimized synthetic logic gates were able to be seamlessly integrated to achieve final arithmetic operations on small molecule inputs in cells. Toehold-switch-based ribocomputing devices may provide a fundamental basis for synthetic RNA-based arithmetic logic units or higher-order systems in cells.

Keywords: toehold switch; arithmetic operation; RNA–RNA interaction; molecular computing; reversible computing

Citation: Choi, S.; Lee, G.; Kim, J. Cellular Computational Logic Using Toehold Switches. *Int. J. Mol. Sci.* **2022**, *23*, 4265. https://doi.org/10.3390/ijms23084265

Academic Editors: Vaclav Brazda and Richard Bowater

Received: 8 March 2022
Accepted: 10 April 2022
Published: 12 April 2022

1. Introduction

Synthetic biology aims to create technologies for designing and building biological systems with programmable and predictable dynamics [1]. Since the demonstration of synthetic biological circuits in living cells over two decades ago [2,3], considerable progress has been made towards more sophisticated artificial cellular functions, such as feedback oscillation [4], combinatorial logic computation [5–7], and information storage [8,9]. In principle, synthetic circuits can be constructed using any biological molecule as a backbone. Natural and engineered protein regulators can provide the framework to implement logic circuits and computing devices [10,11], including de-novo-designed components [12]. Still, the construction of synthetic biological circuits presents numerous challenges, including the lack of composability [13], limited modularity [14], unpredictable cross-reactivity [15], cellular resource usage [16,17], and unexpected idiosyncratic behavior in real-world applications [18]. Therefore, novel approaches for synthetic biological circuits, including the development of readily characterized, standardized, and modular components are required to overcome the innate difficulties in managing and programming cellular behavior towards large, complex synthetic systems [19].

Nucleic-acid-based genetic devices have made remarkable progress in molecular computing and may provide the required platform for scalable synthetic biological systems. Complex logic circuits and advanced computing systems have been implemented using toehold-mediated strand displacement, including a bistable circuit [20], a square-root circuit [21], neural networks for memory [22] and pattern recognition [23], and an arithmetic logic unit [24]. Furthermore, these molecular computing systems are amenable to computational design and analysis [25,26]. While DNA strand-displacement circuits have been demonstrated in live cells [27] and for live-cell imaging [28], DNA logic gates are generally not suitable for in vivo applications due to the challenges in generating single-stranded

DNAs [29] and maintaining stable gate configurations [30] in cells. In comparison, RNA provides several advantageous characteristics for synthetic biological computing devices. RNA strands can be easily programmed to interact in a designed fashion due to the single-stranded nature of RNA. The co-transcriptional folding of RNA allows the formation of stable secondary structures [31], suitable for natural and synthetic riboswitches [32]. Furthermore, RNA signals can be easily modulated in a tunable manner [33] and can also be amplified with several RNA and protein counterparts [34–38]. Therefore, RNA has been exploited as a platform to engineer gene expression programs that operate robustly in vivo [39–44].

Building on the success of synthetic RNA regulatory parts and inspired by natural RNA regulators [45], several de-novo-designed RNA regulators have been utilized for synthetic biological devices with a large library of well-characterized parts [40,41]. As an example, toehold switches control gene expression in trans via well-established Watson-Crick base pairing of switch and trigger RNA molecules (Figure 1a) [40]. Unlike conventional riboregulators [39], the toehold switches remove nearly all the sequence constraints, exhibit a wide dynamic range, and show excellent programmability with a large library of orthogonal parts. The versatility of toehold switches for synthetic genetic circuit construction is exemplified by the recent developments in cellular logic computation [46], translational repressing riboregulators [47], incoherent feed-forward loop circuits [48], synthetic transcription terminators [49], the protein quality control system [50], modulators of riboswitch circuits [51], and regulators of mammalian cells [52]. Beyond cellular circuits, the toehold switches find use in other platforms, such as cell-free systems [53–57] and paper-based diagnostic devices [58–60] for broader applications.

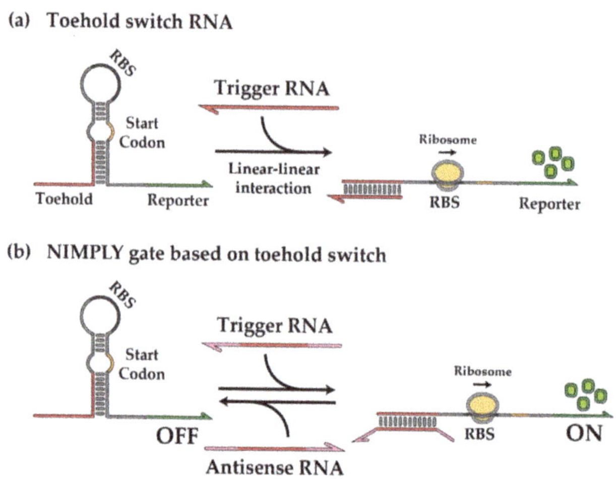

Figure 1. De-novo-designed toehold switch and toehold-switch-based NIMPLY gate. (**a**) Scheme of toehold switch operation. The toehold switch has repressed the translation state through the secondary structure sequestering the RBS and start codon. Linear-linear interaction between the toehold switch and the trigger RNA exposes the RBS and start codon with the strand displacement process; therefore, the translation of the downstream gene is resumed. (**b**) Scheme of NIMPLY gate operation. Antisense RNA has extended overhang sequences at both ends and can inhibit the trigger RNA through sequence displacement or complementary binding. Thus, the toehold switch is reverted to the initial OFF state.

In particular, toehold switches may form the basis for constructing an arithmetic logic unit (ALU) in vivo. A generalized toehold switch architecture, termed ribocomputing devices, concatenated several toehold switch sensor domains and utilized the self-assembly of RNA species to compute multi-input AND/OR/NOT operations [46]. The design flexibility of toehold switches, if effectively utilized, can lead to the streamlined design and construction of a basic form of synthetic biological ALU. Previous work has demonstrated basic ALUs, including a half adder and a half subtractor, in bacteria and mammalian cells [5,61,62]. These binary calculators can perform bitwise calculations across two 1-bit input signals and serve as building blocks for higher-level systems. A half adder takes two bits of information and generates two output bits: one for the sum and one for the carry. The sum bit can be calculated via an XOR gate, and the carry bit can be calculated via an AND gate. Thus, it is straightforward to calculate the carry bit, as previously shown, but a toehold-switch-based XOR gate needs to be engineered. Unlike previous XOR gate implementation, for example, in vitro [63], in prokaryotes [61,62,64,65], and in eukaryotes [66–68], a toehold-switch-based XOR gate is an RNA-only synthetic logic device. Building on previous work in which an antisense RNA that titrates a cognate trigger RNA molecule can be used for a NIMPLY gate operation (Figure 1b), a toehold-switch-based XOR gate with a compact architecture can be obtained by concatenating two orthogonal switches in an OR-gate fashion. A half subtractor can be analogously constructed with an XOR gate and a NIMPLY gate. In addition, a Feynman gate, one of the reversible logic gates that map input and output signals in a one-to-one manner [69], can be obtained using an XOR gate and a BUFFER gate. In summary, we present the binary operation of an XOR gate, cellular arithmetic operations with a half adder and a half subtractor, and a Feynman gate in *E. coli* using a de-novo-designed toehold switch and antisense RNAs. Synthetic RNA-based ALUs could lay the foundation for making sophisticated molecular devices with neural-network-like capabilities for biomedical applications.

2. Results

2.1. XOR Gate

2.1.1. Design of XOR Gate with Toehold Switches

The NIMPLY gate often used in synthetic biology and genetic circuits [61,70] was previously demonstrated using toehold switches [46]. A NIMPLY B is equivalent to A AND (NOT B), and an XOR gate can be constructed using two NIMPLY gates connected via an OR gate as follows: A XOR B = (A NIMPLY B) OR (B NIMPLY A). Thus, we sought to first demonstrate two orthogonal NIMPLY gates. The mechanism for NIMPLY gates is analogous to previous work where the switch RNA is activated by the trigger RNA (A), and the antisense RNA (B) deactivates the trigger RNA via direct hybridization or toehold-mediated strand displacement to separate the trigger RNA bound to the switch RNA. The extended overhang sequences at both ends provide the thermodynamic driving force to shift the equilibrium towards trigger and antisense RNA binding rather than trigger and switch RNA binding.

To implement a NIMPLY gate in *E. coli*, the three circuit components—switch RNA, trigger RNA, and antisense RNA—should be selected from the existing library with appropriate modifications (Figure 2a). We selected a pair of second-generation toehold switches with large dynamic range and strong orthogonality. These two switches are connected with a 9-nt linker sequence to create an OR gate, as previously demonstrated [46]. The overhang sequences of both trigger RNAs and antisense RNAs were designed via the RNA secondary structure prediction software NUPACK [71–75] (Table S1). Fifteen nucleotide overhangs were attached to both ends of the trigger and antisense RNAs, and a single nucleotide bulge was inserted between the overhang region and the switch binding domain to prevent the formation of long double-stranded RNA that could be targeted for degradation by RNase III [76,77]. The design candidates were analyzed for ensemble defect [78], overhang accessibility, and crosstalk in silico to select the best designs to be tested in experiments.

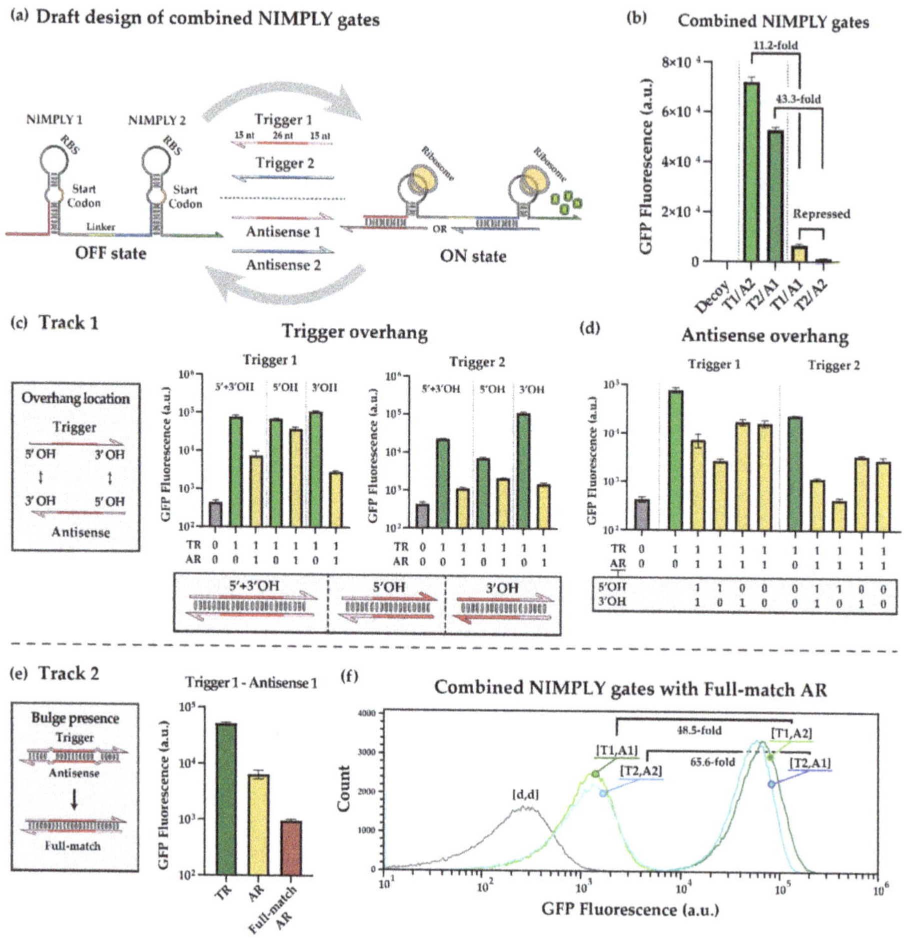

Figure 2. Design of combined NIMPLY gates and optimization strategies. (**a**) Scheme of combined NIMPLY gates composed of two orthogonal NIMPLY gates connected by a 9-nt linker. Trigger and antisense RNAs control the translation states of NIMPLY gates. (**b**) Performance of initial design for combined NIMPLY gates. T1 and T2 denote trigger RNAs that activate NIMPLY gates 1 and 2, respectively, and A1 and A2 denote antisense RNAs that annihilate trigger RNAs 1 and 2, respectively. (**c**,**d**) Effect of the location of extended overhangs on trigger and antisense RNAs. TR and AR indicate trigger RNA and antisense RNA, respectively. Absence of trigger (TR = 0) indicates that only the NIMPLY gate RNA is present. (**e**) Effect of the presence of bulge on antisense RNA. TR and AR indicate trigger RNA and antisense RNA, respectively. Full match means that no bulge was introduced in the antisense RNA design. (**f**) Flow cytometry GFP fluorescence histograms for the NIMPLY complex with full-match antisense RNAs. T1, T2, A1, and A2 indicate trigger and antisense RNAs for switches 1 and 2 as above, and d represents a decoy RNA that does not interact with the switch RNA. T7 RNA polymerase was induced by 1 mM IPTG in *E. coli* BL21 DE3 strain. GFP fluorescence was measured on the flow cytometry (error bars indicate ± SD from three biological replicates). Cellular autofluorescence was subtracted in all cases. Autofluorescence level was measured from cells not bearing a GFP-expressing plasmid.

The NIMPLY gates were tested in *E. coli* BL21 DE3 strain with the switch, trigger, and antisense RNAs expressed from separate low, medium, and high copy plasmids, respectively. The RNA components were under the control of a T7 promoter, and genomically encoded T7 RNA polymerase was induced by Isopropyl β-d-1-thiogalactopyranoside (IPTG). GFP was used to characterize the switch output performance via flow cytometry. First, the OR gate functionality was verified in the absence of antisense RNAs, where the GFP fluorescence was increased at least 100-fold in the presence of either trigger RNAs (Figure S1). Next, the switch, trigger, and antisense RNAs were expressed together in cells to evaluate the function of the NIMPLY gates. The design with the least expected intramolecular and intermolecular structures showed the best performance among the candidates (Figure 2b). The performance was evaluated by dividing the ON state, with a cognate trigger RNA and a non-cognate antisense RNA, by the repressed state, with cognate trigger and antisense RNA pairs. Consequently, we observed increases of 11.2-fold and 43.3-fold for T1-A1 and T2-A2, respectively. Other design candidates with expected secondary structures within trigger RNAs showed relatively poor functionality (Figure S2). Since the design variants on previous NIMPLY gate designs [46] were not extensively characterized, we aimed to further explore and optimize the design choices to enhance the functionality of the NIMPLY gates and hence the performance of the synthetic XOR gate.

2.1.2. Optimization Strategies for Toehold-Switch-Based XOR Gate

For the design variants of the trigger and antisense RNAs within the NIMPLY gate, we mainly adjusted the location of the extended overhangs and the presence of bulge. First, we investigated whether the location of the overhang could affect the functionality of the trigger or antisense RNAs. The GFP fluorescence output for trigger RNA 2 with overhangs showed a stronger reduction than the trigger RNA 2 without the overhang sequences (Figure S3). We hypothesized that the close proximity of RBS within switch 2 and the 5′ overhang of trigger RNA 2 affect the access of RBS through steric hindrance. Therefore, trigger RNA variants with only a 5′ overhang or a 3′ overhang were constructed, and the impact of the overhang location on the switch performance was investigated. Trigger RNAs with only a 5′ extended overhang showed weak repression by antisense RNA for switch 1 and weak activation for switch 2, indicating that the 5′ extended overhang could reduce performance. On the other hand, trigger RNAs with only a 3′ extended overhang showed improved performance for both switches compared to the trigger RNAs with both 5′ and 3′ overhang domains (Figure 2c). The antisense RNAs were similarly modified to test the impact of overhang domains: antisense RNAs with only a 5′ overhang showed improved fold repression, while antisense RNAs with only a 3′ overhang showed weak repression comparable to the antisense RNAs without the extended overhangs (Figure 2d).

Other design candidates were analyzed for the impact of overhang locations on the trigger and antisense RNAs, and a similar trend was observed (Figure S4). Although the 5′ extended overhang can be considered disposable, simply removing the existing 5′ overhang caused crosstalk in some cases because it was not considered during the design phase (Figure S5). Fortunately, the apparent crosstalk was negligible when the 3′ overhang trigger RNAs were paired with antisense RNAs with both overhang sequences (Figure S6). Additionally, an expanded hairpin loop of the switch RNA was explored to help reduce the potential steric hindrance of trigger RNA on the RBS. The increased hairpin size in switch RNA increased the ON level expression but also generally increased the OFF-state leakage (Figure S7). Together, we observed the impact of extended overhang locations on both trigger and antisense RNAs and trade-offs in the switch RNA hairpin loop size on the performance of NIMPLY gates.

To further enhance the functionality of the NIMPLY gates, we investigated the effect of bulges within the trigger–antisense RNA complex on the repression efficiency. The antisense RNA presumably works in one of two ways: (1) dissociating the trigger from the switch or (2) capturing the trigger freely floating in the cytoplasm [46]. Single nucleotide bulges located between the overhang and switch binding domain can act as an energy barrier to the strand displacement pathway that removes the trigger from the switch [79,80]; in that case, the direct hybridization mechanism would be predominant. In order to increase the repression efficiency of antisense RNA, an antisense RNA with extended overhangs but without bulges was designed and tested. The repression efficiency was enhanced nearly 10-fold on trigger 1 (Figure 2e), and the combined NIMPLY gates with optimization exhibited 48.5-fold and 65.6-fold improvements, respectively (Figure 2f). Therefore, we successfully constructed two orthogonal NIMPLY gates with large dynamic ranges using optimization strategies on switch, trigger, and antisense RNA designs. These may serve as useful strategies for other toehold-switch-based logic circuit designs and potentially for other RNA regulatory devices as well.

2.2. *In Vivo Characterization of XOR Gate*

Encouraged by the optimized NIMPLY gates, we then aimed to construct an XOR gate with two chemical inducers as inputs: IPTG and anhydrotetracycline (aTc). An XOR gate provides a true output with an odd number of true inputs (Figure 3a). In the case of the chemically inducible XOR gate that we aimed to construct, the GFP output should be high when either IPTG or aTc is present, but not both. To achieve this, Lac and Tet operators were strategically placed downstream of T7 promoters that drive the expression of trigger and antisense RNAs, such that the trigger RNA of one NIMPLY gate and the antisense RNA of the other NIMPLY gate are simultaneously induced by the same chemical inducer for both inducers (Figure 3b). Specifically, an IPTG induction promotes the expression of trigger RNA 1 and antisense RNA 2, such that the output GFP expression is high. The process works similarly for aTc induction. However, the simultaneous treatment of both inducers results in the expression of both trigger RNAs as well as both antisense RNAs, such that the GFP expression is inefficient. While the Lac and Tet operator sequences are also expressed upon the expression of the trigger and antisense RNAs, the expected secondary structure changes on the core signaling parts of the trigger and antisense RNAs were not noticeable (Figure S8).

At the molecular level, the optimized NIMPLY gates previously characterized were used along with the overhang deletion and bulge deletion strategies. The switch RNA that combines two orthogonal switches in an OR-gate fashion is expressed from a low copy plasmid. To facilitate strong repression by the antisense RNAs, both the trigger RNAs were expressed from a medium copy plasmid, and both the antisense RNAs were expressed from a high copy plasmid. The performance of the XOR gate was evaluated in *E. coli* BL21 AI strain, where genomically encoded T7 RNA polymerase was induced by arabinose. A number of basic molecular interactions were verified for the XOR gate: the crosstalk between the switch and antisense RNAs was negligible (Figure S9); both trigger RNAs, despite the attached additional operator sequences, could turn on the switch RNAs (Figure S10); the antisense RNAs could annihilate the cognate trigger RNA activities, as expected, with little crosstalk (Figure S11). When all the components were put together and the chemical inducers were used, the XOR gate functioned as expected, with a high ON state for either IPTG or aTc input but with a low OFF state for no inducer or both inducer cases (Figure 3c). An XOR gate using trigger RNAs with both 5′ and 3′ overhangs was also shown to be functional, albeit with a reduced ON state for trigger 2 (Figure S12). Furthermore, the GFP output pattern changed sharply as the concentration of inducers was adjusted, indicating that the XOR gate showed a switch-link function suitable for digital circuits (Figure 3d). When incorporated within larger logic circuits, this digital logic ensures an all-or-none response to a variety of inputs and provides a robust output signal regardless of input perturbations [81], thus conveying information with less noise and high accuracy for decision-making processes [82,83].

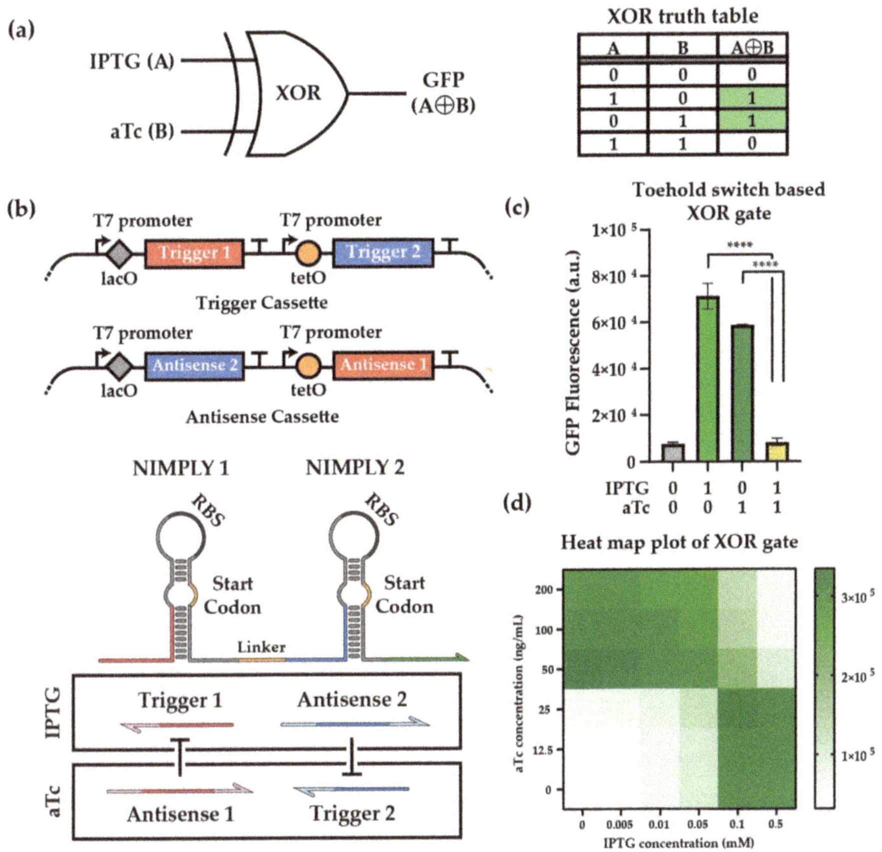

Figure 3. Toehold-switch-based XOR gate. (**a**) Schematics of XOR gate configuration. IPTG and aTc were used as input signals A and B for the XOR gate, and the output signal was visualized through GFP fluorescence. The truth table of the XOR gate indicated the ON and OFF states of the XOR gate depending on the combination of inducer molecules. (**b**) Genetic blueprint of trigger and antisense cassettes and schematics of the XOR gate. Lac operator was placed upstream of T1 and A2, and Tet operator was placed in front of T2 and A1. The optimized extended overhangs were used. (**c**) Performance of toehold-switch-based XOR gate. T7 RNA polymerase was induced with the pretreatment of 1% (*w*/*w*) arabinose in *E. coli* BL21 AI strain. XOR gate components were induced by 0.5 mM IPTG and 100 ng/mL aTc. GFP fluorescence was measured via flow cytometry. Cellular autofluorescence was subtracted in all cases. Autofluorescence level was measured from cells not bearing a GFP-expressing plasmid. Statistical analysis was performed to compare each state of the XOR gate. (Two-tailed Student's *t*-test; **** $p < 0.0001$; Error bars indicate ± SD from three biological replicates) (**d**) Heat map plot of XOR gate output. The color scale was ranged between the minimum and maximum values of the XOR gate output. IPTG and aTc were treated in gradient concentration as described in the table. Each point of the heat map indicates the median value of three replicates.

2.2.1. Cellular Arithmetic Operation of a Half Adder and a Half Subtractor

Building on the RNA devices that were modularized and rigorously characterized earlier, the logical complexity of synthetic RNA circuits can be further increased. As a test case, we focused on basic binary calculators: the half adder and the half subtractor. A half adder takes two input bits and generates two output bits that require an XOR gate for SUM output and an AND gate for CARRY output (Figure 4a). A half subtractor can be analogously constructed, where an XOR gate computes the DIFFERENCE output and a NIMPLY gate computes the BORROW output. Fortunately, a high-performance AND gate was available from the toehold switch library, and another orthogonal NIMPLY gate was constructed with available toehold switches after NUPACK analysis. The functionality of the AND gate and the NIMPLY gate were verified in isolation (Figure S13). Then, these new genetic elements were incorporated into expression cassettes in the same plasmid backbones as before. The XOR gate with GFP output was used to compute the SUM and DIFFERENCE output bits, and the newly introduced AND gate and NIMPLY gate with mCherry output were used to compute the CARRY and BORROW output bits in the half adder and half subtractor, respectively (Figure 4b). To investigate the performance of the binary calculators at the single-cell level, we characterized the system by flow cytometry. For all input combinations, the half adder and half subtractor showed correct ON and OFF states with statistically significant differences (Figures 4c and S14). Nevertheless, as the genetic complexity and the number of heterologous expression cassettes increased, a concomitant decrease in circuit performance was observed when compared to the single XOR gate. Therefore, we checked all combinations of RNA–RNA interactions with NUPACK 4.0.0.25 and confirmed that on-target MFE structures were maintained, albeit with some unintended crosstalk interactions (Table S2). Further improvements in circuit elements, as well as contexts such as promoter arrangements and spacer sequences, may allow for the successful implementation of even more complex circuits such as a full adder.

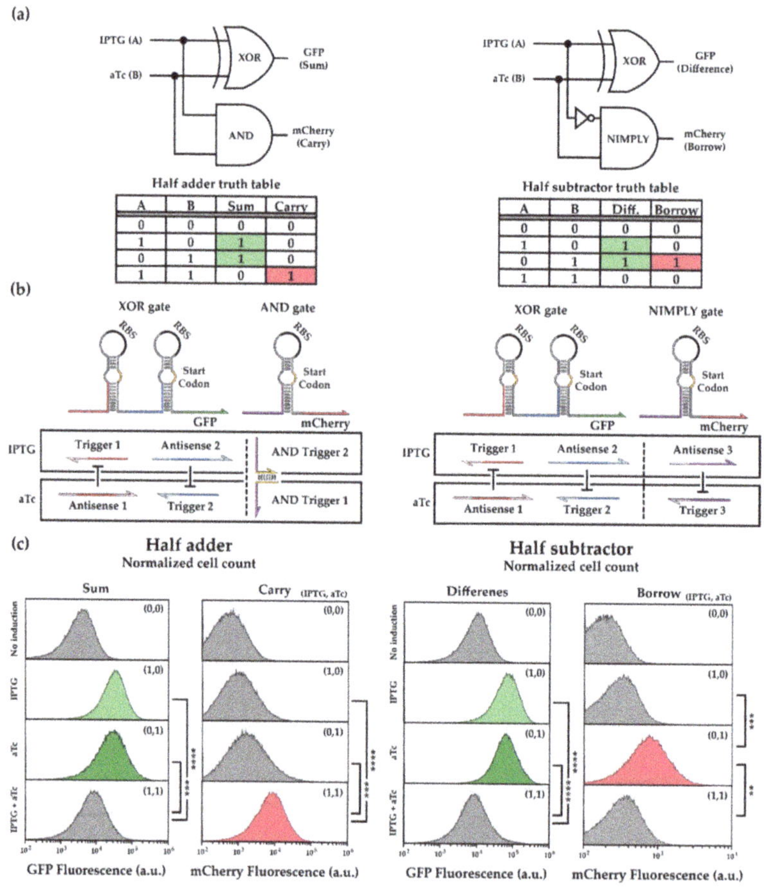

Figure 4. Binary operation of half adder and half subtractor. (**a**) Schematic of half adder and half

subtractor configurations. IPTG and aTc were used as input signals, and the output signals were visualized through GFP and mCherry fluorescence. GFP was assigned to the XOR gate output, and mCherry was assigned to AND or NIMPLY gate output. The truth table of the binary calculators indicated the ON and OFF states of each binary calculator depending on the combination of inducer molecules. Diff. denotes the Difference bit of the half subtractor. (**b**) Schematic of toehold-switch-based half adder and half subtractor. Trigger and antisense RNAs under the same inducer control are shown in boxes. (**c**) Flow cytometry GFP and mCherry fluorescence histograms for the half adder and the half subtractor. The presence of IPTG and aTc was displayed within each panel in brackets. T7 RNA polymerase was induced with the pretreatment of 1% (w/w) arabinose in *E. coli* BL21 AI strain. RNAs of the half adder and the half subtractor were induced by 1 mM IPTG and 200 ng/mL aTc. Statistical analysis was performed for comparing each state of the binary calculators. (Two-tailed Student's *t*-test; ** $p < 0.01$; *** $p < 0.001$; **** $p < 0.0001$).

2.2.2. Cellular Reversible Logic Operation of Feynman Gate

Reversible computing, a nonconventional form of computing with one-to-one mapping of inputs and outputs, may be useful for biomolecular diagnostic and sensing applications. One of the reversible computing devices, a Feynman gate, can also be constructed using a method similar to other binary calculators, using an XOR gate and a BUFFER gate. Due to the unique output patterns, it is also called as a controlled NOT gate when the output signal (Q) changes from BUFFER gate to NOT gate in response to the input signal (Figure 5a). At the molecular level, we executed the same set of sequences as those for the half subtractor, except that a trigger 3 without extended overhangs was used. The RNA-based Feynman gate possessed both a functional switching ability and the capacity for information storage (Figure 5b and Figure S15). The circuit acted as a BUFFER gate for input B in the absence of input A, whereas the circuit functioned as a NOT gate for input B in the presence of input A. Furthermore, information about the input combinations was preserved in the Feynman gate because of the one-to-one manner of input to output mapping. Overall, we demonstrated that toehold-switch-based ribocomputing designs may prove useful for reversible computing in cells.

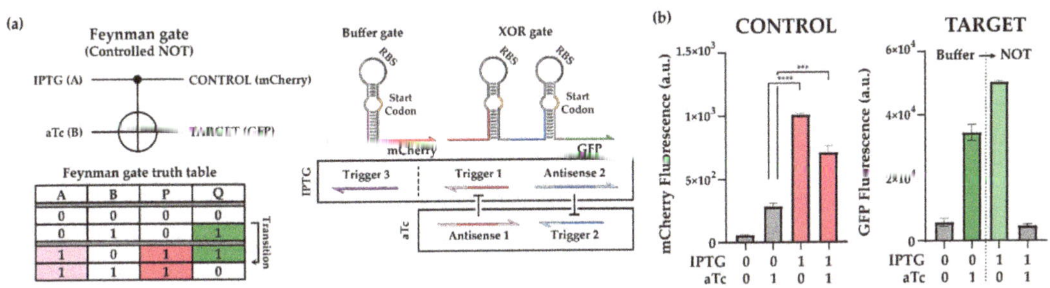

Figure 5. Demonstration of toehold-switch-based Feynman gate. (**a**) Schematic of the Feynman gate. Truth table of the Feynman gate indicated the ON and OFF states for each of the binary calculators depending on the combination of inducer molecules. Transition denotes the functional transition of a Buffer gate to a NOT gate. IPTG and aTc were used as input signals. Trigger and antisense RNAs under the same inducer control are shown in boxes. (**b**) Performance of Feynman gate. T7 RNA polymerase was induced with the pretreatment of 1% (w/w) arabinose in *E. coli* BL21 AI strain. RNAs of Feynman gate were induced by 1 mM IPTG and 200 ng/mL aTc. GFP and mCherry fluorescence were measured on flow cytometry. Cellular autofluorescence was subtracted in all cases. Autofluorescence level was measured from cells not bearing a GFP- or mCherry-expressing plasmid. Statistical analysis was performed for comparing each state of the Feynman gate. (Two-tailed Student's *t*-test; *** $p < 0.001$; **** $p < 0.0001$; Error bars indicate ± SD from three biological replicates).

3. Discussion

In this study, we present the binary operation of XOR gates, cellular arithmetic operation including a half adder and a half subtractor, and a Feynman gate in single-cell *E. coli* using de-novo-designed toehold switches and antisense RNAs. A systematic approach was taken where the basic building block, a NIMPLY gate, was optimized and then subsequently used for an XOR gate, which in turn could be used for basic ALUs. While the NIMPLY gate design was previously demonstrated [46], it was simply used as a proxy for a NOT gate, and further optimization of its performance was limited. Hence, we investigated several design candidates using a number of parameters, including the ensemble defect, the overhang domain accessibility, and the cross-reactivity in silico. The design candidates with accessible overhang domains generally showed better performance (Figures 2b and S2). Crucially, several optimization strategies can improve the performance of NIMPLY gate designs. A thorough analysis of the impact of extended overhang sequences revealed that a 3′ extended overhang on the trigger RNA and a 5′ extended overhang on the antisense RNA improved performance compared to having overhangs on the other location or on both sides. We reasoned that the negative effect of a 5′ extended overhang on the trigger RNA might be due to the potential interference on the ribosome binding to the RBS of the switch. One piece of evidence in support of this hypothesis is that the trigger with a 5′ extended overhang showed improved functionality for toehold switches with increased loop length. Although the mechanistic reasoning on the impact of overhang locations on the antisense RNA is unclear, there still may be physical interference during the initiation stage of trigger and antisense RNA interactions. Recognizing that the single nucleotide bulges located between the overhang and switch binding domains can act as an energy barrier to strand displacement [79,80], we tested the antisense RNA with no bulges and observed improved repression efficiency. These optimization strategies laid the foundation for constructing more complex systems building on the NIMPLY gate designs.

Notably, the antisense RNA designs can be extended to other related synthetic RNA regulators. As an example, a recently reported 3-way junction (3WJ) repressor [47] can be analogously regulated using the antisense RNA design for trigger RNAs (Figure S16). The output characteristics can be considered as an implementation of an IMPLY gate (Figure S17). If applied to the previously reported NOR gate constructed using the 3WJ repressor, an XNOR gate could be constructed similar to the toehold-switch-based XOR gate reported here (Figure S17). Recognizing that NAND gate outputs are distinct from those of XOR gates in the no input case, a NAND gate can be constructed from the current XOR gate by changing the inducible promoters of the trigger RNAs to constitutive promoters (Figure S17). Another important class of de-novo-designed RNA regulator, the small transcription activating RNA (STAR) [41,42], was also subject to antisense RNA-based regulation (Figure S18). Both the T181- and AD1-based STAR designs were successfully regulated using antisense RNA that targets the toehold-binding domain and several bases within the stem-binding domain of the STAR trigger RNAs. Together, these findings indicate that the antisense RNA regulators can be adapted in a straightforward manner to other synthetic RNA regulators and can potentially be used to scale up the complexity of synthetic RNA-based regulatory circuits.

The successful demonstration of a synthetic XOR gate can be seen as a benchmark for systematic synthetic gene circuit construction. Previously, several lines of work demonstrated RNA-based XOR gates, including sRNA [62], miRNA [67], and gRNA [68] that encompass bacterial cells as well as mammalian cell lines. Still, the repression mechanism within the XOR gates relied on protein regulators such as phage-encoded λ repressor protein (CI) [62]. Thus, our demonstration of an RNA-only XOR gate provided a distinct design approach for synthetic XOR gates with performance comparable to the previous work in bacterial cells [62]. More importantly, these RNA-only logic gates can be seamlessly combined for basic ALUs, a half adder and a half subtractor, with performance rivaling previous work [61,62]. Despite thorough in silico analysis and screening for optimized system composition, the performance of basic ALUs showed fold changes less than those of individual gates. There are a number of potential pitfalls in the circuit construction, including the leaky expression of promoters, the unexpected interaction between components, and the cellular burden on synthetic RNA expression. Fortunately, these shortcomings can be mitigated with alternative tightly controlled promoters, such as AraBAD or rhaBAD [84–86] or novel synthetic promoters [87], and by the division of load to different cell populations with multicellular networks [88–90].

Herein, we provided a framework for constructing several synthetic RNA-only basic ALUs with de-novo-designed toehold switches at the single-cell level. This design paradigm offers excellent programmability with simple structural specifications rather than sequence constraints. First, the concatenation of switch RNAs can effectively reduce the encoding space of genetic programs and, therefore, enable the operation of complex systems in *E. coli*. Second, the ALUs can be designed with almost no sequence constraints with in silico screening and optimization. Third, a large library of orthogonal toehold switches provides the required parts for building complex systems. Lastly, the system inherits the general advantages of RNA-based operations, including a fast response time, reduced resource usage, and multiplexing [46,91,92]. Recent developments on degradation-tunable RNAs in combination with toehold switches may provide further design flexibility [93]. Notably, a variety of ALU circuits using DNA strand displacement reactions [24] showcases the power of nucleic-acid-based molecular computations. Still, the demonstrations of ALUs in living cells are limited in complexity and scope. The toehold-switch-based ribocomputing circuits could open a new avenue to exploring the rich design space of synthetic RNA-based ALUs, building up to higher-order systems such as a full adder and a full subtractor, ultimately leading to neural-network-like functions in cells.

4. Materials and Methods

4.1. *E. coli* Strains and Plasmid Construction

The following *E. coli* strains were used in this study: BL21 DE3 (Invitrogen; F^- *ompT hsdS*$_B$ (r_B^- m_B^-) *gal dcm*), BL21 AI (Invitrogen; F^- *ompT hsdS*$_B$ (r_B^- m_B^-) *gal dcm araB*::T7RNAP-*tetA*), and DH5α (Invitrogen; *endA1 recA1 gyrA96 thi-1 glnV44 relA1 hsdR17*(r_K^- m_K^+) λ^-).

The backbones for the plasmids used in this research were taken from the commercial vectors pET15b, pCDFDuet, pCOLADuet, and pACYCDuet (EMD Millipore). The switch RNA of the NIMPLY complex was constructed using ACTS Type II N3 and ACTS Type II N7 from previous research [46] and was constructed in pACYCDuet. All the trigger RNAs and trigger cassettes were constructed in pCDFDuet. All the antisense RNAs and antisense cassettes were constructed in pET15b. The switch RNAs of the AND gate and the NIMPLY gate were constructed in pCOLADuet. All constructs were cloned via blunt end ligation [94], Gibson Assembly [95], circular polymerase extension cloning (CPEC) [96], and/or round-the-horn site-directed mutagenesis [97]. The plasmid architecture and specific part sequences are listed in Tables S3–S11. Plasmids were constructed in *E. coli* DH5α and purified using the EZ-PureTM plasmid Prep Kit. Ver. 2 (Enzynomics). Plasmid sequences were confirmed by DNA sequencing after every cloning step. Plasmids were transformed through chemical transformation [98].

4.2. Cell Culture and Induction Condition

For in vivo experiments, *E. coli* BL21 DE3 and AI strains were used; they contain chromosomally integrated T7 RNA polymerase under the control of IPTG-inducible lacUV5 promoter and arabinose-inducible P_{BAD} promoter, respectively. For the in vivo characterization of the NIMPLY complex in Figure 2, chemically transformed *E. coli* BL21 DE3 cells were cultured on LB agar plates (1.5% agar) with appropriate antibiotics: pACYCDuet (25 μg/mL Chloramphenicol), pCDFDuet (50 μg/mL Spectinomycin), and pET15b (100 μg/mL Ampicillin). Single colonies were grown overnight (~16 h) in 96-well plates with shaking at 800 rpm, 37 °C. Overnight cultures were diluted 1/100-fold into fresh media and returned to shaking (800 rpm, 37 °C). After 80 min, cell cultures were induced with 1 mM IPTG (Promega) and returned to the shaker (800 rpm, 37 °C) until fluorescence measurement after 3 h 30 min. For the experiments on the toehold-switch-based XOR gates, a half adder, a half subtractor, and a Feynman gate, chemically transformed *E. coli* BL21 AI cells were cultured on LB agar plates (BD biosciences; 1.5% agar) with appropriate antibiotics. All antibiotics were purchased from Gold biotechnology: pACYCDuet (25 µg/mL Chloramphenicol), pCOLADuet (50 µg/mL Kanamycin), pCDFDuet (50 µg/mL Spectinomycin), and pET15b (100 µg/mL Ampicillin). Single colonies were grown overnight (~16 h) in 96-well plates with shaking at 800 rpm, 37 °C. Overnight cultures were diluted 1/100-fold into fresh media and returned to shaking (800 rpm, 37 °C). After 80 min, cell cultures were induced with 1% (*w*/*w*) arabinose (Gold biotechnology) to produce T7 RNA polymerase for 1 h with shaking (800 rpm, 37 °C). Then the cell cultures were induced with 0.5 mM IPTG, 100 ng/mL aTc (Takara) for the XOR gates and 1 mM IPTG, 200 ng/mL aTc for a half adder, a half subtractor, and a Feynman gate, and returned to the shaker (800 rpm, 37 °C) until fluorescence measurement after 3 h 30 min.

4.3. Microplate Reader Analysis

For the experiment on the XOR gate with a gradient concentration of chemical inducers (Figure 3d), 200 µL of cell cultures were added per well on a 96-well Black Plate 33,396 (SPL) after 1 mM IPTG induction. GFP fluorescence (excitation: 479 nm, emission: 520 nm), mCherry fluorescence (excitation: 587 nm, emission: 610 nm), and OD600 were measured in a Synergy H1 microplate reader (Biotek) running Gen5 3.08 software. GFP and mCherry fluorescence levels were normalized as follows: the fluorescence of LB blank was subtracted for background normalization, and a measured fluorescence value was divided by its OD600. The number of biological replicates was three for in vivo experiments.

4.4. Fluorescence Measurements Using Flow Cytometry

GFP fluorescence was measured using flow cytometry (CytoFLEX S, Beckman Coulter, Brea, CA, USA) after fixation. The cell pellet was resuspended with 2% (*w/v*) paraformaldehyde solution (Sigma Aldrich) and fixed for 15 min at room temperature. After fixation, samples were washed twice using 1X phosphate-buffered saline (PBS; Enzynomics). Fixed cells were diluted by a factor of ~5 into 1X PBS. Cells were detected using a forward scatter (FSC) trigger, and at least 100,000 events were recorded for each measurement. Cell population was gated according to the FSC and side scatter (SSC) distributions as described previously [40]. To evaluate the circuit output, the fluorescence of GFPmut3b-ASV was measured on a FITC channel, excited with a 488-nm and detected with a 525/40-nm bandpass filter. The fluorescence of mCherry was measured on ECD/mCherry channel, excited with a 561-nm and detected with a 610/20-nm bandpass filter. GFP and mCherry fluorescence histograms yielded unimodal population distributions, and the geometric mean was employed for the average fluorescence across the approximately log-normal fluorescence distribution from three biological replicates. The fold repression of GFP and mCherry were then calculated by taking the average fluorescence from the cognate RNA-expressing case and dividing it by the fluorescence from the antisense RNA-expressing case. Cellular autofluorescence was subtracted in all cases.

5. Conclusions

Expanding the pool of programmable and predictable logic gates is one of the important goals of synthetic biology. Here, we aimed to demonstrate several RNA-only logic gates using toehold switches and antisense RNAs. RNA-only XOR gates, serving as the basic building blocks of arithmetic logic circuits, were constructed using orthogonal NIMPLY gates. Subsequently, the optimized synthetic logic gates were incorporated into arithmetic operations and reversible logic gates via a bottom-up approach in single-cell *E. coli*. In conclusion, toehold-switch-based ribocomputing devices can provide a platform for synthetic RNA-based higher-order circuits in cells.

Supplementary Materials: The following are available online at https://www.mdpi.com/article/10.3390/ijms23084265/s1.

Author Contributions: Conceptualization: S.C. and J.K.; Methodology: S.C. and G.L.; Formal analysis: S.C.; Data curation: S.C. and G.L.; Supervision: J.K.; Writing—original draft preparation: S.C. and G.L.; Writing—review and editing: J.K. All authors have read and agreed to the published version of the manuscript.

Funding: This research was supported by the National Research Foundation of Korea (NRF2019R1A2C1086830) grant funded by the Korean government (MSIT), POSTECH Basic Science Research Institute Grant (NRF-2021R1A6A1A10042944), and Korea Basic Science Institute (National Research Facilities and Equipment Center) grant funded by the Ministry of Education (2021R1A6C101A390).

Data Availability Statement: The data presented in this study are available on request from the corresponding author.

Conflicts of Interest: The authors declare no conflict of interest.

References

1. Cameron, D.E.; Bashor, C.J.; Collins, J.J. A brief history of synthetic biology. *Nat. Rev. Microbiol* **2014**, *12*, 381–390. [CrossRef]
2. Gardner, T.S.; Cantor, C.R.; Collins, J.J. Construction of a genetic toggle switch in Escherichia coli. *Nature* **2000**, *403*, 339–342. [CrossRef] [PubMed]
3. Elowitz, M.B.; Leibler, S.A. synthetic oscillatory network of transcriptional regulators. *Nature* **2000**, *403*, 335–338. [CrossRef] [PubMed]
4. Weitz, M.; Kim, J.; Kapsner, K.; Winfree, E.; Franco, E.; Simmel, F.C. Diversity in the dynamical behaviour of a compartmentalized programmable biochemical oscillator. *Nat. Chem.* **2014**, *6*, 295–302. [CrossRef]
5. Weinberg, B.H.; Pham, N.T.H.; Caraballo, L.D.; Lozanoski, T.; Engel, A.; Bhatia, S.; Wong, W.W. Large-scale design of robust genetic circuits with multiple inputs and outputs for mammalian cells. *Nat. Biotechnol.* **2017**, *35*, 453–462. [CrossRef]
6. Auslander, D.; Auslander, S.; Pierrat, X.; Hellmann, L.; Rachid, L.; Fussenegger, M. Programmable full-adder computations in communicating three-dimensional cell cultures. *Nat. Methods* **2018**, *15*, 57–60. [CrossRef]
7. Sexton, J.T.; Tabor, J.J. Multiplexing cell-cell communication. *Mol. Syst. Biol.* **2020**, *16*, e9618. [CrossRef]
8. Munck, C.; Sheth, R.U.; Freedberg, D.E.; Wang, H.H. Recording mobile DNA in the gut microbiota using an Escherichia coli CRISPR-Cas spacer acquisition platform. *Nat. Commun.* **2020**, *11*, 95. [CrossRef]
9. Shipman, S.L.; Nivala, J.; Macklis, J.D.; Church, G.M. CRISPR-Cas encoding of a digital movie into the genomes of a population of living bacteria. *Nature* **2017**, *547*, 345–349. [CrossRef]
10. Vishweshwaraiah, Y.L.; Chen, J.; Chirasani, V.R.; Tabdanov, E.D.; Dokholyan, N.V. Two-input protein logic gate for computation in living cells. *Nat. Commun.* **2021**, *12*, 6615. [CrossRef]
11. Gao, X.J.; Chong, L.S.; Kim, M.S.; Elowitz, M.B. Programmable protein circuits in living cells. *Science* **2018**, *361*, 1252–1258. [CrossRef] [PubMed]
12. Chen, Z.; Kibler, R.D.; Hunt, A.; Busch, F.; Pearl, J.; Jia, M.; VanAernum, Z.L.; Wicky, B.I.M.; Dods, G.; Liao, H.; et al. De novo design of protein logic gates. *Science* **2020**, *368*, 78–84. [CrossRef] [PubMed]
13. Elbaz, J.; Yin, P.; Voigt, C.A. Genetic encoding of DNA nanostructures and their self-assembly in living bacteria. *Nat. Commun.* **2016**, *7*, 11179. [CrossRef] [PubMed]
14. Chen, Z.; Elowitz, M.B. Programmable protein circuit design. *Cell* **2021**, *184*, 2284–2301. [CrossRef] [PubMed]
15. Stanton, B.C.; Nielsen, A.A.; Tamsir, A.; Clancy, K.; Peterson, T.; Voigt, C.A. Genomic mining of prokaryotic repressors for orthogonal logic gates. *Nat. Chem. Biol.* **2014**, *10*, 99–105. [CrossRef]
16. Bolognesi, B.; Lehner, B. Reaching the limit. *Elife* **2018**, *7*, e39804. [CrossRef]
17. Ceroni, F.; Boo, A.; Furini, S.; Gorochowski, T.E.; Borkowski, O.; Ladak, Y.N.; Awan, A.R.; Gilbert, C.; Stan, G.B.; Ellis, T. Burden-driven feedback control of gene expression. *Nat. Methods* **2018**, *15*, 387–393. [CrossRef]
18. Zhu, L.; Zhu, Y.; Zhang, Y.; Li, Y. Engineering the robustness of industrial microbes through synthetic biology. *Trends Microbiol.* **2012**, *20*, 94–101. [CrossRef]
19. Purnick, P.E.; Weiss, R. The second wave of synthetic biology: From modules to systems. *Nat. Rev. Mol. Cell. Biol.* **2009**, *10*, 410–422. [CrossRef]
20. Kim, J.; White, K.S.; Winfree, E. Construction of an in vitro bistable circuit from synthetic transcriptional switches. *Mol. Syst. Biol.* **2006**, *2*, 68. [CrossRef]
21. Qian, L.; Winfree, E. Scaling up digital circuit computation with DNA strand displacement cascades. *Science* **2011**, *332*, 1196–1201. [CrossRef] [PubMed]
22. Qian, L.; Winfree, E.; Bruck, J. Neural network computation with DNA strand displacement cascades. *Nature* **2011**, *475*, 368–372. [CrossRef] [PubMed]
23. Cherry, K.M.; Qian, L. Scaling up molecular pattern recognition with DNA-based winner-take-all neural networks. *Nature* **2018**, *559*, 370–376. [CrossRef]

24. Su, H.; Xu, J.; Wang, Q.; Wang, F.; Zhou, X. High-efficiency and integrable DNA arithmetic and logic system based on strand displacement synthesis. *Nat. Commun.* **2019**, *10*, 5390. [CrossRef]
25. Yordanov, B.; Kim, J.; Petersen, R.L.; Shudy, A.; Kulkarni, V.V.; Phillips, A. Computational Design of Nucleic Acid Feedback Control Circuits. *ACS Synth. Biol.* **2014**, *3*, 600–616. [CrossRef]
26. Lakin, M.R.; Youssef, S.; Polo, F.; Emmott, S.; Phillips, A. Visual DSD: A design and analysis tool for DNA strand displacement systems. *Bioinformatics* **2011**, *27*, 3211–3213. [CrossRef]
27. Groves, B.; Chen, Y.J.; Zurla, C.; Pochekailov, S.; Kirschman, J.L.; Santangelo, P.J.; Seelig, G. Computing in mammalian cells with nucleic acid strand exchange. *Nat. Nanotechnol.* **2016**, *11*, 287–294. [CrossRef]
28. Choi, H.M.; Chang, J.Y.; Trinh le, A.; Padilla, J.E.; Fraser, S.E.; Pierce, N.A. Programmable in situ amplification for multiplexed imaging of mRNA expression. *Nat. Biotechnol.* **2010**, *28*, 1208–1212. [CrossRef]
29. Zhang, D.Y.; Seelig, G. Dynamic DNA nanotechnology using strand-displacement reactions. *Nat. Chem.* **2011**, *3*, 103–113. [CrossRef] [PubMed]
30. Rothemund, P.W.K. Folding DNA to create nanoscale shapes and patterns. *Nature* **2006**, *440*, 297–302. [CrossRef]
31. Pan, T.; Sosnick, T. RNA folding during transcription. *Annu Rev. Biophys. Biomol. Struct.* **2006**, *35*, 161–175. [CrossRef] [PubMed]
32. Watters, K.E.; Strobel, E.J.; Yu, A.M.; Lis, J.T.; Lucks, J.B. Cotranscriptional folding of a riboswitch at nucleotide resolution. *Nat. Struct. Mol. Biol.* **2016**, *23*, 1124–1131. [CrossRef] [PubMed]
33. Shao, B.; Rammohan, J.; Anderson, D.A.; Alperovich, N.; Ross, D.; Voigt, C.A. Single-cell measurement of plasmid copy number and promoter activity. *Nat. Commun* **2021**, *12*, 1475. [CrossRef] [PubMed]
34. Itzkovitz, S.; van Oudenaarden, A. Validating transcripts with probes and imaging technology. *Nat. Methods* **2011**, *8*, S12–S19. [CrossRef]
35. Coller, J.; Wickens, M. Tethered function assays: An adaptable approach to study RNA regulatory proteins. *Methods Enzymol.* **2007**, *429*, 299–321.
36. Woo, C.H.; Jang, S.; Shin, G.; Jung, G.Y.; Lee, J.W. Sensitive fluorescence detection of SARS-CoV-2 RNA in clinical samples via one-pot isothermal ligation and transcription. *Nat. Biomed. Eng.* **2020**, *4*, 1168–1179. [CrossRef]
37. Santiago-Frangos, A.; Hall, L.N.; Nemudraia, A.; Nemudryi, A.; Krishna, P.; Wiegand, T.; Wilkinson, R.A.; Snyder, D.T.; Hedges, J.F.; Cicha, C.; et al. Intrinsic signal amplification by type III CRISPR-Cas systems provides a sequence-specific SARS-CoV-2 diagnostic. *Cell Rep. Med.* **2021**, *2*, 100319. [CrossRef]
38. Kellner, M.J.; Koob, J.G.; Gootenberg, J.S.; Abudayyeh, O.O.; Zhang, F. SHERLOCK: Nucleic acid detection with CRISPR nucleases. *Nat. Protoc.* **2019**, *14*, 2986–3012. [CrossRef]
39. Isaacs, F.J.; Dwyer, D.J.; Ding, C.; Pervouchine, D.D.; Cantor, C.R.; Collins, J.J. Engineered riboregulators enable post-transcriptional control of gene expression. *Nat. Biotechnol.* **2004**, *22*, 841–847. [CrossRef]
40. Green, A.A.; Silver, P.A.; Collins, J.J.; Yin, P. Toehold switches: De-novo-designed regulators of gene expression. *Cell* **2014**, *159*, 925–939. [CrossRef]
41. Chappell, J.; Takahashi, M.K.; Lucks, J.B. Creating small transcription activating RNAs. *Nat. Chem. Biol.* **2015**, *11*, 214–220. [CrossRef] [PubMed]
42. Chappell, J.; Westbrook, A.; Verosloff, M.; Lucks, J.B. Computational design of small transcription activating RNAs for versatile and dynamic gene regulation. *Nat. Commun.* **2017**, *8*, 1051. [CrossRef]
43. Jang, S.; Jang, S.; Xiu, Y.; Kang, T.J.; Lee, S.-H.; Koffas, M.A.A.G.; Jung, G.Y. Development of Artificial Riboswitches for Monitoring of Naringenin In Vivo. *ACS Synth. Biol.* **2017**, *6*, 2077–2085. [CrossRef] [PubMed]
44. Hanewich-Hollatz, M.H.; Chen, Z.; Hochrein, L.M.; Huang, J.; Pierce, N.A. Conditional Guide RNAs: Programmable Conditional Regulation of CRISPR/Cas Function in Bacterial and Mammalian Cells via Dynamic RNA Nanotechnology. *ACS Cent. Sci.* **2019**, *5*, 1241–1249. [CrossRef] [PubMed]
45. Serganov, A.; Nudler, E. A Decade of Riboswitches. *Cell* **2013**, *152*, 17–24. [CrossRef] [PubMed]
46. Green, A.A.; Kim, J.; Ma, D.; Silver, P.A.; Collins, J.J.; Yin, P. Complex cellular logic computation using ribocomputing devices. *Nature* **2017**, *548*, 117–121. [CrossRef] [PubMed]
47. Kim, J.; Zhou, Y.; Carlson, P.D.; Teichmann, M.; Chaudhary, S.; Simmel, F.C.; Silver, P.A.; Collins, J.J.; Lucks, J.B.; Yin, P.; et al. De novo-designed translation-repressing riboregulators for multi-input cellular logic. *Nat. Chem. Biol.* **2019**, *15*, 1173–1182. [CrossRef]
48. Hong, S.; Jeong, D.; Ryan, J.; Foo, M.; Tang, X.; Kim, J. Design and evaluation of synthetic RNA-based incoherent feed-forward loop circuits. *Biomolecules* **2021**, *11*, 1182. [CrossRef]
49. Hong, S.; Kim, J.; Kim, J. Multilevel Gene Regulation Using Switchable Transcription Terminator and Toehold Switch in Escherichia coli. *Appl. Sci.* **2021**, *11*, 4532. [CrossRef]
50. Yang, J.; Han, Y.H.; Im, J.; Seo, S.W. Synthetic protein quality control to enhance full-length translation in bacteria. *Nat. Chem. Biol.* **2021**, *17*, 421–427. [CrossRef]
51. Hwang, Y.; Kim, S.G.; Jang, S.; Kim, J.; Jung, G.Y. Signal amplification and optimization of riboswitch-based hybrid inputs by modular and titratable toehold switches. *J. Biol. Eng.* **2021**, *15*, 11. [CrossRef]
52. Zhao, E.M.; Mao, A.S.; de Puig, H.; Zhang, K.; Tippens, N.D.; Tan, X.; Ran, F.A.; Han, I.; Nguyen, P.Q.; Chory, E.J.; et al. RNA-responsive elements for eukaryotic translational control. *Nat. Biotechnol.* **2021**. [CrossRef]

53. Huang, A.; Nguyen, P.Q.; Stark, J.C.; Takahashi, M.K.; Donghia, N.; Ferrante, T.; Dy, A.J.; Hsu, K.J.; Dubner, R.S.; Pardee, K.; et al. BioBits™ Explorer: A modular synthetic biology education kit. *Sci. Adv.* **2018**, *4*, eaat5105. [CrossRef] [PubMed]
54. McNerney, M.P.; Zhang, Y.; Steppe, P.; Silverman, A.D.; Jewett, M.C.; Styczynski, M.P. Point-of-care biomarker quantification enabled by sample-specific calibration. *Sci. Adv.* **2019**, *5*, eaax4473. [CrossRef] [PubMed]
55. Sadat Mousavi, P.; Smith, S.J.; Chen, J.B.; Karlikow, M.; Tinafar, A.; Robinson, C.; Liu, W.; Ma, D.; Green, A.A.; Kelley, S.O.; et al. A multiplexed, electrochemical interface for gene-circuit-based sensors. *Nat. Chem.* **2020**, *12*, 48–55. [CrossRef] [PubMed]
56. Nguyen, P.Q.; Soenksen, L.R.; Donghia, N.M.; Angenent-Mari, N.M.; de Puig, H.; Huang, A.; Lee, R.; Slomovic, S.; Galbersanini, T.; Lansberry, G.; et al. Wearable materials with embedded synthetic biology sensors for biomolecule detection. *Nat. Biotechnol.* **2021**, *39*, 1366–1374. [CrossRef]
57. Amalfitano, E.; Karlikow, M.; Norouzi, M.; Jaenes, K.; Cicek, S.; Masum, F.; Sadat Mousavi, P.; Guo, Y.; Tang, L.; Sydor, A.; et al. A glucose meter interface for point-of-care gene circuit-based diagnostics. *Nat. Commun.* **2021**, *12*, 724. [CrossRef] [PubMed]
58. Takahashi, M.K.; Tan, X.; Dy, A.J.; Braff, D.; Akana, R.T.; Furuta, Y.; Donghia, N.; Ananthakrishnan, A.; Collins, J.J. A low-cost paper-based synthetic biology platform for analyzing gut microbiota and host biomarkers. *Nat. Commun.* **2018**, *9*, 3347. [CrossRef]
59. Pardee, K.; Green, A.A.; Takahashi, M.K.; Braff, D.; Lambert, G.; Lee, J.W.; Ferrante, T.; Ma, D.; Donghia, N.; Fan, M.; et al. Rapid, Low-Cost Detection of Zika Virus Using Programmable Biomolecular Components. *Cell* **2016**, *165*, 1255–1266. [CrossRef]
60. Pardee, K.; Green, A.A.; Ferrante, T.; Cameron, D.E.; DaleyKeyser, A.; Yin, P.; Collins, J.J. Paper-Based Synthetic Gene Networks. *Cell* **2014**, *159*, 940–954. [CrossRef]
61. Wong, A.; Wang, H.; Poh, C.L.; Kitney, R.I. Layering genetic circuits to build a single cell, bacterial half adder. *BMC Biol.* **2015**, *13*, 40. [CrossRef] [PubMed]
62. Rosado, A.; Cordero, T.; Rodrigo, G. Binary addition in a living cell based on riboregulation. *PLoS Genet.* **2018**, *14*, e1007548. [CrossRef] [PubMed]
63. Goldsworthy, V.; LaForce, G.; Abels, S.; Khisamutdinov, E.F. Fluorogenic RNA Aptamers: A Nano-platform for Fabrication of Simple and Combinatorial Logic Gates. *Nanomaterials* **2018**, *8*, 984. [CrossRef]
64. Buchler, N.E.; Gerland, U.; Hwa, T. On schemes of combinatorial transcription logic. *Proc. Natl. Acad. Sci. USA* **2003**, *100*, 5136–5141. [CrossRef] [PubMed]
65. Bonnet, J.; Yin, P.; Ortiz, M.E.; Subsoontorn, P.; Endy, D. Amplifying genetic logic gates. *Science* **2013**, *340*, 599–603. [CrossRef] [PubMed]
66. Auslander, S.; Auslander, D.; Muller, M.; Wieland, M.; Fussenegger, M. Programmable single-cell mammalian biocomputers. *Nature* **2012**, *487*, 123–127. [CrossRef] [PubMed]
67. Matsuura, S.; Ono, H.; Kawasaki, S.; Kuang, Y.; Fujita, Y.; Saito, H. Synthetic RNA-based logic computation in mammalian cells. *Nat. Commun.* **2018**, *9*, 4847. [CrossRef]
68. Kim, H.; Bojar, D.; Fussenegger, M. A CRISPR/Cas9-based central processing unit to program complex logic computation in human cells. *Proc. Natl. Acad. Sci. USA* **2019**, *116*, 7214–7219. [CrossRef]
69. Fratto, B.E.; Katz, E. Reversible Logic Gates Based on Enzyme-Biocatalyzed Reactions and Realized in Flow Cells: A Modular Approach. *ChemPhysChem* **2015**, *16*, 1405–1415. [CrossRef]
70. Tan, S.-I.; Ng, I.S. CRISPRi-Mediated NIMPLY Logic Gate for Fine-Tuning the Whole-Cell Sensing toward Simple Urine Glucose Detection. *ACS Synth. Biol.* **2021**, *10*, 412–421. [CrossRef]
71. Fornace, M.E.; Porubsky, N.J.; Pierce, N.A. A Unified Dynamic Programming Framework for the Analysis of Interacting Nucleic Acid Strands: Enhanced Models, Scalability, and Speed. *ACS Synth. Biol.* **2020**, *9*, 2665–2678. [CrossRef] [PubMed]
72. Dirks, R.M.; Bois, J.S.; Schaeffer, J.M.; Winfree, E.; Pierce, N.A. Thermodynamic Analysis of Interacting Nucleic Acid Strands. *SIAM Rev.* **2007**, *49*, 65–88. [CrossRef]
73. Dirks, R.M.; Pierce, N.A. An algorithm for computing nucleic acid base-pairing probabilities including pseudoknots. *J. Comput. Chem.* **2004**, *25*, 1295–1304. [CrossRef] [PubMed]
74. Dirks, R.M.; Pierce, N.A. A partition function algorithm for nucleic acid secondary structure including pseudoknots. *J. Comput. Chem.* **2003**, *24*, 1664–1677. [CrossRef]
75. Wolfe, B.R.; Pierce, N.A. Sequence Design for a Test Tube of Interacting Nucleic Acid Strands. *ACS Synth. Biol.* **2015**, *4*, 1086–1100. [CrossRef]
76. Nicholson, A.W. Ribonuclease III mechanisms of double-stranded RNA cleavage. *Wiley Interdiscip. Rev. RNA* **2014**, *5*, 31–48. [CrossRef] [PubMed]
77. Court, D.L.; Gan, J.; Liang, Y.H.; Shaw, G.X.; Tropea, J.E.; Costantino, N.; Waugh, D.S.; Ji, X. RNase III: Genetics and function; structure and mechanism. *Annu Rev. Genet.* **2013**, *47*, 405–431. [CrossRef]
78. Zadeh, J.N.; Wolfe, B.R.; Pierce, N.A. Nucleic acid sequence design via efficient ensemble defect optimization. *J. Comput. Chem.* **2011**, *32*, 439–452. [CrossRef]
79. Zhang, D.Y.; Winfree, E. Control of DNA Strand Displ.lacement Kinetics Using Toehold Exchange. *J. Am. Chem. Soc.* **2009**, *131*, 17303–17314. [CrossRef]
80. Zhang, D.Y.; Chen, S.X.; Yin, P. Optimizing the specificity of nucleic acid hybridization. *Nat. Chem.* **2012**, *4*, 208–214. [CrossRef]
81. Mancuso, C.P.; Kiriakov, S.; Khalil, A.S. Cellular Advantages to Signaling in a Digital World. *Cell Syst.* **2016**, *3*, 114–115. [CrossRef] [PubMed]
82. Rubens, J.R.; Selvaggio, G.; Lu, T.K. Synthetic mixed-signal computation in living cells. *Nat. Commun.* **2016**, *7*, 11658. [CrossRef]

83. Balazsi, G.; van Oudenaarden, A.; Collins, J.J. Cellular decision making and biological noise: From microbes to mammals. *Cell* **2011**, *144*, 910–925. [CrossRef] [PubMed]
84. Guzman, L.M.; Belin, D.; Carson, M.J.; Beckwith, J. Tight regulation, modulation, and high-level expression by vectors containing the arabinose PBAD promoter. *J. Bacteriol.* **1995**, *177*, 4121–4130. [CrossRef] [PubMed]
85. Egan, S.M.; Schleif, R.F. A Regulatory Cascade in the Induction of rhaBAD. *J. Mol. Biol.* **1993**, *234*, 87–98. [CrossRef] [PubMed]
86. Terpe, K. Overview of bacterial expression systems for heterologous protein production: From molecular and biochemical fundamentals to commercial systems. *Appl. Microbiol. Biotechnol.* **2006**, *72*, 211–222. [CrossRef]
87. Ruegg, T.L.; Pereira, J.H.; Chen, J.C.; DeGiovanni, A.; Novichkov, P.; Mutalik, V.K.; Tomaleri, G.P.; Singer, S.W.; Hillson, N.J.; Simmons, B.A.; et al. Jungle Express is a versatile repressor system for tight transcriptional control. *Nat. Commun.* **2018**, *9*, 3617. [CrossRef]
88. Tamsir, A.; Tabor, J.J.; Voigt, C.A. Robust multicellular computing using genetically encoded NOR gates and chemical 'wires'. *Nature* **2011**, *469*, 212–215. [CrossRef]
89. Regot, S.; Macia, J.; Conde, N.; Furukawa, K.; Kjellen, J.; Peeters, T.; Hohmann, S.; de Nadal, E.; Posas, F.; Sole, R. Distributed biological computation with multicellular engineered networks. *Nature* **2011**, *469*, 207–211. [CrossRef]
90. Osmekhina, E.; Jonkergouw, C.; Schmidt, G.; Jahangiri, F.; Jokinen, V.; Franssila, S.; Linder, M.B. Controlled communication between physically separated bacterial populations in a microfluidic device. *Commun. Biol.* **2018**, *1*, 97. [CrossRef]
91. Callura, J.M.; Dwyer, D.J.; Isaacs, F.J.; Cantor, C.R.; Collins, J.J. Tracking, tuning, and terminating microbial physiology using synthetic riboregulators. *Proc. Natl. Acad. Sci. USA* **2010**, *107*, 15898. [CrossRef] [PubMed]
92. Ceroni, F.; Algar, R.; Stan, G.-B.; Ellis, T. Quantifying cellular capacity identifies gene expression designs with reduced burden. *Nat. Methods* **2015**, *12*, 415–418. [CrossRef] [PubMed]
93. Zhang, Q.; Ma, D.; Wu, F.; Standage-Beier, K.; Chen, X.; Wu, K.; Green, A.A.; Wang, X. Predictable control of RNA lifetime using engineered degradation-tuning RNAs. *Nat. Chem. Biol.* **2021**, *17*, 828–836. [CrossRef]
94. Bhat, G.J.; Lodes, M.J.; Myler, P.J.; Stuart, K.D. A simple method for cloning blunt ended DNA fragments. *Nucleic Acids. Res.* **1991**, *19*, 398. [CrossRef] [PubMed]
95. Gibson, D.G.; Young, L.; Chuang, R.Y.; Venter, J.C.; Hutchison, C.A., III; Smith, H.O. Enzymatic assembly of DNA molecules up to several hundred kilobases. *Nat. Methods* **2009**, *6*, 343–345. [CrossRef] [PubMed]
96. Quan, J.; Tian, J. Circular polymerase extension cloning for high-throughput cloning of complex and combinatorial DNA libraries. *Nat. Protoc.* **2011**, *6*, 242–251. [CrossRef]
97. Liu, H.; Naismith, J.H. An efficient one-step site-directed deletion, insertion, single and multiple-site plasmid mutagenesis protocol. *BMC Biotechnol* **2008**, *8*, 91. [CrossRef]
98. Green, M.R.; Sambrook, J. The Inoue Method for Preparation and Transformation of Competent Escherichia coli: "Ultracompetent" Cells. *Cold Spring Harb. Protoc.* **2020**, *2020*, 101196. [CrossRef]

International Journal of
Molecular Sciences

MDPI

Communication

Hidden Information Revealed Using the Orthogonal System of Nucleic Acids

Viktor Víglaský

Department of Biochemistry, Institute of Chemistry, Faculty of Sciences, Pavol Jozef Šafárik University, 04001 Košice, Slovakia; viktor.viglasky@upjs.sk; Tel.: +421-55-2341262

Abstract: In this study, the organization of genetic information in nucleic acids is defined using a novel orthogonal representation. Clearly defined base pairing in DNA allows the linear base chain and sequence to be mathematically transformed into an orthogonal representation where the G–C and A–T pairs are displayed in different planes that are perpendicular to each other. This form of base allocation enables the evaluation of any nucleic acid and predicts the likelihood of a particular region to form non-canonical motifs. The G4Hunter algorithm is currently a popular method of identifying G-quadruplex forming sequences in nucleic acids, and offers promising scores despite its lack of a substantial rational basis. The orthogonal representation described here is an effort to address this incongruity. In addition, the orthogonal display facilitates the search for other sequences that are capable of adopting non-canonical motifs, such as direct and palindromic repeats. The technique can also be used for various RNAs, including any aptamers. This powerful tool based on an orthogonal system offers considerable potential for a wide range of applications.

Keywords: orthogonal representation; G-quadruplex; G4Hunter; algorithm

Citation: Víglaský, V. Hidden Information Revealed Using the Orthogonal System of Nucleic Acids. *Int. J. Mol. Sci.* **2022**, *23*, 1804. https://doi.org/10.3390/ijms23031804

Academic Editors: Vaclav Brazda and Richard Bowater

Received: 29 December 2021
Accepted: 2 February 2022
Published: 4 February 2022

1. Introduction

As is well known, DNA molecules often occur in an antiparallel double-stranded structure due to Watson–Crick (WC) base pairing, with adenine and guanine bases pairing with thymine and cytosine, respectively. A unique feature of these molecules is their ability to pair not only through WC pairing, but also through Hoogsteen bonds. Non-canonical structures can be stabilized by a combination of both types of hydrogen bonds and can also contain several unpaired bases, such as G quadruplex, i-motif, triplexes, hairpin, and cruciform [1,2]. Non-canonical structures exist in cells and play important roles in gene expression regulation [1].

Nucleic acid consists of building blocks of nucleotides that are arranged in different permutations, with the order of the nucleotides determining the sequence of DNA or RNA molecules (Figure 1A). The nucleic acid sequence is crucial for the arrangement of amino acids in proteins and the 3D structure of RNA, which does not necessarily translate into protein. These sequences are not coincidental. DNA consists of characteristic sequence motifs typical of any organism, usually untranslated, that play key roles at various levels of gene expression [3–5]. For example, they separate coding and non-coding regions, control the efficiency of promoter sequences, segment chromosomes, and signal for transcription and translation machineries. Countless other examples are known where specific sequence motifs play a key role in regulating the gene expression and the cell signaling system [2].

The identification of sequence clusters and their mutations is particularly useful for understanding the expression of structural genes, which are responsible for various pathological manifestations. An awareness of the DNA sequence alone is not sufficient to provide a full understanding of these processes, and therefore a number of diverse bioinformatic approaches have been developed that enable the identification of so-called non-standard sequences in the genome. The dramatic increase in the accumulation of genomic data over the last decade poses a considerable challenge in terms of processing and provides an opportunity to develop computational analyzes that are capable of sophisticated screening processes of unknown genomes, including their graphical representation [6].

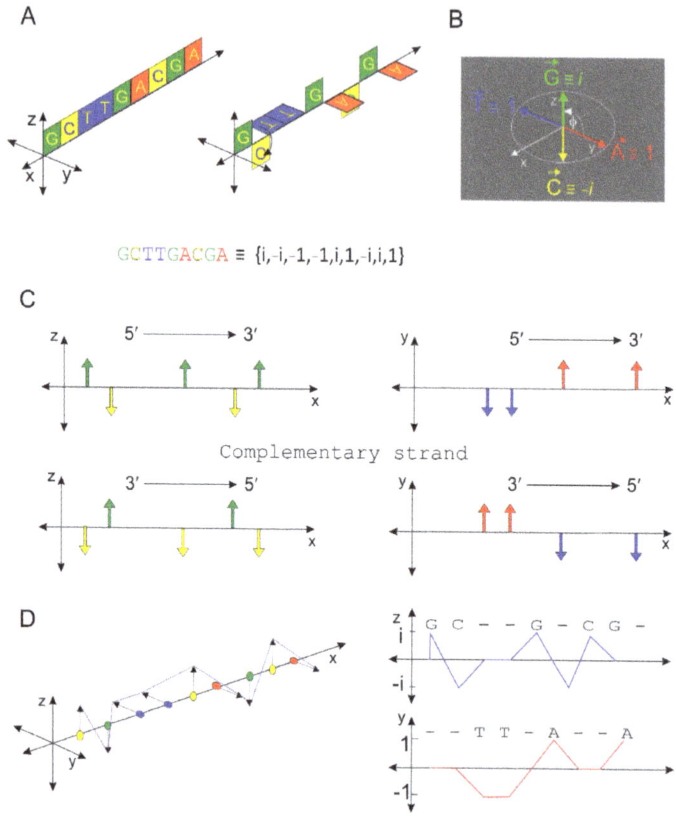

Figure 1. Basic properties of an orthogonal system. Standard sequence visualization is performed on two perpendicular planes, where nucleotides A + T are on the xy planes and C + G are on the xz planes. The nucleotide order is expressed by an integer value on the *x*-axis (**A**). There is a close analogy with the representation of complex integers (**B**), and a unit circle is used for this purpose. In the complex space, any oligonucleotide in the DNA sequence can be expressed instead of A, T, C, and G by four values: −1, 1, −i, and i, respectively. The complementary strand of DNA is a mirror image of the original sequence on a given plane of display (**C**). The sequence can be displayed in a complex space using vectors that can be projected into a real or complex plane (**D**).

The approach known as "digital signal processing" has seen increasing use in genomic DNA research as a means of revealing genome structures and identifying hidden periodicities and features that cannot be determined using conventional DNA symbolic and graphical representation techniques [6]. Various numerical, vector, color, and different graphical representation of nucleobases in DNA have already been described in earlier studies [6–12]. For canonical putative sequences adopting cruciform or G-quadruplex structures, it is more appropriate to use an application specially tailored for this purpose, for example, computational approaches, which study these motifs to allow for a detailed analysis of the genomes [13–17].

Interestingly, the G4Hunter algorithm offers one of the highest search scores for identifying sequences that form G-quadruplexes, but there is still a lack of a rational explanation for this success rate. The G4Hunter algorithm considers the G-richness and

G-skewness of a given sequence, and provides a quadruplex propensity score as an output. The searching strategy is simple; each position in a sequence is given a score between −4 and 4. Scores of 0 indicate A and T, while positive scores indicate G and negative scores C. A single G achieves a score of 1, and two, three, and four neighboring Gs scores of 2, 3, and 4, respectively; a score of 4 also suggests the presence of higher numbers of Gs. The C bases are scored similarly, but all of the values are negative [18]. The G4Hunter algorithm also retains some G–C pairing features; the G score has the opposite value of C, but not in the case of A–T pairing. This study will present an alternative to the G4Hunter approach. In this system, the basic attribute related to base pairing is preserved for both WC base-pairs. Although the basic principle of the system is very simple, it does not appear to have been described before.

2. Results and Discussion

2.1. Principle of the Orthogonal Algorithm

The principle of the algorithm is shown in Figure 1. Complementary oligonucleotides are shown in the following colors: A—red; T—blue; G—green; C—yellow. A + T bases occur only on the xy planes and C + G bases only on the xz planes. The representative sequence is d(GCTTGACGA) (panel A). There is a close analogy with the representation of complex numbers, and it is therefore possible to state that A + T are projected in the real plane and G + C in the imaginary plane (panel B). Based on this analogy, the values 1, −1, i, and −i can be assigned to the individual nucleotides A, T, G, and C, respectively. If the size of vectors A, T, G, and C are equivalent and equal to 1, then the endpoint of each vector lies on the unit circle, and it is possible to express a representative sequence using a linear string {i, −i, −1, −1, i, 1, −i, i, 1}. Any DNA sequence can be divided into real and imaginary components, but both categories are coupled. The definition of the axes is variable but due to the symmetry of this view, similar results would also be obtained with a different choice of planes and axes. In principle, only a single condition is required to be met; C must be opposite G, and A must be opposite T. An antiparallel strand represents a mirror image for both components (panel C). The vector representation and projection into real and imaginary planes are shown in panel D. In situations when it cannot be ruled out that the individual endpoints of vectors A, T, C, and G lie on an ellipse and that the angle φ is not exactly 90 degrees, the quantitative results will offer an even more reliable score than a purely orthogonal system for sequences forming a specific non-canonical motif (see below).

The profile of projection into the plane is given by the sequence, and an example of this is shown in Figure 2. The projection shows the following two sets of sequences: ATA(G/C)T(G/C)AATTTT(G/C) and GCG(A/T)C(A/T)GGCCCC(A/T). The area is not solely dependent on a given nucleotide, but is also influenced to some extent by the neighboring nucleotides. For example, the area in the xy plane given by the CAC sequence is equal to 1, the TAT is equal to 0.5, and the CAT is 0.75. The total area of a given sequence in Figure 2 in one of the projection planes, which achieves a negative value of −2.5. An important parameter is obtained if this value is divided by the number of nucleotides.

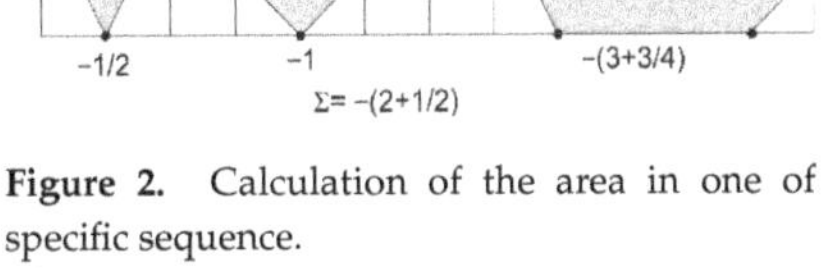

Figure 2. Calculation of the area in one of the planes determined by the projection of a specific sequence.

2.2. G-Quadruplex Forming Sequences and Non-Canonical Motifs

The orthogonal system was applied to a series of sequences that are known to be capable of forming a G-quadruplex motif. Five examples of G-quadruplex sequences, human telomeric repeats (HTR), c-myc promoter sequence, thrombin binding aptamer (TBA), d[(C(G_4C_2)$_3G_4$C], and d[T(G_4T_2)$_3G_4$T] are shown in Figure 3A. Each of the DNA sequences is capable of forming a relatively stable G-quadruplex structure in the presence of a potassium ion [19–25]. This set of sequences is displayed in the xz-projection.

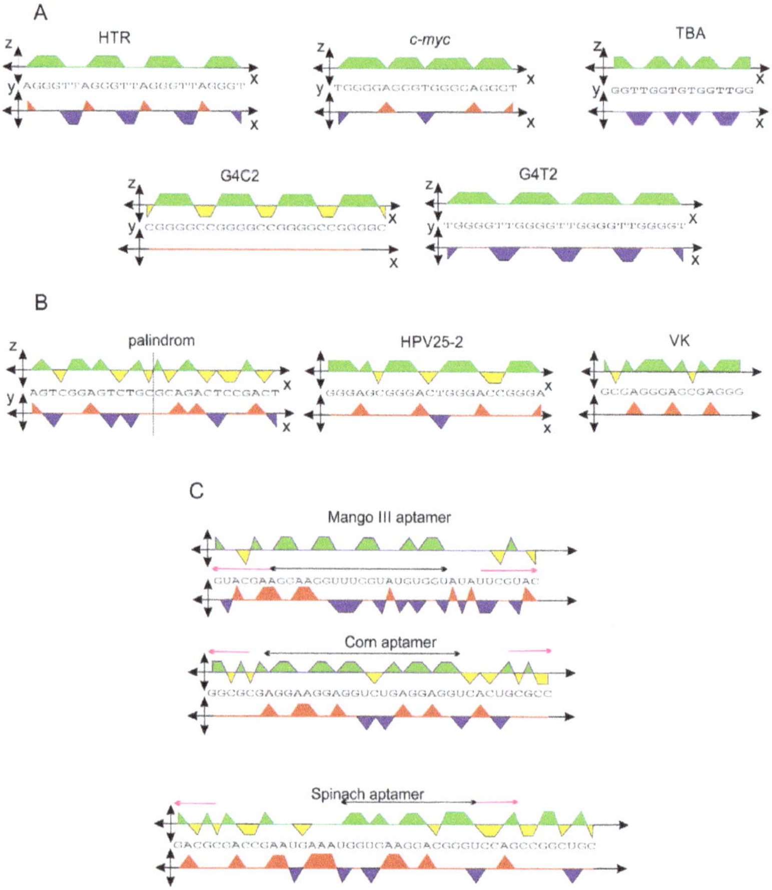

Figure 3. (**A**) Orthogonal projection of sequences adopting G-quadruplexes: human telomeric repeats (HTR), *c-myc* promoter sequence, and thrombin binding aptamer (TBA). (**B**) Example of a palindromic sequence adopting a hairpin and sequences adopting other non-canonical structures. (**C**) Sequence of RNA aptamers Mango III and Corn adopting a structure consisting of both G-quadruplex and dsRNA.

The areas of green projection for HTR, *c-myc*, and TBA are 12, 14, and 8, respectively. The orthogonal system provides the following scores: 0.52, 0.74, and 0.53, respectively. In contrast, the G4Hunter scores are as follows: 1.57, 2.11, and 2.2, respectively. However, if the radius "r" of the circle is equal to 3, as shown in Figure 1, then the scores multiplied by a factor of 3 provide values of 1.56, 2.22, and 1.59, respectively, with the first two values being very close to those obtained using the G4Hunter algorithm (Table 1). The scores for G4C2 and G4T2 give values of 1.13 and 2.0, while those obtained from G4Hunter are 2.08 and 2.67. The deviation between these types of algorithms is a result of the overly strong parameterization in G4Hunter in cases of two, four, or more adjacent Gs. The orthogonal projection and G4Hunter algorithm provide similar results for sequences consisting of less than four contiguous Gs.

Table 1. Scores obtained with the orthogonal presentation and G4Hunter algorithm.

Sequence	xz-Projection Area, r = 3	xy-Projection Area, r = 3	G4hunter Score	xz-Projection Area, r = 3, ψ = 15° #	xz-Projection Area, r = 3, ψ = 30° #	G4 Formation
HTR	1.56	−0.39	1.57	1.66	1.75	yes
c-myc	2.22	0	2.11	2.22	2.22	yes
TBA	1.59	−1.2	2.20	1.90	2.19	yes
G4C2	1.13	0	2.08	1.13	1.13	yes
G4T2	2.00	−0.87	2.67	2.23	2.43	yes
HPV25-2	1.16	+0.34	1.68	1.07	0.99	no
VK	1.1	+0.60	1.4	0.94	0.80	no
palindrom *	0	0	0	0	0	no
cs-Mango III	1.35	−0.41	0.85	1.46	1.55	yes
cs-Corn	1.41	+0.18	0.94	1.36	1.32	yes

*—any sequences where the number of As and Ts is equal and the number of Gs and Cs then the score is 0; special cases include any perfect palindrom; see also Figure 8; #—ψ-correction for quasi-orthogonal system.

The HPV25-2 and VK (pdb ID: 2MJJ) sequences are known not to form G-quadruplexes [26,27], and the scores for these sequences are 1.16 and 1.10 for HPV25-2 and VK, respectively. However, the G4hunter algorithm gives a false positive score, indicating that the sequences have the capacity to form a G-quadruplex structure. If the score obtained by the orthogonal system falls within the range of 1.1–1.2, the prediction of G-quadruplex formation can be somewhat ambiguous.

However, if the score obtained from the xy projection does not show higher positive values, then the sequence still has the potential to adopt a G-quadruplex structure, but experimental verification would be recommended to confirm the formation of a G-quadruplex from the sequence in such a case. In essence, an increasing number of As in a sequence reduces the inclination to adopt G-quadruplex, mainly if the xz-score is less than 1.2. Therefore, the G4C2 sequence listed in Table 1 does not lose the potential to form a G-quadruplex, even at a lower xz-score of 1.13, but this is not the case for the VK and HPV25-2 sequences. For example, while the d(G_3A_2)G_3 sequence still has the potential to form a G-quadruplex with xz- and xy-scores of 1.83 and 1.00, respectively, the CD spectrum results (not shown in this study) do not confirm the formation of the G-quadruplex structure of the sequence d(G_3A_3)G_3 with xz- and xy scores of 1.57 and 1.29, respectively, These findings would suggest that the xz-score alone may not be a sufficient indicator to confirm the actual presence of G quadruplexes.

Even more interesting results were obtained in the case of the two RNA aptamers Mango III and Corn [28,29]. The orthogonal system is not only applicable for DNA sequences, but it can also be expanded for use with RNA molecules, with the U being used instead of T with the same value. The central sequence scores (cs) obtained for these aptamers are highlighted by black double-arrows in Figure 3C, and the values are shown in Table 1. The G4Hunter algorithm failed for both aptamers, with no G-quadruplex formation predicted, but the orthogonal system did predict G-quadruplex formation, with a xz-score higher than 1.2. In addition, clear palindromic regions were identified, highlighted with the purple arrows in Figure 3. Such a complex view of a given sequence clearly suggests that a G-quadruplex could form in the central region and that the terminal sequences would also be paired. The 3D structures of these aptamers only confirm these predicted results (pdb ID: 6E80 and 6E8T).

We accept that the orthogonal system is not a completely perfect method, but the accuracy can be increased if the orthogonality is slightly disturbed, resulting in a reduction in the number of false positives. The generalization of the system shown in Figure 1B is such that no nucleotide needs to be defined as a purely real or imaginary number; their coordinates lie on a circle or ellipse, depending on constants r_1 and r_2. For the sake of simplicity, these constants were equal to 1. If the condition of complementarity is maintained, the coordinates [y, z] for A, T, C, and G vectors can generally be expressed as follows:

$$A = r_1.[\cos(\alpha); i\sin(\alpha)],$$

$$T = r_1.[\cos(\alpha + \pi); i\sin(\alpha + \pi)],$$

$$G = r_2.[\cos(\beta); i\sin(\beta)],$$

$$C = r_2.[\cos(\beta + \pi); i\sin(\beta + \pi)],$$

where r_1 and r_2 are variable constants (radius), and the difference α–β expresses the angle φ between vectors A and G or C and T. If the angular difference is greater than 90° than angle ψ, then the contribution of the imaginary components for A directly reduces the score in the imaginary plane (xz), Figure 4.

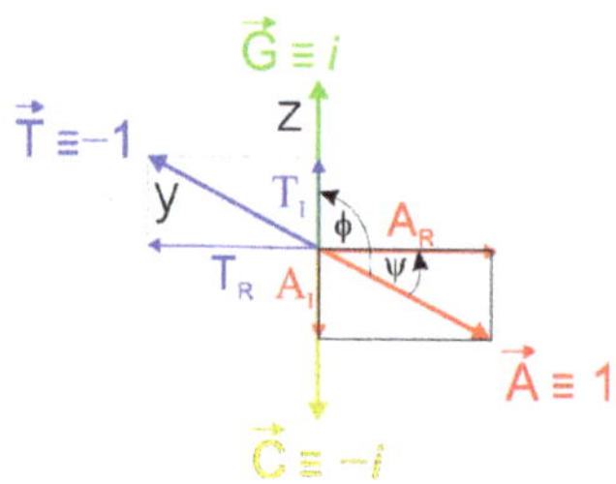

Figure 4. A quasi-orthogonal system in which the xy plane rotates around the x-axis at angle ψ. The projection of the vector A and T into imaginary and real components is also shown. Imaginary components contribute in the xz planes to the C + G score.

The result is a decrease in the probability of G-quadruplex formation. As has been shown previously, the HPV25-2 and VK sequences show a significant signal from A-nucleotides [26,27]. On the other hand, the presence of Ts increases the probability of G-quadruplex formation. The value of angle ψ can be estimated from the experimentally confirmed sequences forming a G-quadruplex in which the orthogonal scores are ambiguous. The scores recalculated for two different values of angle ψ, 15° and 30°, are also shown in Table 1. Implementing this correction results in a significant reduction in ambiguity. The ψ around 30° seems to be more ideal, with the threshold for G-quadruplex formation approaching 1.1. This so-called ψ-correction has been applied to more than 100 experimentally validated sequences that have adopted the G-quadruplex structure, but no exception has been found to date.

2.3. Genetic Code in Orthogonal Presentation

The system presented here can be applied to all sizes of nucleic acids, including short oligonucleotides. Recent research has revealed that short sequence regions often play a key role; for example, they are a target for many proteins and they are recognized by various restriction enzymes, transcription factors, and ribosomes. It is clear that short trinucleotide sequences are sufficient to encode amino acids in the form of a genetic code. The numerical transformation of the genetic code into an orthogonal system is shown in Table 2.

Table 2. Genetic code in numeric representation, radius r equals 1.

Ala/A	GCU {i, −i, −1}	GCC {i, −i, −i}	GCA {i, −i, 1}	GCG {i, −i, i}		
Arg/R	CGU {−i, i, −1}	CGC {i, i, −i}	CGA {−i, i, 1}	CGG {−i, i, i}	AGA {1, i, 1}	AGG {1, i, i}
Asn/N	AAU {1, 1, −1}	AAC {1, 1, −i}				
Asp/D	GAU {i, 1, −1}	GAC {i, 1, −i}				
Cys/C	UGU {−1, i, −1}	UGC {−1, i, −i}				
Gln/Q	CAA {−i, 1, 1}	CAG {−i, 1, i}				
Glu/E	GAA {i, 1, 1}	GAG {i, 1, i}				
Gly/G	GGU {i, i, −1}	GGC {i, i, −i}	GGA {i, i, 1}	GGG {i, i, i}		
His/H	CAU {−i, 1, −1}	CAC {−i, 1, −i}				
Ile/I	AUU {1, −1, −1}	AUC {1, −1, -i}	AUA {1, −1, 1}			
Leu/L	UUA {−1, −1, 1}	UUG {−1, −1, i}	CUU {−i, −1, −1}	CUC {−i, −1, −i}	CUA {−i, −1, 1}	CUG {−i, −1, i}
Lys/K	AAA {1, 1, 1}	AAG {1, 1, i}				
Met/M	AUG {1, −1, i}					
Phe/F	UUU {−1, −1, −1}	UUC {−1, −1, −i}				
Pro/P	CCU {−i, −i, −1}	CCC {−i, −i, −i}	CCA {−i, −i, 1}	CCG {−i, −i, i}		
Ser/S	UCU {−1, −i, −1}	UCC {−1, −i, −i}	UCA {−1, −i, 1}	UCG {−1, −i, i}	AGU {1, i, −1}	AGC {1, i, −i}
Thr/T	ACU {1, −i, −1}	ACC {1, −i, −i}	ACA {1, −i, 1}	ACG {1, −i, i}		
Trp/W	UGG {−1, i, i}					
Tyr/Y	UAU {−1, 1, −1}	UAC {−1, 1, −i}				
Val/V	GUU {i, −1, −1}	GUC {i, −1, −i}	GUA {i, −1, 1}	GUG {i, −1, i}		
STOP	UAG {−1, 1, i}	UGA {−1, i, 1}	UAA {−1, 1, 1}			

The 3D examples of the two mirror codons, the start codon-methionine and isoleucine are shown in Figure 5. Each pair of graphical representations is equivalent, the only difference being that they are shown from a different angle. Any triplet-nucleotide sequence can be represented by a single line (dashed lines). These types of graphical and numerical representations could be of considerable use in bioinformatic analyses [30].

An even more interesting representation, analogous to the previous application for the DNA and RNA sequences, is shown in Figure 6. There is no ambiguity concerning which color is dominant for a particular group of codons. The different color coding of the codon tetrahedral representation has also been performed and described in a previous study, although the strategy used in that case was based on a slightly different but still complex basis [7]. Nevertheless, the orthogonal representation method is a simpler technique and can also be transformed into a tetrahedral representation.

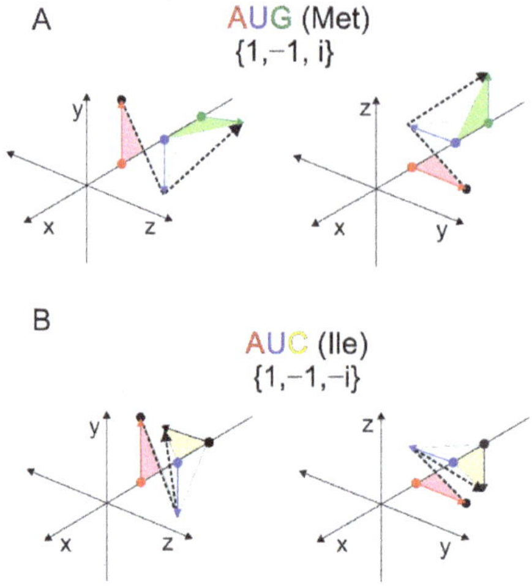

Figure 5. Orthogonal projection of two codons: Met (**A**) and Ile (**B**). The difference is found in the third base. The corresponding polylines are mirrored in the orthogonal projection (dashed lines).

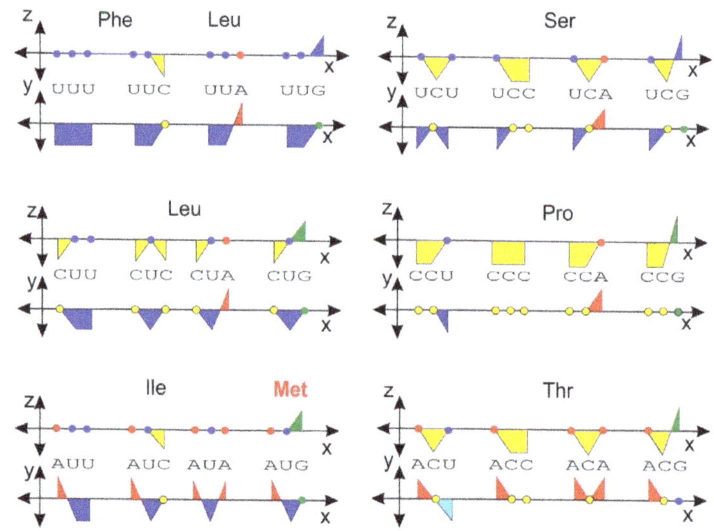

Figure 6. *Cont.*

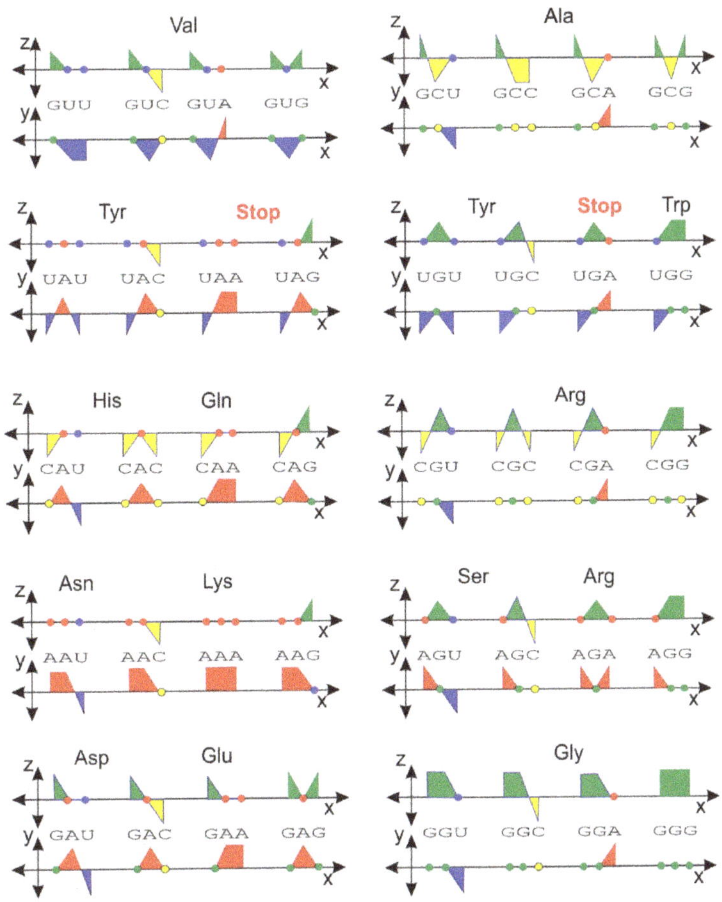

Figure 6. Genetic code represented in the orthogonal system. Each codon is projected into both the xy- and xz-planes.

The vector representation derived from the orthogonal system offers an alternative view on the genetic code (Figure 7). Interestingly, some combinations of double degeneracy in the third codon base for a single amino acid, specifically a combination of CG ($-i$, i) or UA (-1, 1), are not permitted. No amino acid is specified by these combinations, except those that are more degenerate than Gly, Ser, Leu, Pro, Arg, Ile, Thr, Val, and Ala.

Analogically, the vector representation is also applicable for longer sequences. The sequences used for the projection in the xy- and xz-planes shown in Figure 3 are displayed in the vector representation in Figure 8. Again, the fact that G-quadruplexes show some features is confirmed. The sequences adopting biologically relevant G-quadruplexes also show a tendency not to turn right, a feature that may suggest that many As can exert some destabilization effect on G-quadruplex formation. If the start and end points in this presentation of the trajectory are identical, then the sequence consists of the same number of As and Ts and the same number of Gs and Cs. If the second half of the trajectory is identical to the first, then the sequence is a perfect palindrome, e.g., Pal1: d (GAGTCTGCAGACTC). However, the start and end points of imperfect palindromic sequences are not identical. Irrespective of the central sequence, which is not part of the palindromic region (black lines), the trajectory consists of two antiparallel sections, e.g., Pal2: d(GAGTCTGgggCAGACTC), Pal3: d(GAGTCTGtgaagCAGACTC) and Pal4: d(GAGGGaCCCTC).

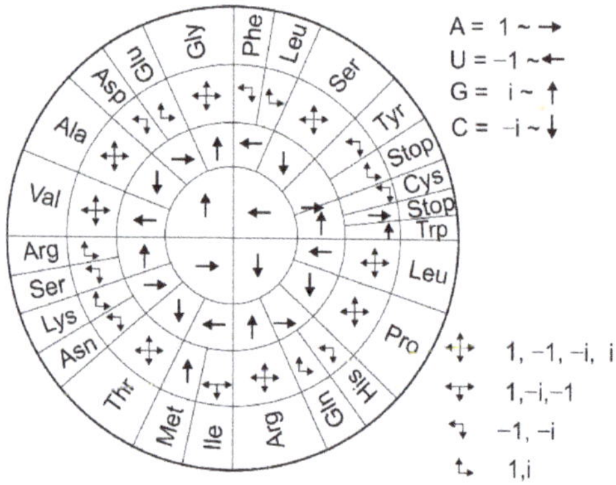

Figure 7. Vector representation of a genetic code.

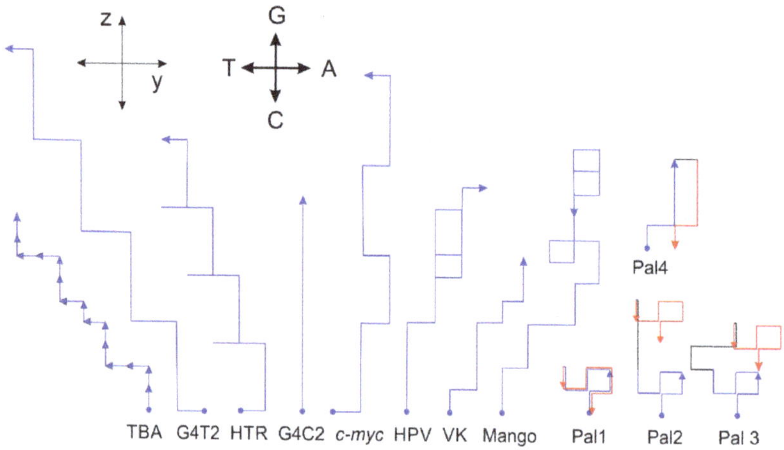

Figure 8. Selected sequences represented by orthogonal vector analysis. The direction of each oligonucleotide is determined by the vector, analogous to that defined in Figure 7. Each oligonucleotide is represented by arrows, analogical to that used in Figure 7. Traces corresponding to sequences forming G-quadruplexes do not tend to point more significantly to the right. The first half trajectory (blue) of the palindromic sequence (Pal) is identical to that of the second (red). Spacers are shown in black lines.

3. Concluding Remarks

The orthogonal system can easily be used for all types and sizes of nucleic acids. It can be adapted to search for tandem forward and inverse repeats, and is, of course, ideal for sequences featuring non-canonical motifs. An indirect side effect of the method is that this presentation offers a rational explanation of why the G4Hunter algorithm provides such promising scores for i-motifs and G-quadruplexes. In addition, the system also explains the weaknesses of the G4Hunter algorithm. An orthogonal system allows any nucleic acid sequences to be presented in numerical, color, and vector representations. The system is particularly efficient at identifying sequential domains responsible for a wide range of biological functions. Nevertheless, a deviation from orthogonality offers a significant improvement in the prediction of G-quadruplex adoption from a specific sequence. Although the quasi-orthogonal system loses its perfect symmetry, it allows for the possibility of distinguishing between G-quadruplexes consisting of loops featuring pure As or Ts nucleotides, a feature which is not possible with the G4hunter algorithm and the orthogonal system. For example, this system would explain why the presence of As reduces the likelihood of G-quadruplex formation.

Funding: This work was supported by the Slovak Grant Agency (1/0138/20).

Institutional Review Board Statement: Not applicable.

Informed Consent Statement: Not applicable.

Data Availability Statement: The data presented in this study are available in the article.

Conflicts of Interest: The author declares no conflict of interest. The funders had no role in the design of the study; in the collection, analyses, or interpretation of data; in the writing of the manuscript; or in the decision to publish the results. The data presented in this study are available in the article.

References

1. Takahashi, S.; Sugimoto, N. Stability prediction of canonical and non-canonical structures of nucleic acids in various molecular environments and cells. *Chem. Soc. Rev.* **2020**, *49*, 8439–8468. [CrossRef] [PubMed]
2. Takahashi, S.; Sugimoto, N. Roles of non-canonical structures of nucleic acids in cancer and neurodegenerative diseases. *Nucleic Acids Res.* **2021**, *49*, 7839–7855.
3. Weiner, A.M. SINEs and LINEs: The art of biting the hand that feeds you. *Curr. Opin. Cell Biol.* **2002**, *14*, 343–350. [CrossRef]
4. Siggers, T.; Gordân, R. Protein-DNA binding: Complexities and multi-protein codes. *Nucleic Acids Res.* **2014**, *42*, 2099–2111. [CrossRef]
5. Balasubramaniyam, T.; Oh, K.I.; Jin, H.S.; Ahn, H.B.; Kim, B.S.; Lee, J.H. Non-Canonical Helical Structure of Nucleic Acids Containing Base-Modified Nucleotides. *Int. J. Mol. Sci.* **2021**, *22*, 9552. [CrossRef]
6. Roy, A.; Raychaudhury, C.; Nandy, A. Novel techniques of graphical representation and analysis of DNA sequences—A review. *J. Biosci.* **1998**, *23*, 55–71. [CrossRef]
7. Cristea, P.D. Conversion of nucleotides sequences into genomic signals. *J. Cell. Mol. Med.* **2002**, *6*, 279–303. [CrossRef]
8. Cristea, P.D. Representation and analysis of DNA sequences. In *Genomic Signal Processing and Statistics: EURASIP Book Series in Signal Processing and Communications*; Dougherty, E.R., Ed.; Hindawi Pub. Corp.: New York, NY, USA, 2005; Volume 2, pp. 15–66.
9. Mendizabal-Ruiz, G.; Román-Godínez, I.; Torres-Ramos, S.; Salido-Ruiz, R.A.; Morales, J.A. On DNA numerical representations for genomic similarity computation. *PLoS ONE* **2017**, *12*, e0173288. [CrossRef]
10. Voss, R.F. Evolution of long-range fractal correlations and 1/f noise in DNA base sequences. *Phys. Rev. Lett.* **1992**, *68*, 3805–3808. [CrossRef]
11. Kwan, H.K.; Arniker, S.B. Numerical representation of DNA sequences. In Proceedings of the 2009 IEEE International Conference on Electro/Information Technology, Windsor, ON, Canada, 7–9 June 2009; pp. 307–310. [CrossRef]
12. Benson, G. Tandem repeats finder: A program to analyze DNA sequences. *Nucleic Acids Res.* **1999**, *27*, 573–580. [CrossRef]
13. Brázda, V.; Kolomazník, J.; Lýsek, J.; Hároníková, L.; Coufal, J.; Št'astný, J. Palindrome analyser—A new web-based server for predicting and evaluating inverted repeats in nucleotide sequences. *Biochem. Biophys. Res. Commun.* **2016**, *478*, 1739–1745. [CrossRef] [PubMed]
14. Brazda, V.; Kolomaznik, J.; Mergny, J.L.; Stastny, J. G4Killer web application: A tool to design G-quadruplex mutations. *Bioinformatics* **2020**, *36*, 3246–3247. [CrossRef]
15. Brázda, V.; Kolomazník, J.; Lýsek, J.; Bartas, M.; Fojta, M.; Šťastný, J.; Mergny, J.L. G4Hunter web application: A web server for G-quadruplex prediction. *Bioinformatics* **2019**, *35*, 3493–3495. [CrossRef] [PubMed]
16. Kikin, O.; D'Antonio, L.; Bagga, P.S. QGRS Mapper: A web-based server for predicting G-quadruplexes in nucleotide sequences. *Nucleic Acids Res.* **2006**, *34*, W676–W682. [CrossRef] [PubMed]
17. Oh, K.I.; Kim, J.; Park, C.J.; Lee, J.H. Dynamics Studies of DNA with Non-canonical Structure Using NMR Spectroscopy. *Int. J. Mol. Sci.* **2020**, *21*, 2673. [CrossRef]
18. Bedrat, A.; Lacroix, L.; Mergny, J.L. Re-evaluation of G-quadruplex propensity with G4Hunter. *Nucleic Acids Res.* **2016**, *44*, 1746–1759. [CrossRef]
19. Macaya, R.F.; Schultze, P.; Smith, F.W.; Roe, J.A.; Feigon, J. Thrombin-binding DNA aptamer forms a unimolecular quadruplex structure in solution. *Proc. Natl. Acad. Sci. USA* **1993**, *90*, 3745–3749. [CrossRef]
20. Demkovičová, E.; Bauer, L'.; Krafčíková, P.; Tlučková, K.; Tóthova, P.; Halaganová, A.; Valušová, E.; Víglaský, V. Telomeric G-Quadruplexes: From Human to Tetrahymena Repeats. *J. Nucleic Acids* **2017**, *2017*, 9170371. [CrossRef]
21. Greider, C.W.; Blackburn, E.H. A telomeric sequence in the RNA of Tetrahymena telomerase required for telomere repeat synthesis. *Nature* **1989**, *337*, 331–337. [CrossRef]
22. Wang, Y.; Patel, D.J. Solution structure of the Tetrahymena telomeric repeat d(T2G4)4 G-tetraplex. *Structure* **1994**, *2*, 1141–1156. [CrossRef]
23. Ambrus, A.; Chen, D.; Dai, J.; Bialis, T.; Jones, R.A.; Yang, D. Human telomeric sequence forms a hybrid-type intramolecular G-quadruplex structure with mixed parallel/antiparallel strands in potassium solution. *Nucleic Acids Res.* **2006**, *34*, 2723–2735. [CrossRef]

24. Mathad, R.I.; Hatzakis, E.; Dai, J.; Yang, D. c-MYC promoter G-quadruplex formed at the 5′-end of NHE III1 element: Insights into biological relevance and parallel-stranded G-quadruplex stability. *Nucleic Acids Res.* **2011**, *39*, 9023–9033. [CrossRef] [PubMed]
25. Brcic, J.; Plavec, J. NMR structure of a G-quadruplex formed by four d(G4C2) repeats: Insights into structural polymorphism. *Nucleic Acids Res.* **2018**, *46*, 11605–11617. [PubMed]
26. Tlučková, K.; Marušič, M.; Tóthová, P.; Bauer, L.; Šket, P.; Plavec, J.; Viglasky, V. Human papillomavirus G-quadruplexes. *Biochemistry* **2013**, *52*, 7207–7216. [CrossRef]
27. Kocman, V.; Plavec, J. A tetrahelical DNA fold adopted by tandem repeats of alternating GGG and GCG tracts. *Nat. Commun.* **2014**, *5*, 5831. [CrossRef]
28. Trachman, R.J., 3rd; Autour, A.; Jeng, S.C.Y.; Abdolahzadeh, A.; Andreoni, A.; Cojocaru, R.; Garipov, R.; Dolgosheina, E.V.; Knutson, J.R.; Ryckelynck, M.; et al. Structure and functional reselection of the Mango-III fluorogenic RNA aptamer. *Nat. Chem. Biol.* **2019**, *15*, 472–479. [CrossRef] [PubMed]
29. Sjekloća, L.; Ferré-D'Amaré, A.R. Binding between G Quadruplexes at the Homodimer Interface of the Corn RNA Aptamer Strongly Activates Thioflavin T Fluorescence. *Cell Chem. Biol.* **2019**, *26*, 1159–1168.e4. [CrossRef]
30. Anastassiou, D. Frequency-domain analysis of biomolecular sequences. *Bioinformatics* **2000**, *16*, 1073–1081. [CrossRef]

International Journal of *Molecular Sciences*

MDPI

Review

(Dys)function Follows Form: Nucleic Acid Structure, Repeat Expansion, and Disease Pathology in *FMR1* Disorders

Xiaonan Zhao * and Karen Usdin *

Laboratory of Cell and Molecular Biology, National Institute of Diabetes, Digestive and Kidney Diseases, National Institutes of Health, Bethesda, MD 20892, USA

* Correspondence: xiaonan.zhao@nih.gov (X.Z.); karenu@nih.gov (K.U.); Tel.: +1-301-451-6322 (X.Z.); +1-301-496-2189 (K.U.)

Abstract: Fragile X-related disorders (FXDs), also known as *FMR1* disorders, are examples of repeat expansion diseases (REDs), clinical conditions that arise from an increase in the number of repeats in a disease-specific microsatellite. In the case of FXDs, the repeat unit is CGG/CCG and the repeat tract is located in the 5′ UTR of the X-linked *FMR1* gene. Expansion can result in neurodegeneration, ovarian dysfunction, or intellectual disability depending on the number of repeats in the expanded allele. A growing body of evidence suggests that the mutational mechanisms responsible for many REDs share several common features. It is also increasingly apparent that in some of these diseases the pathologic consequences of expansion may arise in similar ways. It has long been known that many of the disease-associated repeats form unusual DNA and RNA structures. This review will focus on what is known about these structures, the proteins with which they interact, and how they may be related to the causative mutation and disease pathology in the *FMR1* disorders.

Keywords: fragile X-associated primary ovarian insufficiency (FXPOI); fragile X-associated tremor/ataxia syndrome (FXTAS); fragile X syndrome (FXS); repeat instability; repeat expansion; chromosome fragility; RNA gain-of-function; repeat-associated non-AUG (RAN) translation; repeat-mediated gene silencing

Citation: Zhao, X.; Usdin, K. (Dys)function Follows Form: Nucleic Acid Structure, Repeat Expansion, and Disease Pathology in *FMR1* Disorders. *Int. J. Mol. Sci.* **2021**, *22*, 9167. https://doi.org/10.3390/ijms22179167

Academic Editors: Vaclav Brazda and Richard Bowater

Received: 30 July 2021
Accepted: 18 August 2021
Published: 25 August 2021

1. Introduction

Repeat expansion diseases (REDs) are a group of human diseases caused by the presence of a large number of repeats in a microsatellite or short tandem repeat (STR) [1]. Unlike the microsatellite instability caused by a mismatch repair (MMR) deficiency that affects STRs genome-wide, each of these diseases results from expansion at a single disease-specific locus. While contractions of the repeat are occasionally seen, expansions predominate in both somatic and germline cells. The propensity to expand becomes apparent when the repeat number exceeds a certain critical threshold, with expansions increasing in frequency as the repeat number increases. These expansions occur in both intergenerational transmission and in the somatic cells during the lifetime of the individual. In general, for many of the diseases that are not congenital, the age at onset decreases and disease severity or disease penetrance increase with increasing repeat number [1]. As will be discussed in more detail later in this review, the characteristic features and genetic requirements for expansion in many of these diseases suggest that they may arise in similar ways. Furthermore, the pathology in many of these diseases may also arise from similar consequences of the expansion process.

More than 40 REDs have been identified to date, including Huntington's disease (HD), myotonic dystrophy type 1 (DM1), *C9orf72*-associated amyotrophic lateral sclerosis/frontotemporal dementia (ALS/FTD), and the *FMR1* disorders, also known as the fragile X (FX)-related disorders (FXDs). In the case of the FXDs, the repeat unit is CGG/CCG and the repeat tract is located in the 5′ UTR of *FMR1*, a gene located on the long arm of the X chromosome (reviewed in [2]). Normal alleles have 15–45 repeats, with alleles with

~30 repeats being the most common. In this context the repeat tract is thought to play a role in the regulation of synthesis of the *FMR1* gene product, FMRP, via the production of a protein generated from an upstream open reading frame using a near-cognate AUG codon [3]. Normal alleles are relatively stable. However, larger alleles tend to expand both in germline [4] and somatic cells [5]. Most of the historical focus has been on germline expansion, and while somatic expansion does play a role in other REDs [6,7], its role in the FXDs is unknown. Expanded repeats have paradoxical effects on expression of the *FMR1* gene, with alleles with 55–200 repeats (known as premutation (PM) alleles [2,8]) being hyper-expressed, and alleles with >200 repeats (known as full mutation (FM) alleles [9,10]) being epigenetically silenced. The net result is that females with PM alleles are at risk of a form of female infertility known as fragile X-associated primary ovarian insufficiency (FXPOI), and both PM males and females are at risk for a neurodegenerative condition known as fragile X-associated tremor/ataxia syndrome (FXTAS). The PM is seen in 1:200 females and 1:400 males [11]. Penetrance for FXTAS increases with age and repeat number, with >60% of male PM carriers showing symptoms by age 70, as compared to ~16% of females [11]. FXPOI affects ~20–30% of PM carriers [12] and there is a non-linear relationship between repeat number and FXPOI risk that is not well understood [13,14]. Cells from PM carriers show splicing abnormalities [15], lamin A/C dysregulation [16], mitochondrial disfunction, and the presence of intranuclear inclusions in the brain and ovary [17,18]. Female PM carriers are at risk of transmitting FM alleles to their children, with the risk of doing so being related to their repeat number, the number of AGG interruptions seen at the 5′ end of the repeat tract, and maternal age [19]. The risk of maternal transmission of a FM allele approaches 100% when the repeat number exceeds 90, irrespective of age or interruptions [19]. In contrast, male PM carriers do not transmit FM alleles, likely due to the tendency of long repeat tracts to contract in sperm [20]. FM alleles are seen at a frequency of ~1 in 2000 to 1 in 7000 in the general population, with a variation in prevalence seen in different populations [21]. Most males who inherit FM alleles have fragile X syndrome (FXS), the most common monogenic cause of intellectual disability and autism [22]. Females tend to be less severely affected due to the protective effect of their second X chromosome. Silencing of the FM allele results in the loss of FMRP, a multi-functional protein best known for its role in negatively regulating the translation of genes important for learning and memory [23]. FM alleles are also associated with a folate-sensitive fragile site, a gap or constriction of the chromosome, coincident with the repeat [10]. Female FM fetuses also show a high frequency loss of the affected X chromosome, resulting in Turner syndrome [24].

As with other expansion-prone repeats, the CGG/CCG repeats responsible for the *FMR1* disorders form a variety of nucleic acid secondary structures (Figure 1). These structures have the potential to interfere with many biological processes. As such, they have the potential not only to cause the mutation responsible for the FXDs, but they may also be responsible for some of the pathological consequences of the mutation. Interestingly, many targets of FMRP form G4 structures to which the protein binds [25], and FMRP has also been implicated in R-loop processing [26], thus representing other ways that non-canonical nucleic acid structures and proteins intersect in these disorders. However, in this review we will focus primarily on what is known about the DNA and RNA structures formed by the FX repeats themselves and their biological effects in the context of both expansion and disease pathology in the *FMR1* disorders.

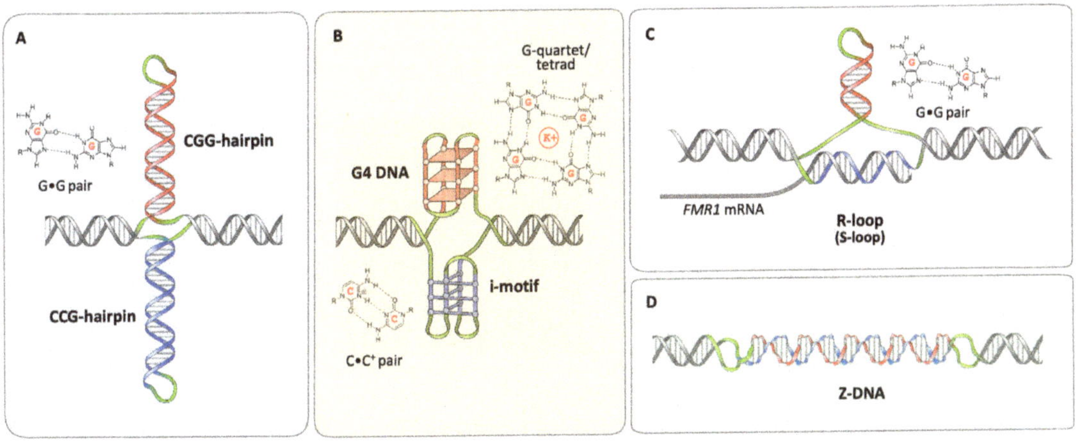

Figure 1. Generic representation of the types of DNA and RNA structures formed by FX repeats. Structures shown include hairpins formed by each strand of the repeat (**A**), a quadruplex or G4 DNA structure and an i-motif structure (**B**), an R-loop with associated hairpin formed by the non-template strand resulting in an S-loop (**C**) and Z-DNA (**D**). The CGG strand is shown in red and the CCG strand in blue. Unpaired loops regions are shown in green and the non-repetitive flanking DNA is shown in grey. Note that in addition to unpaired loop bases, some of these structures also contain non-Watson Crick base pairs or mismatches. The structures of the constituent non-canonical base interactions are shown alongside each structure.

2. Secondary Structures Formed by FX Repeats

Like the repeats responsible for many of the other REDs, individual DNA strands of the FX repeat can form stable hairpins containing a mixture of Watson–Crick and non-Watson–Crick base pairs or mismatches [27–32]. CGG-DNA hairpins are the most stable of the hairpins formed by different trinucleotide repeats, with a $(CGG)_{15}$ hairpin having a Tm of 75 °C in physiologically reasonable buffers [33]. In contrast, similarly sized CCG hairpins have a Tm of 30–37 °C depending on pH, and are less stable than CGG, CTG, and CAG repeats [33]. While similar experiments have not been performed for CGG and CCG repeats, evidence from cleavage by zinc finger nucleases specific for CAG and CTG repeats provides evidence for the formation of such hairpins in mammalian cells [34]. In principle, hairpin formation by both strands of the repeat could result in a cruciform-like structure, as illustrated in Figure 1A. CGG repeats also form stable hairpins in RNA [35–37]. In addition to hairpins, the formation of intramolecular and intermolecular G4 quadruplex structures by both CGG repeat-containing DNA and RNA have been reported in some studies [27,38–45] (Figure 1B). These structures are sometimes overlooked because CGG hairpins form readily and once formed are very stable, whilst the G4 structures are only seen in the presence of K^+ [27]. Nonetheless, once formed these structures are stable at temperatures of >85 °C with physiologically reasonable K^+ concentrations [27]. The CCG strand of the repeat has also been shown to form a variety of intramolecular and intermolecular four-stranded structures, including i-motif structures containing intercalated C•C^+ base pairs [46–48] as illustrated in Figure 1B.

In addition to intrastrand DNA and RNA structures, the 5′ end of the *FMR1* gene forms a stable R-loop in vivo, as illustrated in Figure 1C [49–52]. In these structures, the G-rich transcript forms a hybrid with the C-rich template strand, likely during transcription. This results in a three-stranded structure involving an RNA:DNA hybrid and a displaced DNA strand. The *FMR1* R-loop extends well into the 5′ and 3′ flanking regions [49,51], regions that also have a strong GC skew [53]. Non-denaturing bisulfite mapping shows that most of the cytosines on the non-template strand are resistant to bisulfite modification [49], consistent with the formation of intrastrand folded structures by the non-template strand. An R-loop containing a non-template-strand hairpin, sometimes referred to as an S-loop (for slipped hairpin R-loops), is illustrated in Figure 1C, but an R-loop with a G4 structure, a

G-loop, is also possible. In either case the occasional modified cytosines seen on the bisulfite-treated non-template strand [49] would correspond to bases in the loops of these structures. Structures formed by the non-template strand may in turn help stabilize the R-loop [54]. Since the CGG/CCG repeats at the *FMR1* locus are bidirectionally transcribed, they can also form double R-loops [55]. In addition to these inter- and intra-strand structures, there is evidence that even the CGG•CCG duplex is atypical, adopting a left-handed Z-DNA conformation as illustrated in Figure 1D [56].

3. Repeat Expansion

One important clue to the process that causes repeat expansion in the REDs has emerged from recent genome-wide association studies (GWAS) in different RED patient cohorts. These studies have implicated the MMR proteins MSH3, MLH1, and MLH3 as important modifiers of somatic expansion risk and/or age at onset/disease severity in many REDs [6,7,57–63]. MSH3 forms a heterodimer with MSH2 in the MutSβ complex, one of the two mismatch recognition complexes involved in MMR in mammals, while MLH1 and MLH3 form the heterodimer MutLγ, a complex that acts downstream of MutSβ in the MMR pathway [64]. Notably, single nucleotide polymorphisms associated with increased MSH3 expression are associated with increased somatic expansion in an HD patient cohort [7], suggesting that, unlike the microsatellite instability associated with certain cancers, functional MMR proteins are required for expansion. A requirement of these same proteins for repeat expansion is seen in a mouse model of FXDs as well as other mouse models of REDs (reviewed in [65,66]). A role for MMR in repeat expansion is consistent with the fact that many of the unusual structures formed by the repeats contain mismatches or regions of single-strandedness that can be bound by MutSβ and the related protein MutSα, a heterodimer of MSH2 and MSH6 [64,67]. While GWAS studies of factors that affect germline expansion risk have not yet been performed for REDs, in the FXD mouse model it is known that the same factors that affect somatic expansion risk also affect germline expansion risk (reviewed in [65]).

However, how the MMR substrates arise is unclear. It may be that they form during strand slippage or strand displacement during replication or repair. Since expansion in many REDs can occur in non-dividing cells like oocytes and neurons [19,68], repair may be a more likely source of these substrates, at least in disease-relevant cell types. One model for expansion invokes a role of base excision repair (BER) of 7,8-dihydro-8-oxoguanine (8-oxoG), the most common oxidation product in DNA, with strand slippage or strand displacement during BER generating hairpin loop-outs that are bound by the MutS proteins [69]. Hairpin formation may trigger multiple rounds of BER since guanines in the loop of hairpins are susceptible to DNA damage and are less likely to be repaired [70]. A role for BER would be consistent with the fact that loss of the 7,8-dihydro-8-oxoguanine glycosylase (OGG1) leads to reduced expansion in the liver (but not in the brain or gametes) of an HD mouse model [69]. Loss of NEIL1, the other major DNA glycosylase able to remove 8-oxoG, also led to a decline in expansion in HD mouse brain [71]. GWAS studies in other REDs have not as yet identified a role for BER proteins in the expansion process [6,7,57–63]. However, this does not definitively rule out a role for BER. A role for oxidative damage in repeat expansion is supported by the observation that oxidizing agents increase repeat expansion in a mouse model of FXDs [72] and in cell models of HD [73]. However, antioxidants have no effect on an FXD mouse cell model (Miller and Usdin, unpublished observations), and only a modest effect on repeat expansions in HD mouse models [74,75]. Thus, spontaneous oxidative damage may not be a major contributor to expansion under normal circumstances.

Furthermore, expansions in human PM carriers require transcription of the *FMR1* gene or at least for the allele to be in a region of transcriptionally competent chromatin [76]. Canonical BER has no such strict transcriptional requirement, although it is possible that transcription provides the opportunity for secondary structures to form that in turn would be predisposed to oxidative damage [70]. An alternative source of MMR substrates may be transcription itself, which can result in the formation of an S-loop as illustrated in Figure 1C. The S-loop may be the MMR target. It is also possible that resolution of the R-loop would then leave the template strand unable to bind its complementary strand and since the CCG-rich strand can also form hairpins, this could result in the cruciform-like double loop-out structure shown in Figures 1A and 2A that could also be a target for MMR.

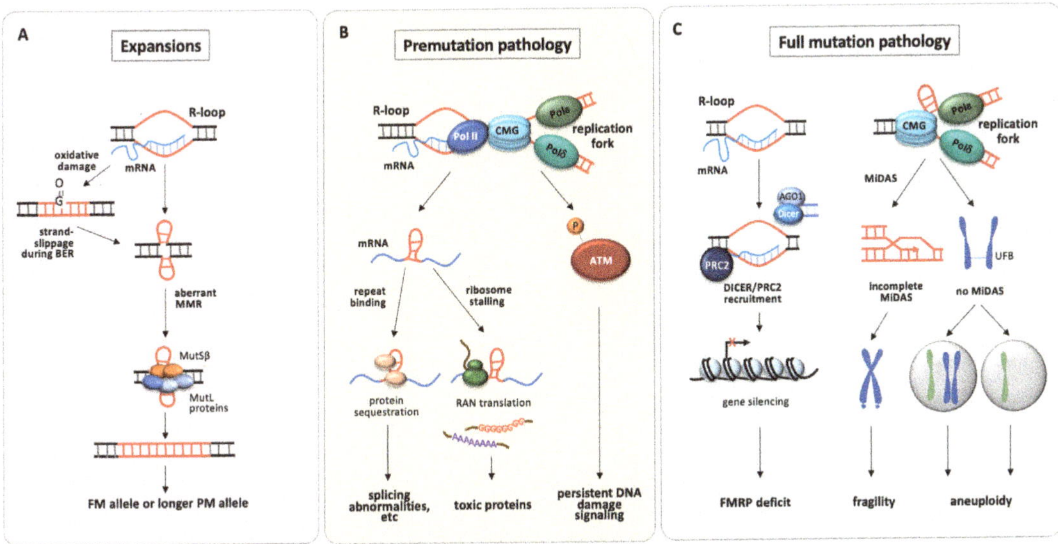

Figure 2. Models for the roles of non-canonical DNA and RNA structures in the etiology of the repeat expansion mutation and the resultant pathology seen in individuals with PM and FM alleles. (**A**) R-loops formed during transcription contain a region of single-stranded DNA that would be prone to oxidative damage. Base excision repair of such damage could generate loop-outs by strand-slippage or strand-displacement during repair synthesis [71,77,78]. R-loops may also facilitate the direct formation of loop-outs, first by the unpaired non-template strand when the template strand is involved in the RNA:DNA hybrid, and subsequently by the template strand after the R-loop is resolved. The loop-outs are bound by mismatch repair factors like MutSβ and MutLγ [79–82] and are processed via a DSB [83] to generate expansions. (**B**) CGG-hairpins in the *FMR1* transcript can bind and sequester proteins [84,85] or trigger RAN translation of toxic proteins [86,87]. Persistent R-loops, perhaps exacerbated by replication-transcription collisions may result in DSBs that cause persistent DNA damage signaling [49]. (**C**) R-loop formation allows the recruitment of PRC2 to the *FMR1* gene [88]. DICER complexes associated with dsRNA produced from the *FMR1* locus [36] may also contribute to silencing by facilitating recruitment of SUV39H [89]. Secondary structures may cause stalling of the replication fork that triggers MiDAS [90]. Failure to complete MiDAS results in chromosome fragility, while failure to initiate MiDAS results in the formation of UFBs and ultimately the gain or loss of the affected X chromosome [90].

Work on a mouse model of the FXDs shows a dependence on both MutSβ and MutLγ for repeat expansion [79–82,91], consistent with GWAS of REDs. However, other genetic modifiers of expansion risk in this mouse model suggest that the MMR protein-dependent expansion pathway differs in key ways from canonical MMR. For example, in addition to MutSβ, MutSα also plays an important role in expansion [64], as do MutLα and MutLβ, two other MLH1 containing complexes found in mammals [80]. MutSβ and MutSα are not known to act together in MMR. Neither are MutLγ and MutLα, while the contribution of MutLβ to MMR is unclear. Furthermore, DNA ligase IV, which is required for non-homologous end-joining (NHEJ), a form of double-strand break (DSB) repair, protects against expansion in a mouse model of FXDs [83]. This suggests that expansion involves a DSB intermediate. It may be that a DSB results from cleavage of a double loop-out by MutLγ which normally cuts the strand opposite a mismatch [92]. However, the details of this process and the downstream events that result in the generation of an expansion are still unknown.

4. Consequences of Repeat Expansion

4.1. Pathology in PM Carriers

Most work on PM pathology has focused on FXTAS rather than FXPOI. While relatively little is known about which cells are most vulnerable in these disorders, it could be that similar mechanisms act to reduce cell viability in both cases. The fact that FM carriers who make little, if any, *FMR1* mRNA and FMRP, do not show FXTAS or FXPOI symptoms suggests that the CGG-repeat-containing RNA produced from PM alleles is responsible, rather than any decline in the amount of FMRP. An RNA-based pathology is supported by the demonstration that ectopic expression of the CGG-tract causes reduced cell viability [72,93–97], the production of inclusions [94,98,99], disruption of the nuclear lamin A/C architecture in neuronal cell lines [16], and neurodegeneration in both flies [94] and mice [96]. It also alters the ovarian response to gonadotropins and results in reduced fertility in mice {Shelly, 2021}. Interestingly, PM alleles show elevated levels of *FMR1* transcription initiation [8]. R-loop formation could potentially contribute to this via its effects on chromatin decondensation [100], inhibition of binding of DNA methyltransferases [101], or the recruitment of activators including the ten-eleven translocation (TET) DNA demethylases [102]. It is also possible that the formation of hairpins or G4 DNA by the non-template strand predisposes these regions to oxidative damage, in turn increasing transcription, as has been described for the *PCNA* gene [103].

Several different models that invoke RNA hairpins formed by CGG-repeats have been proposed to explain PM pathology, as illustrated in Figure 2B. One such model proposes that binding of specific proteins to the CGG-repeat-containing RNA hairpins results in them being sequestered and unable to carry out their normal activities [84,104]. Numbered amongst these proteins are the splicing factor src-associated in mitosis of 68 kDa (Sam68) [104], and the DiGeorge syndrome critical region gene 8 (DGCR8) protein [84], a double-stranded RNA-binding protein involved in the microRNA (miRNA)-processing pathway. Consistent with a role for sequestration of these proteins, Sam68-mediated splicing abnormalities are seen in FXTAS patient cells [104], and decreased levels of mature miRNAs are seen in the brains of FXTAS patients. This is associated with decreased dendritic complexity and reduced viability of neuronal cells in culture that can be reversed by overexpression of DGCR8 [84].

Repeat-associated non-AUG (RAN) translation, a form of translation that initiates at near cognate codons upstream of or within the repeat, has also been suggested to account for PM pathology [85,86,105–107], as previously proposed for other REDs [87]. RAN translation is thought to be triggered by the stalling of the ribosome by RNA hairpins, consistent with work suggesting that kinetic barriers to the ribosome favor initiation at otherwise suboptimal initiation codons located upstream of the true initiation codon [108]. In reporter constructs with PM-sized repeat tracts, RAN translation can occur in both the sense strand producing polyglycine (FMRpolyG), polyalanine (FMRpolyA), and polyarginine (FMRpolyR)-containing proteins, and the antisense strand producing polyproline (ASFMRpolyP), polyalanine (ASFMRpolyA), and polyarginine (ASFMRpolyR)-containing proteins. FMRpolyG and FMRpolyA can be seen in intranuclear neuronal inclusions in FXTAS patients using immunochemical detection methods [109–112], and overexpression of FMRpolyG in particular is toxic in various model systems [86,107].

Interestingly, there are two other potential intersections of RNA structure and protein interactions in RAN translation. The first is related to the fact that many repeat-containing transcripts activate the double-stranded RNA-dependent protein kinase PKR [113,114], presumably due to their ability to form hairpins. This results in an increase in the phosphorylation of eukaryotic translation initiation factor 2 subunit alpha (eIF2α) which in turn exacerbates RAN translation [115]. Supporting the role of PKR in RED pathology is the fact that its inhibition reduces RAN protein expression and improves disease symptoms in a mouse model of *C9orf72* ALS/FTD [114]. Whether PKR plays a similar role in the context of CGG-repeat expansion remains to be seen. The repeats did not cause significant PKR activation in a tissue culture model [36]; however, whether this is due to the cell type used or the level of CGG-RNA produced is unclear. The second intersection with RNA structure is the demonstration that FMRpolyG binds CGG-RNA quadruplex structures in vitro, with evidence of G4 RNA promoting the liquid-to-solid transition and aggregate formation of FMRpolyG in a FXTAS mouse model [45]. However, overexpression of FMRpolyG is not always associated with FXTAS pathology in mice [116]. Furthermore, FMRpolyG is not detected by mass spectroscopy of brain extracts of FXTAS patients [117] and is only present at very low levels in inclusions isolated from such patients [17]. This raises the possibility that despite the immunological detection of these proteins in patient samples, their concentration may be too low to account for the pathology observed in PM carriers.

In addition to PKR activation by the repeat-containing RNA hairpins, elevated type 1 interferon (IFN) signaling is seen in *C9orf72* ALS/FTD [118]. This process, like PKR activation, is part of the normal cellular response to double-stranded RNAs. In ALS/FTD it is associated with sterile inflammation and neuronal death. Cell death can be suppressed by inhibitors of Janus kinase, a key component of the major signaling pathway activated by IFNs but not by PKR inhibitors [118]. Whether a similar effect is seen for the CGG-RNA hairpins in PM carriers remains to be seen.

R-loop formation at the *FMR1* locus has also been proposed as a source of pathology in PM carriers [49,119] as illustrated in Figure 2B. R-loops are prone to single-stranded breaks and DSBs resulting from clustered single-stranded breaks [120]. Hyperphosphorylation of ataxia-telangiectasia mutated kinase (ATM), a consequence of DSBs, is seen in FXTAS cell and animal models, and γH2AX, a marker of double-strand breaks, is present in nuclear inclusions in FXTAS patient tissue [17,97]. However, while mutations that affect R-loop levels genome-wide are associated with a variety of neurodegenerative diseases [121], given the prevalence of R-loops in the genome it is unclear whether the addition of a single, albeit a large and stable, R-loop at a PM allele would be sufficient to trigger neuronal cell death.

In addition to pathology characteristic of PM carriers, many carriers of large PM alleles, or rare FM alleles that do not become silenced, show reduced levels of FMRP that could contribute to some of the symptoms seen in this population [122,123]. The reduced FMRP levels are thought to be due to the stalling of the 40S ribosomal subunit by the hairpin formed by the repeats in the 5′ UTR of the *FMR1* transcript [122,124].

4.2. Pathology in FM Carriers

4.2.1. *FMR1* Gene Silencing

The 5′ end of the *FMR1* gene in FM carriers is epigenetically modified, resulting in gene silencing and an absence or deficiency in FMRP. In FM carriers the DNA in this region of the gene is hypermethylated and associated with modified histones typical of heterochromatin, including histone H3 trimethylated at lysine 27 (H3K27Me3) [125]. H3K27Me3 is deposited by the polycomb repressive complex 2 (PRC2). R-loops are important for PRC2-mediated gene silencing at several loci [88]. PRC2 binds to R-loops directly and drives R-loop production in Drosophila [126]. PRC2 has also been reported to bind to G-rich RNA and to G4-forming RNA sequences in particular [127]. R-loops have also been implicated in silencing in both FXS and a related RED, Friedreich ataxia [52]. The *FMR1* transcript is important for recruiting PRC2 to the 5′ end of FM alleles that have been reactivated with 5-deazadeoxycytidine (AZA), a DNA methyltransferase inhibitor [125]. Inhibition of PRC2 or blocking its recruitment to the *FMR1* 5′ UTR prevents H3K27 trimethylation at this locus [50]. This in turn prevents the remethylation and resilencing of FM alleles that typically occur after AZA is withdrawn [50,125]. These data would be consistent with a model in which PRC2 binds to the 5′ end of the *FMR1* transcript, while the transcript is also simultaneously bound to the 5′ end of the *FMR1* gene via an R-loop. This would tether PRC2 in the vicinity of the *FMR1* promoter, as illustrated in Figure 2C. PRC2-mediated H3K27 trimethylation is favored by loss of marks of active chromatin [128–130]. This loss could be triggered by R-loop formation itself via increased transcription termination [53,131], or as a downstream consequence of the induction of DNA damage at R-loops [132,133]. Silencing has traditionally been considered to occur when the repeat number exceeds 200 based on data from Southern blotting; however, higher-resolution techniques like capillary electrophoresis suggest that the threshold may be higher than this [134]. What triggers the transition from the hyper-expressed state to the silenced state is unknown and the role of an R-loop in gene silencing of FM alleles seems paradoxical given its proposed roles in hyperexpression of PM alleles. However, there are many reports of similar paradoxical effects of R-loops in the literature (see [135] for a good recent discussion). The R-loop formed by an FM allele while it was still transcriptionally active would be more stable than an R-loop formed on a PM allele. As such the R-loops formed on FM alleles may form a more effective block to transcription elongation. This would result in a larger drop in H3K36me3 levels, which in turn would favor H3K27 trimethylation.

Members of the argonaute protein family and the endoribonuclease DICER1, proteins that are important for RNA-induced gene silencing via the small interfering RNA (siRNA) pathway, have also been suggested to play a role in *FMR1* gene resilencing [136]. This presumably reflects a role for double-stranded RNA in the silencing process. However, whether the source of double-stranded RNA is the RNA hairpin formed by the FX repeats or the product of the annealing of the *FMR1* transcript and an antisense transcript from this locus [137] is unclear. DICER-mediated gene silencing is thought to be accomplished via SUV39H-mediated trimethylation of H3K9 [89]. Since inhibitors of H3K9 methylation [138] and H3K27 trimethylation [50] delay resilencing after AZA treatment, methylation at both residues might be involved in restoring DNA methylation at this locus.

4.2.2. Chromosome Fragility

Fragile sites (FSs) are breaks or gaps that are visible in otherwise condensed chromosomes in metaphase spreads of cells treated with different classes of replication inhibitors [139]. They are thought to represent regions of the genome that are difficult to replicate. In the case of the *FMR1* locus, expression of the fragile site, FRAXA, is induced by folate-stress that causes nucleotide pool imbalances [140]. CGG repeats are known to be difficult to replicate both in vitro [27] and in vivo [141], and replication stalling is seen at the 5′ end of the endogenous *FMR1* gene [142]. Given the ability of CGG-repeat structures to block DNA synthesis in vitro [27], these structures could account for the replication difficulty shown in Figure 2C. The formation of a block to the replication fork is consistent with the fact that FM alleles are prone to mitotic DNA synthesis (MiDAS) when subjected to folate-stress. MiDAS is thought to be a form of break-induced replication (BIR), a salvage pathway involved in the processing of stalled replication forks to allow replication of the chromosome to be completed [90]. Suppression of MiDAS prevents chromosome fragility, but alleles that fail to initiate BIR at all are associated with high levels of ultrafine bridges (UFBs), anaphase bridges involving single-stranded regions of DNA that are histone-free [46]. Failure to resolve these UFBs results in non-disjunction of the chromosomes and subsequent aneuploidy [90] that may account for the high frequency of Turner syndrome observed in female carriers of FM alleles [24].

Replication difficulties may also account for the fact that male PM carriers do not transmit FM alleles to their children since, unlike oocytes which are post-mitotic, male gametes undergo multiple rounds of replication prior to fertilization. As such, there may be selective pressure for smaller alleles in males that is not seen in females.

5. Concluding Remarks

While the ability of the FX repeats to form secondary structures of various sorts has been known for some time, work in recent years has begun to identify ways to target these structures or the downstream consequences of these structures, so as to ameliorate their effects. For example, CCG-repeat-containing antisense oligonucleotides (ASOs) reduce

R-loop formation and ameliorate some of the downstream consequences of the formation of RNA hairpins [143]. Small molecules that target CGG-RNA hairpins have also been shown to have beneficial effects in cell and mouse models of the PM [144–146]. Additionally, the ability of PKR to promote RAN translation can be inhibited by metformin [114], a widely used oral hypoglycemic agent used to treat type 2 diabetes. Thus, an understanding of the secondary structures formed by disease-associated repeats and their downstream consequences is beginning to reveal therapeutic opportunities that may be useful for treating these disorders.

Funding: The APC was made possible by funding to KU (1ZIADK057808) from the Intramural program of the NIDDK, National Institutes of Health, USA.

Conflicts of Interest: The authors declare no conflict of interest.

References

1. Paulson, H. Repeat expansion diseases. *Handb. Clin. Neurol.* **2018**, *147*, 105–123. [CrossRef] [PubMed]
2. Loesch, D.; Hagerman, R. Unstable mutations in the FMR1 gene and the phenotypes. *Adv. Exp. Med. Biol.* **2012**, *769*, 78–114. [CrossRef] [PubMed]
3. Rodriguez, C.M.; Wright, S.E.; Kearse, M.G.; Haenfler, J.M.; Flores, B.N.; Liu, Y.; Ifrim, M.F.; Glineburg, M.R.; Krans, A.; Jafar-Nejad, P.; et al. A native function for RAN translation and CGG repeats in regulating fragile X protein synthesis. *Nat. Neurosci.* **2020**, *23*, 386–397. [CrossRef] [PubMed]
4. Nolin, S.L.; Glicksman, A.; Tortora, N.; Allen, E.; Macpherson, J.; Mila, M.; Vianna-Morgante, A.M.; Sherman, S.L.; Dobkin, C.; Latham, G.J.; et al. Expansions and contractions of the FMR1 CGG repeat in 5508 transmissions of normal, intermediate, and premutation alleles. *Am. J. Med. Genet. A* **2019**, *179*, 1148–1156. [CrossRef] [PubMed]
5. Lokanga, R.A.; Entezam, A.; Kumari, D.; Yudkin, D.; Qin, M.; Smith, C.B.; Usdin, K. Somatic expansion in mouse and human carriers of fragile X premutation alleles. *Hum. Mutat.* **2013**, *34*, 157–166. [CrossRef]
6. Ciosi, M.; Maxwell, A.; Cumming, S.A.; Hensman Moss, D.J.; Alshammari, A.M.; Flower, M.D.; Durr, A.; Leavitt, B.R.; Roos, R.A.C.; Team, T.-H.; et al. A genetic association study of glutamine-encoding DNA sequence structures, somatic CAG expansion, and DNA repair gene variants, with Huntington disease clinical outcomes. *EBioMedicine* **2019**, *48*, 568–580. [CrossRef]
7. Genetic Modifiers of Huntington's Disease Consortium. CAG Repeat Not Polyglutamine Length Determines Timing of Huntington's Disease Onset. *Cell* **2019**, *178*, 887–900.e814. [CrossRef]
8. Tassone, F.; Hagerman, R.J.; Taylor, A.K.; Gane, L.W.; Godfrey, T.E.; Hagerman, P.J. Elevated levels of FMR1 mRNA in carrier males: A new mechanism of involvement in the fragile-X syndrome. *Am. J. Hum. Genet.* **2000**, *66*, 6–15. [CrossRef] [PubMed]
9. Fu, Y.H.; Kuhl, D.P.; Pizzuti, A.; Pieretti, M.; Sutcliffe, J.S.; Richards, S.; Verkerk, A.J.; Holden, J.J.; Fenwick, R.G., Jr.; Warren, S.T.; et al. Variation of the CGG repeat at the fragile X site results in genetic instability: Resolution of the Sherman paradox. *Cell* **1991**, *67*, 1047–1058. [CrossRef]
10. Verkerk, A.J.; Pieretti, M.; Sutcliffe, J.S.; Fu, Y.H.; Kuhl, D.P.; Pizzuti, A.; Reiner, O.; Richards, S.; Victoria, M.F.; Zhang, F.P.; et al. Identification of a gene (FMR-1) containing a CGG repeat coincident with a breakpoint cluster region exhibiting length variation in fragile X syndrome. *Cell* **1991**, *65*, 905–914. [CrossRef]
11. Hagerman, R.; Hagerman, P. Fragile X-associated tremor/ataxia syndrome: Pathophysiology and management. *Curr. Opin. Neurol.* **2021**, *34*, 541–546. [CrossRef] [PubMed]
12. Allen, E.G.; Charen, K.; Hipp, H.S.; Shubeck, L.; Amin, A.; He, W.; Hunter, J.E.; Sherman, S.L. Clustering of comorbid conditions among women who carry an FMR1 premutation. *Genet. Med.* **2020**, *22*, 758–766. [CrossRef]
13. Ennis, S.; Ward, D.; Murray, A. Nonlinear association between CGG repeat number and age of menopause in FMR1 premutation carriers. *Eur. J. Hum. Genet.* **2006**, *14*, 253–255. [CrossRef] [PubMed]
14. Sullivan, A.K.; Marcus, M.; Epstein, M.P.; Allen, E.G.; Anido, A.E.; Paquin, J.J.; Yadav-Shah, M.; Sherman, S.L. Association of FMR1 repeat size with ovarian dysfunction. *Hum. Reprod.* **2005**, *20*, 402–412. [CrossRef] [PubMed]
15. Tseng, E.; Tang, H.T.; AlOlaby, R.R.; Hickey, L.; Tassone, F. Altered expression of the FMR1 splicing variants landscape in premutation carriers. *Biochim. Biophys. Acta Gene Regul. Mech.* **2017**, *1860*, 1117–1126. [CrossRef]
16. Arocena, D.G.; Iwahashi, C.K.; Won, N.; Beilina, A.; Ludwig, A.L.; Tassone, F.; Schwartz, P.H.; Hagerman, P.J. Induction of inclusion formation and disruption of lamin A/C structure by premutation CGG-repeat RNA in human cultured neural cells. *Hum. Mol. Genet.* **2005**, *14*, 3661–3671. [CrossRef]
17. Iwahashi, C.K.; Yasui, D.H.; An, H.J.; Greco, C.M.; Tassone, F.; Nannen, K.; Babineau, B.; Lebrilla, C.B.; Hagerman, R.J.; Hagerman, P.J. Protein composition of the intranuclear inclusions of FXTAS. *Brain* **2006**, *129*, 256–271. [CrossRef]
18. Ma, L.; Herren, A.W.; Espinal, G.; Randol, J.; McLaughlin, B.; Martinez-Cerdeno, V.; Pessah, I.N.; Hagerman, R.J.; Hagerman, P.J. Composition of the Intranuclear Inclusions of Fragile X-associated Tremor/Ataxia Syndrome. *Acta Neuropathol. Commun.* **2019**, *7*, 143. [CrossRef]

19. Yrigollen, C.M.; Martorell, L.; Durbin-Johnson, B.; Naudo, M.; Genoves, J.; Murgia, A.; Polli, R.; Zhou, L.; Barbouth, D.; Rupchock, A.; et al. AGG interruptions and maternal age affect FMR1 CGG repeat allele stability during transmission. *J. Neurodev. Disord.* **2014**, *6*, 24. [CrossRef]
20. Reyniers, E.; Vits, L.; De Boulle, K.; Van Roy, B.; Van Velzen, D.; de Graaff, E.; Verkerk, A.J.; Jorens, H.Z.; Darby, J.K.; Oostra, B.; et al. The full mutation in the FMR-1 gene of male fragile X patients is absent in their sperm. *Nat. Genet.* **1993**, *4*, 143–146. [CrossRef] [PubMed]
21. Lozano, R.; Azarang, A.; Wilaisakditipakorn, T.; Hagerman, R.J. Fragile X syndrome: A review of clinical management. *Intractable Rare Dis. Res.* **2016**, *5*, 145–157. [CrossRef] [PubMed]
22. Hagerman, R.J.; Berry-Kravis, E.; Hazlett, H.C.; Bailey, D.B., Jr.; Moine, H.; Kooy, R.F.; Tassone, F.; Gantois, I.; Sonenberg, N.; Mandel, J.L.; et al. Fragile X syndrome. *Nat. Rev. Dis. Primers* **2017**, *3*, 17065. [CrossRef]
23. Richter, J.D.; Zhao, X. The molecular biology of FMRP: New insights into fragile X syndrome. *Nat. Rev. Neurosci.* **2021**, *22*, 209–222. [CrossRef] [PubMed]
24. Dobkin, C.; Radu, G.; Ding, X.H.; Brown, W.T.; Nolin, S.L. Fragile X prenatal analyses show full mutation females at high risk for mosaic Turner syndrome: Fragile X leads to chromosome loss. *Am. J. Med. Genet. A* **2009**, *149A*, 2152–2157. [CrossRef] [PubMed]
25. Darnell, J.C.; Jensen, K.B.; Jin, P.; Brown, V.; Warren, S.T.; Darnell, R.B. Fragile X mental retardation protein targets G quartet mRNAs important for neuronal function. *Cell* **2001**, *107*, 489–499. [CrossRef]
26. Chakraborty, A.; Jenjaroenpun, P.; Li, J.; El Hilali, S.; McCulley, A.; Haarer, B.; Hoffman, E.A.; Belak, A.; Thorland, A.; Hehnly, H.; et al. Replication stress induces global chromosome breakage in the fragile X genome. *Cell Rep.* **2021**, *34*, 108838. [CrossRef] [PubMed]
27. Usdin, K.; Woodford, K.J. CGG repeats associated with DNA instability and chromosome fragility form structures that block DNA synthesis in vitro. *Nucleic Acids Res.* **1995**, *23*, 4202–4209. [CrossRef]
28. Mitas, M.; Yu, A.; Dill, J.; Haworth, I.S. The trinucleotide repeat sequence d(CGG)15 forms a heat-stable hairpin containing Gsyn. Ganti base pairs. *Biochemistry* **1995**, *34*, 12803–12811. [CrossRef]
29. Yu, A.; Barron, M.D.; Romero, R.M.; Christy, M.; Gold, B.; Dai, J.; Gray, D.M.; Haworth, I.S.; Mitas, M. At physiological pH, d(CCG)15 forms a hairpin containing protonated cytosines and a distorted helix. *Biochemistry* **1997**, *36*, 3687–3699. [CrossRef] [PubMed]
30. Nadel, Y.; Weisman-Shomer, P.; Fry, M. The fragile X syndrome single strand d(CGG)n nucleotide repeats readily fold back to form unimolecular hairpin structures. *J. Biol. Chem.* **1995**, *270*, 28970–28977. [CrossRef] [PubMed]
31. Chen, X.; Mariappan, S.V.; Catasti, P.; Ratliff, R.; Moyzis, R.K.; Laayoun, A.; Smith, S.S.; Bradbury, E.M.; Gupta, G. Hairpins are formed by the single DNA strands of the fragile X triplet repeats: Structure and biological implications. *Proc. Natl. Acad. Sci. USA* **1995**, *92*, 5199–5203. [CrossRef]
32. Gacy, A.M.; Goellner, G.; Juranic, N.; Macura, S.; McMurray, C.T. Trinucleotide repeats that expand in human disease form hairpin structures in vitro. *Cell* **1995**, *81*, 533–540. [CrossRef]
33. Mitas, M. Trinucleotide repeats associated with human disease. *Nucleic Acids Res.* **1997**, *25*, 2245–2254. [CrossRef] [PubMed]
34. Liu, G.; Chen, X.; Bissler, J.J.; Sinden, R.R.; Leffak, M. Replication-dependent instability at (CTG) x (CAG) repeat hairpins in human cells. *Nat. Chem. Biol.* **2010**, *6*, 652–659. [CrossRef]
35. Zumwalt, M.; Ludwig, A.; Hagerman, P.J.; Dieckmann, T. Secondary structure and dynamics of the r(CGG) repeat in the mRNA of the fragile X mental retardation 1 (FMR1) gene. *RNA Biol.* **2007**, *4*, 93–100. [CrossRef] [PubMed]
36. Handa, V.; Saha, T.; Usdin, K. The fragile X syndrome repeats form RNA hairpins that do not activate the interferon-inducible protein kinase, PKR, but are cut by Dicer. *Nucleic Acids Res.* **2003**, *31*, 6243–6248. [CrossRef]
37. Sobczak, K.; de Mezer, M.; Michlewski, G.; Krol, J.; Krzyzosiak, W.J. RNA structure of trinucleotide repeats associated with human neurological diseases. *Nucleic Acids Res.* **2003**, *31*, 5469–5482. [CrossRef] [PubMed]
38. Khateb, S.; Weisman-Shomer, P.; Hershco, I.; Loeb, L.A.; Fry, M. Destabilization of tetraplex structures of the fragile X repeat sequence (CGG)n is mediated by homolog-conserved domains in three members of the hnRNP family. *Nucleic Acids Res.* **2004**, *32*, 4145–4154. [CrossRef] [PubMed]
39. Khateb, S.; Weisman-Shomer, P.; Hershco-Shani, I.; Ludwig, A.L.; Fry, M. The tetraplex (CGG)n destabilizing proteins hnRNP A2 and CBF-A enhance the in vivo translation of fragile X premutation mRNA. *Nucleic Acids Res.* **2007**, *35*, 5775–5788. [CrossRef]
40. Weisman-Shomer, P.; Cohen, E.; Fry, M. Distinct domains in the CArG-box binding factor A destabilize tetraplex forms of the fragile X expanded sequence d(CGG)n. *Nucleic Acids Res.* **2002**, *30*, 3672–3681. [CrossRef] [PubMed]
41. Weisman-Shomer, P.; Cohen, E.; Hershco, I.; Khateb, S.; Wolfovitz-Barchad, O.; Hurley, L.H.; Fry, M. The cationic porphyrin TMPyP4 destabilizes the tetraplex form of the fragile X syndrome expanded sequence d(CGG)n. *Nucleic Acids Res.* **2003**, *31*, 3963–3970. [CrossRef] [PubMed]
42. Kettani, A.; Kumar, R.A.; Patel, D.J. Solution structure of a DNA quadruplex containing the fragile X syndrome triplet repeat. *J. Mol. Biol.* **1995**, *254*, 638–656. [CrossRef] [PubMed]
43. Malgowska, M.; Gudanis, D.; Kierzek, R.; Wyszko, E.; Gabelica, V.; Gdaniec, Z. Distinctive structural motifs of RNA G-quadruplexes composed of AGG, CGG and UGG trinucleotide repeats. *Nucleic Acids Res.* **2014**, *42*, 10196–10207. [CrossRef] [PubMed]
44. Binas, O.; Bessi, I.; Schwalbe, H. Structure Validation of G-Rich RNAs in Noncoding Regions of the Human Genome. *ChemBioChem* **2020**, *21*, 1656–1663. [CrossRef]

45. Asamitsu, S.; Yabuki, Y.; Ikenoshita, S.; Kawakubo, K.; Kawasaki, M.; Usuki, S.; Nakayama, Y.; Adachi, K.; Kugoh, H.; Ishii, K.; et al. CGG repeat RNA G-quadruplexes interact with FMRpolyG to cause neuronal dysfunction in fragile X-related tremor/ataxia syndrome. *Sci. Adv.* **2021**, *7*. [CrossRef]
46. Chen, Y.W.; Satange, R.; Wu, P.C.; Jhan, C.R.; Chang, C.K.; Chung, K.R.; Waring, M.J.; Lin, S.W.; Hsieh, L.C.; Hou, M.H. Co(II)(Chromomycin)(2) Complex Induces a Conformational Change of CCG Repeats from i-Motif to Base-Extruded DNA Duplex. *Int. J. Mol. Sci.* **2018**, *19*, 2796. [CrossRef] [PubMed]
47. Yang, B.; Rodgers, M.T. Base-pairing energies of proton-bound heterodimers of cytosine and modified cytosines: Implications for the stability of DNA i-motif conformations. *J. Am. Chem. Soc.* **2014**, *136*, 282–290. [CrossRef]
48. Fojtik, P.; Vorlickova, M. The fragile X chromosome (GCC) repeat folds into a DNA tetraplex at neutral pH. *Nucleic Acids Res.* **2001**, *29*, 4684–4690. [CrossRef]
49. Loomis, E.W.; Sanz, L.A.; Chedin, F.; Hagerman, P.J. Transcription-associated R-loop formation across the human FMR1 CGG-repeat region. *PLoS Genet.* **2014**, *10*, e1004294. [CrossRef]
50. Kumari, D.; Usdin, K. Sustained expression of FMR1 mRNA from reactivated fragile X syndrome alleles after treatment with small molecules that prevent trimethylation of H3K27. *Hum. Mol. Genet.* **2016**, *25*, 3689–3698. [CrossRef]
51. Abu Diab, M.; Mor-Shaked, H.; Cohen, E.; Cohen-Hadad, Y.; Ram, O.; Epsztejn-Litman, S.; Eiges, R. The G-rich Repeats in FMR1 and C9orf72 Loci Are Hotspots for Local Unpairing of DNA. *Genetics* **2018**, *210*, 1239–1252. [CrossRef]
52. Groh, M.; Lufino, M.M.; Wade-Martins, R.; Gromak, N. R-loops associated with triplet repeat expansions promote gene silencing in Friedreich ataxia and fragile X syndrome. *PLoS Genet.* **2014**, *10*, e1004318. [CrossRef]
53. Ginno, P.A.; Lim, Y.W.; Lott, P.L.; Korf, I.; Chedin, F. GC skew at the 5′ and 3′ ends of human genes links R-loop formation to epigenetic regulation and transcription termination. *Genome Res.* **2013**, *23*, 1590–1600. [CrossRef]
54. De Magis, A.; Manzo, S.G.; Russo, M.; Marinello, J.; Morigi, R.; Sordet, O.; Capranico, G. DNA damage and genome instability by G-quadruplex ligands are mediated by R loops in human cancer cells. *Proc. Natl. Acad. Sci. USA* **2019**, *116*, 816–825. [CrossRef] [PubMed]
55. Reddy, K.; Tam, M.; Bowater, R.P.; Barber, M.; Tomlinson, M.; Nichol Edamura, K.; Wang, Y.H.; Pearson, C.E. Determinants of R-loop formation at convergent bidirectionally transcribed trinucleotide repeats. *Nucleic Acids Res.* **2011**, *39*, 1749–1762. [CrossRef] [PubMed]
56. Renciuk, D.; Kypr, J.; Vorlickova, M. CGG repeats associated with fragile X chromosome form left-handed Z-DNA structure. *Biopolymers* **2011**, *95*, 174–181. [CrossRef] [PubMed]
57. Genetic Modifiers of Huntington's Disease Consortium. Identification of Genetic Factors that Modify Clinical Onset of Huntington's Disease. *Cell* **2015**, *162*, 516–526. [CrossRef]
58. Moss, D.J.H.; Pardinas, A.F.; Langbehn, D.; Lo, K.; Leavitt, B.R.; Roos, R.; Durr, A.; Mead, S.; TRACK-HD Investigators; REGISTRY Investigators; et al. Identification of genetic variants associated with Huntington's disease progression: A genome-wide association study. *Lancet Neurol.* **2017**, *16*, 701–711. [CrossRef]
59. Lee, J.M.; Chao, M.J.; Harold, D.; Abu Elneel, K.; Gillis, T.; Holmans, P.; Jones, L.; Orth, M.; Myers, R.H.; Kwak, S.; et al. A modifier of Huntington's disease onset at the MLH1 locus. *Hum. Mol. Genet.* **2017**, *26*, 3859–3867. [CrossRef]
60. Bettencourt, C.; Hensman-Moss, D.; Flower, M.; Wiethoff, S.; Brice, A.; Goizet, C.; Stevanin, G.; Koutsis, G.; Karadima, G.; Panas, M.; et al. DNA repair pathways underlie a common genetic mechanism modulating onset in polyglutamine diseases. *Ann. Neurol.* **2016**, *79*, 983–990. [CrossRef]
61. Goold, R.; Flower, M.; Moss, D.H.; Medway, C.; Wood-Kaczmar, A.; Andre, R.; Farshim, P.; Bates, G.P.; Holmans, P.; Jones, L.; et al. FAN1 modifies Huntington's disease progression by stabilizing the expanded HTT CAG repeat. *Hum. Mol. Genet.* **2019**, *28*, 650–661. [CrossRef] [PubMed]
62. Morales, F.; Vasquez, M.; Santamaria, C.; Cuenca, P.; Corrales, E.; Monckton, D.G. A polymorphism in the MSH3 mismatch repair gene is associated with the levels of somatic instability of the expanded CTG repeat in the blood DNA of myotonic dystrophy type 1 patients. *DNA Repair (Amst)* **2016**, *40*, 57–66. [CrossRef] [PubMed]
63. Holmans, P.A.; Massey, T.H.; Jones, L. Genetic modifiers of Mendelian disease: Huntington's disease and the trinucleotide repeat disorders. *Hum. Mol. Genet.* **2017**, *26*, R83–R90. [CrossRef] [PubMed]
64. Zhao, X.N.; Lokanga, R.; Allette, K.; Gazy, I.; Wu, D.; Usdin, K. A MutSbeta-Dependent Contribution of MutSalpha to Repeat Expansions in Fragile X Premutation Mice? *PLoS Genet.* **2016**, *12*, e1006190. [CrossRef] [PubMed]
65. Zhao, X.; Kumari, D.; Miller, C.J.; Kim, G.Y.; Hayward, B.; Vitalo, A.G.; Pinto, R.M.; Usdin, K. Modifiers of Somatic Repeat Instability in Mouse Models of Friedreich Ataxia and the Fragile X-Related Disorders: Implications for the Mechanism of Somatic Expansion in Huntington's Disease. *J. Huntingt. Dis.* **2021**, *10*, 149–163. [CrossRef] [PubMed]
66. Wheeler, V.C.; Dion, V. Modifiers of CAG/CTG Repeat Instability: Insights from Mammalian Models. *J. Huntingt. Dis.* **2021**, *10*, 123–148. [CrossRef] [PubMed]
67. Owen, B.A.; Yang, Z.; Lai, M.; Gajec, M.; Badger, J.D., 2nd; Hayes, J.J.; Edelmann, W.; Kucherlapati, R.; Wilson, T.M.; McMurray, C.T. (CAG)(n)-hairpin DNA binds to Msh2-Msh3 and changes properties of mismatch recognition. *Nat. Struct. Mol. Biol.* **2005**, *12*, 663–670. [CrossRef]
68. Shelbourne, P.F.; Keller-McGandy, C.; Bi, W.L.; Yoon, S.R.; Dubeau, L.; Veitch, N.J.; Vonsattel, J.P.; Wexler, N.S.; Group, U.S.-V.C.R.; Arnheim, N.; et al. Triplet repeat mutation length gains correlate with cell-type specific vulnerability in Huntington disease brain. *Hum. Mol. Genet.* **2007**, *16*, 1133–1142. [CrossRef]

69. Kovtun, I.V.; Johnson, K.O.; McMurray, C.T. Cockayne syndrome B protein antagonizes OGG1 in modulating CAG repeat length in vivo. *Aging* **2011**, *3*, 509–514. [CrossRef]
70. Jarem, D.A.; Wilson, N.R.; Delaney, S. Structure-dependent DNA damage and repair in a trinucleotide repeat sequence. *Biochemistry* **2009**, *48*, 6655–6663. [CrossRef]
71. Mollersen, L.; Rowe, A.D.; Illuzzi, J.L.; Hildrestrand, G.A.; Gerhold, K.J.; Tveteras, L.; Bjolgerud, A.; Wilson, D.M., 3rd; Bjoras, M.; Klungland, A. Neil1 is a genetic modifier of somatic and germline CAG trinucleotide repeat instability in R6/1 mice. *Hum. Mol. Genet.* **2012**, *21*, 4939–4947. [CrossRef]
72. Entezam, A.; Biacsi, R.; Orrison, B.; Saha, T.; Hoffman, G.E.; Grabczyk, E.; Nussbaum, R.L.; Usdin, K. Regional FMRP deficits and large repeat expansions into the full mutation range in a new Fragile X premutation mouse model. *Gene* **2007**, *395*, 125–134. [CrossRef]
73. Jonson, I.; Ougland, R.; Klungland, A.; Larsen, E. Oxidative stress causes DNA triplet expansion in Huntington's disease mouse embryonic stem cells. *Stem Cell Res.* **2013**, *11*, 1264–1271. [CrossRef] [PubMed]
74. Budworth, H.; Harris, F.R.; Williams, P.; Lee, D.Y.; Holt, A.; Pahnke, J.; Szczesny, B.; Acevedo-Torres, K.; Ayala-Pena, S.; McMurray, C.T. Suppression of Somatic Expansion Delays the Onset of Pathophysiology in a Mouse Model of Huntington's Disease. *PLoS Genet.* **2015**, *11*, e1005267. [CrossRef] [PubMed]
75. Mollersen, L.; Moldestad, O.; Rowe, A.D.; Bjolgerud, A.; Holm, I.; Tveteras, L.; Klungland, A.; Retterstol, L. Effects of Anthocyanins on CAG Repeat Instability and Behaviour in Huntington's Disease R6/1 Mice. *PLoS Curr.* **2016**, *8*. [CrossRef] [PubMed]
76. Lokanga, R.A.; Zhao, X.N.; Entezam, A.; Usdin, K. X inactivation plays a major role in the gender bias in somatic expansion in a mouse model of the fragile X-related disorders: Implications for the mechanism of repeat expansion. *Hum. Mol. Genet.* **2014**, *23*, 4985–4994. [CrossRef]
77. Kovtun, I.V.; Liu, Y.; Bjoras, M.; Klungland, A.; Wilson, S.H.; McMurray, C.T. OGG1 initiates age-dependent CAG trinucleotide expansion in somatic cells. *Nature* **2007**, *447*, 447–452. [CrossRef]
78. Lokanga, R.A.; Senejani, A.G.; Sweasy, J.B.; Usdin, K. Heterozygosity for a hypomorphic Polbeta mutation reduces the expansion frequency in a mouse model of the Fragile X-related disorders. *PLoS Genet.* **2015**, *11*, e1005181. [CrossRef]
79. Lokanga, R.A.; Zhao, X.N.; Usdin, K. The mismatch repair protein MSH2 is rate limiting for repeat expansion in a fragile X premutation mouse model. *Hum. Mutat.* **2014**, *35*, 129–136. [CrossRef]
80. Miller, C.J.; Kim, G.Y.; Zhao, X.; Usdin, K. All three mammalian MutL complexes are required for repeat expansion in a mouse cell model of the Fragile X-related disorders. *PLoS Genet* **2020**, *16*, e1008902. [CrossRef]
81. Zhao, X.; Zhang, Y.; Wilkins, K.; Edelmann, W.; Usdin, K. MutLgamma promotes repeat expansion in a Fragile X mouse model while EXO1 is protective. *PLoS Genet.* **2018**, *14*, e1007719. [CrossRef]
82. Zhao, X.N.; Kumari, D.; Gupta, S.; Wu, D.; Evanitsky, M.; Yang, W.; Usdin, K. Mutsbeta generates both expansions and contractions in a mouse model of the Fragile X-associated disorders. *Hum. Mol. Genet.* **2015**, *24*, 7087–7096. [CrossRef]
83. Gazy, I.; Hayward, B.; Potapova, S.; Zhao, X.; Usdin, K. Double-strand break repair plays a role in repeat instability in a fragile X mouse model. *DNA Repair (Amst)* **2019**, *74*, 63–69. [CrossRef]
84. Sellier, C.; Freyermuth, F.; Tabet, R.; Tran, T.; He, F.; Ruffenach, F.; Alunni, V.; Moine, H.; Thibault, C.; Page, A.; et al. Sequestration of DROSHA and DGCR8 by expanded CGG RNA repeats alters microRNA processing in fragile X-associated tremor/ataxia syndrome. *Cell Rep.* **2013**, *3*, 869–880. [CrossRef] [PubMed]
85. Todd, P.K.; Oh, S.Y.; Krans, A.; He, F.; Sellier, C.; Frazer, M.; Renoux, A.J.; Chen, K.C.; Scaglione, K.M.; Basrur, V.; et al. CGG repeat-associated translation mediates neurodegeneration in fragile X tremor ataxia syndrome. *Neuron* **2013**, *78*, 440–455. [CrossRef]
86. Oh, S.Y.; He, F.; Krans, A.; Frazer, M.; Taylor, J.P.; Paulson, H.L.; Todd, P.K. RAN translation at CGG repeats induces ubiquitin proteasome system impairment in models of fragile X-associated tremor ataxia syndrome. *Hum. Mol. Genet.* **2015**, *24*, 4317–4326. [CrossRef]
87. Nguyen, L.; Cleary, J.D.; Ranum, L.P.W. Repeat-Associated Non-ATG Translation: Molecular Mechanisms and Contribution to Neurological Disease. *Annu. Rev. Neurosci.* **2019**, *42*, 227–247. [CrossRef]
88. Skourti-Stathaki, K.; Torlai Triglia, E.; Warburton, M.; Voigt, P.; Bird, A.; Pombo, A. R-Loops Enhance Polycomb Repression at a Subset of Developmental Regulator Genes. *Mol. Cell* **2019**, *73*, 930–945.e934. [CrossRef] [PubMed]
89. Skourti-Stathaki, K.; Kamieniarz-Gdula, K.; Proudfoot, N.J. R-loops induce repressive chromatin marks over mammalian gene terminators. *Nature* **2014**, *516*, 436–439. [CrossRef] [PubMed]
90. Garribba, L.; Bjerregaard, V.A.; Goncalves Dinis, M.M.; Ozer, O.; Wu, W.; Sakellariou, D.; Pena-Diaz, J.; Hickson, I.D.; Liu, Y. Folate stress induces SLX1- and RAD51-dependent mitotic DNA synthesis at the fragile X locus in human cells. *Proc. Natl. Acad. Sci. USA* **2020**, *117*, 16527–16536. [CrossRef]
91. Hayward, B.E.; Steinbach, P.J.; Usdin, K. A point mutation in the nuclease domain of MLH3 eliminates repeat expansions in a mouse stem cell model of the Fragile X-related disorders. *Nucleic Acids Res.* **2020**, *48*, 7856–7863. [CrossRef]
92. Kadyrova, L.Y.; Gujar, V.; Burdett, V.; Modrich, P.L.; Kadyrov, F.A. Human MutLgamma, the MLH1-MLH3 heterodimer, is an endonuclease that promotes DNA expansion. *Proc. Natl. Acad. Sci. USA* **2020**, *117*, 3535–3542. [CrossRef]
93. Handa, V.; Goldwater, D.; Stiles, D.; Cam, M.; Poy, G.; Kumari, D.; Usdin, K. Long CGG-repeat tracts are toxic to human cells: Implications for carriers of Fragile X premutation alleles. *FEBS Lett.* **2005**, *579*, 2702–2708. [CrossRef] [PubMed]

94. Jin, P.; Zarnescu, D.C.; Zhang, F.; Pearson, C.E.; Lucchesi, J.C.; Moses, K.; Warren, S.T. RNA-mediated neurodegeneration caused by the fragile X premutation rCGG repeats in Drosophila. *Neuron* **2003**, *39*, 739–747. [CrossRef]
95. Willemsen, R.; Hoogeveen-Westerveld, M.; Reis, S.; Holstege, J.; Severijnen, L.A.; Nieuwenhuizen, I.M.; Schrier, M.; van Unen, L.; Tassone, F.; Hoogeveen, A.T.; et al. The FMR1 CGG repeat mouse displays ubiquitin-positive intranuclear neuronal inclusions; implications for the cerebellar tremor/ataxia syndrome. *Hum. Mol. Genet.* **2003**, *12*, 949–959. [CrossRef] [PubMed]
96. Hashem, V.; Galloway, J.N.; Mori, M.; Willemsen, R.; Oostra, B.A.; Paylor, R.; Nelson, D.L. Ectopic expression of CGG containing mRNA is neurotoxic in mammals. *Hum. Mol. Genet.* **2009**, *18*, 2443–2451. [CrossRef]
97. Hoem, G.; Raske, C.R.; Garcia-Arocena, D.; Tassone, F.; Sanchez, E.; Ludwig, A.L.; Iwahashi, C.K.; Kumar, M.; Yang, J.E.; Hagerman, P.J. CGG-repeat length threshold for FMR1 RNA pathogenesis in a cellular model for FXTAS. *Hum. Mol. Genet.* **2011**, *20*, 2161–2170. [CrossRef]
98. Greco, C.M.; Hagerman, R.J.; Tassone, F.; Chudley, A.E.; Del Bigio, M.R.; Jacquemont, S.; Leehey, M.; Hagerman, P.J. Neuronal intranuclear inclusions in a new cerebellar tremor/ataxia syndrome among fragile X carriers. *Brain* **2002**, *125*, 1760–1771. [CrossRef]
99. Tassone, F.; Iwahashi, C.; Hagerman, P.J. FMR1 RNA within the intranuclear inclusions of fragile X-associated tremor/ataxia syndrome (FXTAS). *RNA Biol.* **2004**, *1*, 103–105. [CrossRef]
100. Powell, W.T.; Coulson, R.L.; Gonzales, M.L.; Crary, F.K.; Wong, S.S.; Adams, S.; Ach, R.A.; Tsang, P.; Yamada, N.A.; Yasui, D.H.; et al. R-loop formation at Snord116 mediates topotecan inhibition of Ube3a-antisense and allele-specific chromatin decondensation. *Proc. Natl. Acad. Sci. USA* **2013**, *110*, 13938–13943. [CrossRef]
101. Ginno, P.A.; Lott, P.L.; Christensen, H.C.; Korf, I.; Chedin, F. R-loop formation is a distinctive characteristic of unmethylated human CpG island promoters. *Mol. Cell* **2012**, *45*, 814–825. [CrossRef]
102. Arab, K.; Karaulanov, E.; Musheev, M.; Trnka, P.; Schafer, A.; Grummt, I.; Niehrs, C. GADD45A binds R-loops and recruits TET1 to CpG island promoters. *Nat. Genet.* **2019**, *51*, 217–223. [CrossRef] [PubMed]
103. Redstone, S.C.J.; Fleming, A.M.; Burrows, C.J. Oxidative Modification of the Potential G-Quadruplex Sequence in the PCNA Gene Promoter Can Turn on Transcription. *Chem. Res. Toxicol.* **2019**, *32*, 437–446. [CrossRef]
104. Sellier, C.; Rau, F.; Liu, Y.; Tassone, F.; Hukema, R.K.; Gattoni, R.; Schneider, A.; Richard, S.; Willemsen, R.; Elliott, D.J.; et al. Sam68 sequestration and partial loss of function are associated with splicing alterations in FXTAS patients. *EMBO J.* **2010**, *29*, 1248–1261. [CrossRef] [PubMed]
105. Kearse, M.G.; Green, K.M.; Krans, A.; Rodriguez, C.M.; Linsalata, A.E.; Goldstrohm, A.C.; Todd, P.K. CGG Repeat-Associated Non-AUG Translation Utilizes a Cap-Dependent Scanning Mechanism of Initiation to Produce Toxic Proteins. *Mol. Cell* **2016**, *62*, 314–322. [CrossRef]
106. Krans, A.; Kearse, M.G.; Todd, P.K. Repeat-associated non-AUG translation from antisense CCG repeats in fragile X tremor/ataxia syndrome. *Ann. Neurol.* **2016**, *80*, 871–881. [CrossRef] [PubMed]
107. Sellier, C.; Buijsen, R.A.M.; He, F.; Natla, S.; Jung, L.; Tropel, P.; Gaucherot, A.; Jacobs, H.; Meziane, H.; Vincent, A.; et al. Translation of Expanded CGG Repeats into FMRpolyG Is Pathogenic and May Contribute to Fragile X Tremor Ataxia Syndrome. *Neuron* **2017**, *93*, 331–347. [CrossRef]
108. Kozak, M. Evaluation of the fidelity of initiation of translation in reticulocyte lysates from commercial sources. *Nucleic Acids Res.* **1990**, *18*, 2828. [CrossRef]
109. Krans, A.; Skariah, G.; Zhang, Y.; Bayly, B.; Todd, P.K. Neuropathology of RAN translation proteins in fragile X-associated tremor/ataxia syndrome. *Acta Neuropathol. Commun.* **2019**, *7*, 152. [CrossRef]
110. Bonapace, G.; Gullace, R.; Concolino, D.; Iannello, G.; Procopio, R.; Gagliardi, M.; Arabia, G.; Barbagallo, G.; Lupo, A.; Manfredini, L.I.; et al. Intracellular FMRpolyG-Hsp70 complex in fibroblast cells from a patient affected by fragile X tremor ataxia syndrome. *Heliyon* **2019**, *5*, e01954. [CrossRef] [PubMed]
111. Buijsen, R.A.; Sellier, C.; Severijnen, L.A.; Oulad-Abdelghani, M.; Verhagen, R.F.; Berman, R.F.; Charlet-Berguerand, N.; Willemsen, R.; Hukema, R.K. FMRpolyG-positive inclusions in CNS and non-CNS organs of a fragile X premutation carrier with fragile X-associated tremor/ataxia syndrome. *Acta Neuropathol. Commun.* **2014**, *2*, 162. [CrossRef]
112. Friedman-Gohas, M.; Elizur, S.E.; Dratviman-Storobinsky, O.; Aizer, A.; Haas, J.; Raanani, H.; Orvieto, R.; Cohen, Y. FMRpolyG accumulates in FMR1 premutation granulosa cells. *J. Ovarian Res.* **2020**, *13*, 22. [CrossRef]
113. Tian, B.; White, R.J.; Xia, T.; Welle, S.; Turner, D.H.; Mathews, M.B.; Thornton, C.A. Expanded CUG repeat RNAs form hairpins that activate the double-stranded RNA-dependent protein kinase PKR. *RNA* **2000**, *6*, 79–87. [CrossRef] [PubMed]
114. Zu, T.; Guo, S.; Bardhi, O.; Ryskamp, D.A.; Li, J.; Khoramian Tusi, S.; Engelbrecht, A.; Klippel, K.; Chakrabarty, P.; Nguyen, L.; et al. Metformin inhibits RAN translation through PKR pathway and mitigates disease in C9orf72 ALS/FTD mice. *Proc. Natl. Acad. Sci. USA* **2020**, *117*, 18591–18599. [CrossRef] [PubMed]
115. Green, K.M.; Glineburg, M.R.; Kearse, M.G.; Flores, B.N.; Linsalata, A.E.; Fedak, S.J.; Goldstrohm, A.C.; Barmada, S.J.; Todd, P.K. RAN translation at C9orf72-associated repeat expansions is selectively enhanced by the integrated stress response. *Nat. Commun.* **2017**, *8*, 2005. [CrossRef] [PubMed]
116. Haify, S.N.; Mankoe, R.S.D.; Boumeester, V.; van der Toorn, E.C.; Verhagen, R.F.M.; Willemsen, R.; Hukema, R.K.; Bosman, L.W.J. Lack of a Clear Behavioral Phenotype in an Inducible FXTAS Mouse Model Despite the Presence of Neuronal FMRpolyG-Positive Aggregates. *Front. Mol. Biosci.* **2020**, *7*, 599101. [CrossRef]

117. Holm, K.N.; Herren, A.W.; Taylor, S.L.; Randol, J.L.; Kim, K.; Espinal, G.; Martiinez-Cerdeno, V.; Pessah, I.N.; Hagerman, R.J.; Hagerman, P.J. Human Cerebral Cortex Proteome of Fragile X-Associated Tremor/Ataxia Syndrome. *Front. Mol. Biosci.* **2020**, *7*, 600840. [CrossRef]
118. Rodriguez, S.; Sahin, A.; Schrank, B.R.; Al-Lawati, H.; Costantino, I.; Benz, E.; Fard, D.; Albers, A.D.; Cao, L.; Gomez, A.C.; et al. Genome-encoded cytoplasmic double-stranded RNAs, found in C9ORF72 ALS-FTD brain, propagate neuronal loss. *Sci. Transl. Med.* **2021**, *13*. [CrossRef]
119. Cabal-Herrera, A.M.; Tassanakijpanich, N.; Salcedo-Arellano, M.J.; Hagerman, R.J. Fragile X-Associated Tremor/Ataxia Syndrome (FXTAS): Pathophysiology and Clinical Implications. *Int. J. Mol. Sci.* **2020**, *21*, 4391. [CrossRef]
120. Cristini, A.; Ricci, G.; Britton, S.; Salimbeni, S.; Huang, S.N.; Marinello, J.; Calsou, P.; Pommier, Y.; Favre, G.; Capranico, G.; et al. Dual Processing of R-Loops and Topoisomerase I Induces Transcription-Dependent DNA Double-Strand Breaks. *Cell Rep.* **2019**, *28*, 3167–3181.e3166. [CrossRef]
121. Crossley, M.P.; Bocek, M.; Cimprich, K.A. R-Loops as Cellular Regulators and Genomic Threats. *Mol. Cell* **2019**, *73*, 398–411. [CrossRef] [PubMed]
122. Primerano, B.; Tassone, F.; Hagerman, R.J.; Hagerman, P.; Amaldi, F.; Bagni, C. Reduced FMR1 mRNA translation efficiency in fragile X patients with premutations. *RNA* **2002**, *8*, 1482–1488. [PubMed]
123. Schneider, A.; Winarni, T.I.; Cabal-Herrera, A.M.; Bacalman, S.; Gane, L.; Hagerman, P.; Tassone, F.; Hagerman, R. Elevated FMR1-mRNA and lowered FMRP-A double-hit mechanism for psychiatric features in men with FMR1 premutations. *Transl. Psychiatry* **2020**, *10*, 205. [CrossRef] [PubMed]
124. Feng, Y.; Zhang, F.; Lokey, L.K.; Chastain, J.L.; Lakkis, L.; Eberhart, D.; Warren, S.T. Translational suppression by trinucleotide repeat expansion at FMR1. *Science* **1995**, *268*, 731–734. [CrossRef] [PubMed]
125. Kumari, D.; Usdin, K. Polycomb group complexes are recruited to reactivated FMR1 alleles in Fragile X syndrome in response to FMR1 transcription. *Hum. Mol. Genet.* **2014**, *23*, 6575–6583. [CrossRef]
126. Alecki, C.; Chiwara, V.; Sanz, L.A.; Grau, D.; Arias Perez, O.; Boulier, E.L.; Armache, K.J.; Chedin, F.; Francis, N.J. RNA-DNA strand exchange by the Drosophila Polycomb complex PRC2. *Nat. Commun.* **2020**, *11*, 1781. [CrossRef]
127. Wang, X.; Goodrich, K.J.; Gooding, A.R.; Naeem, H.; Archer, S.; Paucek, R.D.; Youmans, D.T.; Cech, T.R.; Davidovich, C. Targeting of Polycomb Repressive Complex 2 to RNA by Short Repeats of Consecutive Guanines. *Mol. Cell* **2017**, *65*, 1056–1067.e1055. [CrossRef]
128. Jani, K.S.; Jain, S.U.; Ge, E.J.; Diehl, K.L.; Lundgren, S.M.; Muller, M.M.; Lewis, P.W.; Muir, T.W. Histone H3 tail binds a unique sensing pocket in EZH2 to activate the PRC2 methyltransferase. *Proc. Natl. Acad. Sci. USA* **2019**, *116*, 8295–8300. [CrossRef]
129. Yuan, W.; Xu, M.; Huang, C.; Liu, N.; Chen, S.; Zhu, B. H3K36 methylation antagonizes PRC2-mediated H3K27 methylation. *J. Biol. Chem.* **2011**, *286*, 7983–7989. [CrossRef]
130. Schmitges, F.W.; Prusty, A.B.; Faty, M.; Stutzer, A.; Lingaraju, G.M.; Aiwazian, J.; Sack, R.; Hess, D.; Li, L.; Zhou, S.; et al. Histone methylation by PRC2 is inhibited by active chromatin marks. *Mol. Cell* **2011**, *42*, 330–341. [CrossRef]
131. Huertas, P.; Aguilera, A. Cotranscriptionally formed DNA:RNA hybrids mediate transcription elongation impairment and transcription-associated recombination. *Mol. Cell* **2003**, *12*, 711–721. [CrossRef]
132. Shanbhag, N.M.; Rafalska-Metcalf, I.U.; Balane-Bolivar, C.; Janicki, S.M.; Greenberg, R.A. ATM-dependent chromatin changes silence transcription in cis to DNA double-strand breaks. *Cell* **2010**, *141*, 970–981. [CrossRef]
133. Pankotai, T.; Bonhomme, C.; Chen, D.; Soutoglou, E. DNAPKcs-dependent arrest of RNA polymerase II transcription in the presence of DNA breaks. *Nat. Struct. Mol. Biol.* **2012**, *19*, 276–282. [CrossRef]
134. Zhou, Y.; Kumari, D.; Sciascia, N.; Usdin, K. CGG-repeat dynamics and FMR1 gene silencing in fragile X syndrome stem cells and stem cell-derived neurons. *Mol. Autism* **2016**, *7*, 42. [CrossRef]
135. Niehrs, C.; Luke, B. Regulatory R-loops as facilitators of gene expression and genome stability. *Nat. Rev. Mol. Cell Biol.* **2020**, *21*, 167–178. [CrossRef] [PubMed]
136. Hecht, M.; Tabib, A.; Kahan, T.; Orlanski, S.; Gropp, M.; Tabach, Y.; Yanuka, O.; Benvenisty, N.; Keshet, I.; Cedar, H. Epigenetic mechanism of FMR1 inactivation in Fragile X syndrome. *Int. J. Dev. Biol.* **2017**, *61*, 285–292. [CrossRef] [PubMed]
137. Ladd, P.D.; Smith, L.E.; Rabaia, N.A.; Moore, J.M.; Georges, S.A.; Hansen, R.S.; Hagerman, R.J.; Tassone, F.; Tapscott, S.J.; Filippova, G.N. An antisense transcript spanning the CGG repeat region of FMR1 is upregulated in premutation carriers but silenced in full mutation individuals. *Hum. Mol. Genet.* **2007**, *16*, 3174–3187. [CrossRef] [PubMed]
138. Kumari, D.; Sciascia, N.; Usdin, K. Small Molecules Targeting H3K9 Methylation Prevent Silencing of Reactivated FMR1 Alleles in Fragile X Syndrome Patient Derived Cells. *Genes* **2020**, *11*, 356. [CrossRef] [PubMed]
139. Lukusa, T.; Fryns, J.P. Human chromosome fragility. *Biochim. Biophys. Acta* **2008**, *1779*, 3–16. [CrossRef]
140. Sutherland, G.R. Heritable fragile sites on human chromosomes. III. Detection of fra(X)(q27) in males with X-linked mental retardation and in their female relatives. *Hum. Genet.* **1979**, *53*, 23–27. [CrossRef]
141. Voineagu, I.; Surka, C.F.; Shishkin, A.A.; Krasilnikova, M.M.; Mirkin, S.M. Replisome stalling and stabilization at CGG repeats, which are responsible for chromosomal fragility. *Nat. Struct. Mol. Biol.* **2009**, *16*, 226–228. [CrossRef] [PubMed]
142. Gerhardt, J.; Tomishima, M.J.; Zaninovic, N.; Colak, D.; Yan, Z.; Zhan, Q.; Rosenwaks, Z.; Jaffrey, S.R.; Schildkraut, C.L. The DNA replication program is altered at the FMR1 locus in fragile X embryonic stem cells. *Mol. Cell* **2014**, *53*, 19–31. [CrossRef]

143. Derbis, M.; Kul, E.; Niewiadomska, D.; Sekrecki, M.; Piasecka, A.; Taylor, K.; Hukema, R.K.; Stork, O.; Sobczak, K. Short antisense oligonucleotides alleviate the pleiotropic toxicity of RNA harboring expanded CGG repeats. *Nat. Commun.* **2021**, *12*, 1265. [CrossRef] [PubMed]
144. Disney, M.D.; Liu, B.; Yang, W.Y.; Sellier, C.; Tran, T.; Charlet-Berguerand, N.; Childs-Disney, J.L. A small molecule that targets r(CGG)(exp) and improves defects in fragile X-associated tremor ataxia syndrome. *ACS Chem. Biol.* **2012**, *7*, 1711–1718. [CrossRef] [PubMed]
145. Verma, A.K.; Khan, E.; Mishra, S.K.; Mishra, A.; Charlet-Berguerand, N.; Kumar, A. Curcumin Regulates the r(CGG)(exp) RNA Hairpin Structure and Ameliorate Defects in Fragile X-Associated Tremor Ataxia Syndrome. *Front. Neurosci.* **2020**, *14*, 295. [CrossRef]
146. Haify, S.N.; Buijsen, R.A.M.; Verwegen, L.; Severijnen, L.; de Boer, H.; Boumeester, V.; Monshouwer, R.; Yang, W.Y.; Cameron, M.D.; Willemsen, R.; et al. Small molecule 1a reduces FMRpolyG-mediated toxicity in in vitro and in vivo models for FMR1 premutation. *Hum. Mol. Genet.* **2021**. [CrossRef]

International Journal of *Molecular Sciences*

MDPI

Communication

Searching for New Z-DNA/Z-RNA Binding Proteins Based on Structural Similarity to Experimentally Validated Zα Domain

Martin Bartas [1], Kristyna Slychko [1], Václav Brázda [2], Jiří Červeň [1], Christopher A. Beaudoin [3], Tom L. Blundell [3] and Petr Pečinka [1,*]

1 Department of Biology and Ecology, Faculty of Science, University of Ostrava, 710 00 Ostrava, Czech Republic; martin.bartas@osu.cz (M.B.); P21097@student.osu.cz (K.S.); jiri.cerven@osu.cz (J.Č.)
2 Department of Biophysical Chemistry and Molecular Oncology, Institute of Biophysics of the Czech Academy of Sciences, 612 65 Brno, Czech Republic; vaclav@ibp.cz
3 Department of Biochemistry, Sanger Building, University of Cambridge, Tennis Court Rd., Cambridge CB2 1GA, UK; cab233@cam.ac.uk (C.A.B.); tlb20@cam.ac.uk (T.L.B.)
* Correspondence: petr.pecinka@osu.cz

Abstract: Z-DNA and Z-RNA are functionally important left-handed structures of nucleic acids, which play a significant role in several molecular and biological processes including DNA replication, gene expression regulation and viral nucleic acid sensing. Most proteins that have been proven to interact with Z-DNA/Z-RNA contain the so-called Zα domain, which is structurally well conserved. To date, only eight proteins with Zα domain have been described within a few organisms (including human, mouse, *Danio rerio*, *Trypanosoma brucei* and some viruses). Therefore, this paper aimed to search for new Z-DNA/Z-RNA binding proteins in the complete PDB structures database and from the AlphaFold2 protein models. A structure-based similarity search found 14 proteins with highly similar Zα domain structure in experimentally-defined proteins and 185 proteins with a putative Zα domain using the AlphaFold2 models. Structure-based alignment and molecular docking confirmed high functional conservation of amino acids involved in Z-DNA/Z-RNA, suggesting that Z-DNA/Z-RNA recognition may play an important role in a variety of cellular processes.

Keywords: Z-DNA; Z-RNA; Zα domain; protein binding; bioinformatics

Citation: Bartas, M.; Slychko, K.; Brázda, V.; Červeň, J.; Beaudoin, C.A.; Blundell, T.L.; Pečinka, P. Searching for New Z-DNA/Z-RNA Binding Proteins Based on Structural Similarity to Experimentally Validated Zα Domain. *Int. J. Mol. Sci.* **2022**, 23, 768. https://doi.org/10.3390/ijms23020768

Academic Editors: Maurizio Battino, Vaclav Brazda and Richard Bowater

Received: 30 November 2021
Accepted: 5 January 2022
Published: 11 January 2022

1. Introduction

Local DNA structures, also called 'non-B' DNA structures, have been recognised as important regulators of many fundamental regulatory processes, including replication [1], transcription [2], translation [3], epigenetics [4], DNA damage repair [5–7], genome evolution and rearrangement [8]. Negative supercoiling of DNA and protein binding can increase the stability of local DNA conformation and/or induce conformational changes that give rise to various alternative DNA structures, the best-described being cruciforms [7], Z-DNA/Z-RNA [9,10], triplexes [11] and quadruplexes [12]. Recently, a large number of proteins that recognise especially G-quadruplexes [13] and cruciforms [7,14] were characterised. Surprisingly, only a few Z-DNA/Z-RNA binding proteins have been characterised to date [15–23]. Z-DNA is a left-handed form of deoxyribonucleic acid, and its name was derived from the typical 'zig-zag' pattern (Figure 1). This DNA structure was first proposed by Robert Wells and his colleagues in 1970, during their physical and enzymatic studies on d(I–C) polymers (consisting of altered inosine and cytosine units) [24]. The first structure of Z-DNA was subsequently solved by Andrew H. Wang et al. in 1979 using complementary hexamers of $d(CG)_3$ [25]. The next development was the crystallographic structure of the so-called B-Z junction (DNA loci where right-handed B-DNA passes to a left-handed Z-DNA conformation, or vice versa) [26]. Many biochemical and biophysical in vitro experiments have been conducted to better characterise Z-DNA behaviour at close to physiological conditions [27,28] and also to better understand switching between B and Z-DNA [29,30]. Furthermore, several bioinformatic searches have been performed to predict Z-DNA-prone sequence motifs in the genomic DNA of some model organisms, including humans [31,32]. Z-DNA structures can be formed only in specific double-stranded sequences with alternating purine–pyrimidine tracks, which has been determined by crystallography in various nucleotide repeats, where specifically Z-DNA containing GC repeats have been shown to have increased stability [33]. These sequences with a high potential to form Z-DNA were observed in the genomes of organisms across all domains of life, and their particular importance has been shown: e.g., in transposable ALU elements [34], and gene promoters [35]. In 2009, the first human map of experimentally-obtained Z-DNA forming sites was released [36], followed by the ChIP-seq map in 2016, where they associated Z-DNA forming sites with actively transcribed regions in the human genome [37]. Since these discoveries, it is clear that Z-DNA structures arise under physiological conditions. However, compared to classical B-DNA conformations, Z-DNA structures are energetically unfavourable and, therefore, the structure formation requires energy (usually in the form of negative supercoiling), which results in less structural stability [38].

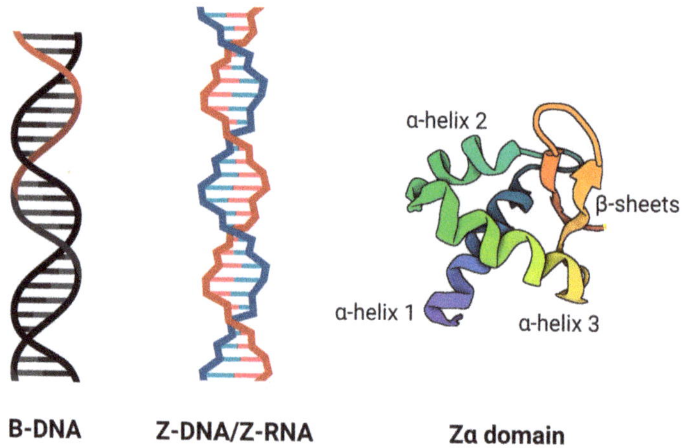

Figure 1. Schematic diagram of classical right-handed B-DNA, left-handed Z-DNA/Z-RNA, and Zα domain consisting of three α-helices and two β-strands. This domain is known to specifically interact with left-handed nucleic acids, mainly through its α-helix 3 and some amino acid residues of beta-strands.

In addition to Z-DNA, there is an analogous structure called Z-RNA (i.e., double-stranded left-handed RNA) that was firstly described in detail in 1984 by Kathleen Hall et al. [39]. Using a combination of spectroscopic techniques, they found that poly(GC)·poly(GC) undergoes a transition from the classical A-form to a left-handed Z-form. Z-RNA has also been found in viral genomes, For example, the influenza virus has been shown to produce Z-RNA during replication, which can induce ZBP1-mediated necroptosis [40]. Additionally, SARS-CoV-2 has been reported to contain loci that theoretically form Z-RNAs (not published, analysed in house using the Non-B DB webserver [41]) [33–35,40,41].

It is assumed that Z-DNA/Z-RNA structures often need 'special' binding proteins for their stabilisation. Most known Z-DNA binding proteins bind to left-handed nucleic acids through the so-called Z-DNA binding domain Zα (Figure 1). One of the first discovered human Z-DNA binding proteins was double-stranded RNA adenosine deaminase (now designated as ADAR1) in 1995 by Herbert et al. [42]. The Zα domain was also discovered in DAI, PKZ, E3L, and ORF112 proteins [21], and a recent study found that this domain is present in RBP7910 protein [43]. The structure of the Zα domain has a specific β-sheet-helix-turn-helix motif (βHTH), which is a subgroup of the winged HTH motif (wHTH). The Zα domain usually consists of three α-helices and sheets of two or three β-strands (αβααββ). The β-wing motif is formed by two antiparallel β-sheets composed of β2 and β3. The resulting β-wing and third α-helix play an important role in recognition and binding to Z-DNA [21,44].

During the past 40 years of research, only about ten Z-DNA (or Z-RNA) binding proteins have been identified in different organisms. All known Z-DNA/Z-RNA proteins that contain Zα domains have been demonstrated to be involved in the immune response (ADAR1, ZBP1, PKZ) [19,45–48] and/or virus-host interactions (E3L protein from *Vaccinia virus*, ORF112 protein from *Cyprinid herpesvirus* 3) [21,49–51]. Some studies have also shown that the binding of the Zα domain to Z-RNA is responsible for the localisation of Z-DNA/Z-RNA binding proteins into cytoplasmic stress granules [52–54]. One of the most well-characterised Z-DNA/Z-DNA binding proteins, ADAR 1, is, in fact, a moonlighting protein [55], and its Z-DNA/Z-RNA binding function was discovered [56] after it was originally described as an adenosine deaminase [57]. This led us to the hypothesis that some functionally characterised proteins may still possess an unidentified Z-DNA/Z-RNA binding function. Therefore, this paper aims to identify new Z-DNA/RNA binding proteins based on structural similarity to an experimentally well-defined Zα domain.

2. Results and Discussion

2.1. Prediction of New Z-DNA/Z-RNA Binding Proteins Based on Structural Similarity to the Experimentally Validated Zα Domain

At the beginning of our study, we made a list of experimentally solved Zα (and Zβ) domain structures (Table 1). After careful consideration (based mainly on the atomic-resolution and selection of a well-characterised human protein), we chose the crystal structure of the Zα domain from the human protein ADAR1 in complex with non-CG-repeat Z-DNA, obtained by Sung Chul Ha et al. in 2009 at a resolution of 2.20 Å [58]. Using this experimental Zα domain structure (PDB: 3f21, chain A), we carried out structural similarity searches using the PDBeFold web server (https://www.ebi.ac.uk/msd-srv/ssm/, (accessed on 10 September 2021)) and RUPEE web server (https://ayoubresearch.com/, (accessed on 21 October 2021)). The PDBeFold algorithm allows examination of a given protein structure for similarity with the whole PDB archive containing nearly 200k of experimentally solved protein structures from a variety of model and nonmodel organisms, whereas RUPEE allows the querying of protein structures predicted by AlphaFold2 [59].

Table 1. Known Z-DNA/RNA binding proteins containing experimentally solved Zα or Zβ domain(s) (PDB IDs are provided). UniProtKB IDs of all proteins are provided as well.

Protein Symbol/ID	Protein Name	Organism	Protein Length	Function	PDB ID	Method/ Resolution	Domain	Ref.
ADAR (P55265)	Double-stranded RNA-specific adenosine deaminase	*Homo sapiens*	1226	Hydrolytic deamination of adenosine to inosine in dsRNA (A-to-I RNA editing)	1XMK	XRC/0.97 Å	Zβ	[60]
					1QGP	NMR	Zα	[61]
					3F21	XRC/2.20 Å	Zα	[58]
					3F22	XRC/2.50 Å	Zα	
					3F23	XRC/2.70 Å	Zα	
					2GXB	XRC/2.25 Å	Zα	[16]
ZBP1 (Q9H171)	Z-DNA-binding protein 1	*Homo sapiens*	429	Innate sensor recognising viral Z-RNA	2L4M	NMR	Zβ	[62]
Zbp1/DAI	Z-DNA-binding protein 1	*Mus musculus*	411		1J75	XRC/1.85 Å	Zα	[18]

Table 1. *Cont.*

Protein Symbol/ID	Protein Name	Organism	Protein Length	Function	PDB ID	Method/ Resolution	Domain	Ref.
PKZ (Q5NE12)	Protein kinase-containing Z-DNA-binding domains	*Danio rerio*	511	Defence response to virus	4LB5	XRC/2.00 Å	Zα	[20]
					4LB6	XRC/1.80 Å		
ORF112 (A4FTK7)	Protein ORF112	*Cyprinid herpesvirus* 3	278	Double-stranded RNA adenosine deaminase activity; RNA binding	4WCG	XRC/1.50 Å	Zα	[21]
E3L (P21605)	Protein E3	*Vaccinia virus*	190	Double-stranded RNA adenosine deaminase activity; inhibition of multiple cellular antiviral responses activated by dsRNA	7C0I	XRC/2.40 Å	Zα	[63]
34L (Q9DHS8)	34L protein	*Yaba-like disease virus*	185	Same as E3L	1SFU	XRC/2.00 Å	Zα	[22]

In Table 2, all non-redundant hits with a Q-score higher than a predefined threshold are shown. The Q-score represents the quality function of the Cα alignment, maximised by the secondary structure matching (SSM) alignment algorithm [64]. The Q-score is reported in an interval from 0 to 1, where the Q-score reaches 1 in the case of identical structures and decreases with an increasing RMSD or a smaller alignment length. A Q-score of 0 indicates completely dissimilar structures. A Q-score higher than 0.1 can indicate some possibly significant level of structural similarity. Nonetheless, in this research, we set a more stringent Q-score threshold of 0.55. This value seemed to be meaningful as there were known structures of Z-DNA/Z-RNA binding proteins that scored below the newly reported domains (i.e., structures where the Z-DNA/Z-RNA binding function has not been described so far).

Table 2. Predicted Z-DNA/RNA binding proteins based on structural similarity to the experimentally validated Zα domain (3f21). Proteins are sorted according to their decreasing similarity score (Q-score); HOP2 is the best hit. UniProtKB IDs of all proteins are provided as well.

Protein Symbol/ID	Protein Name	Organism	Domain	Protein Length	Cellular Localisation/Known Function
HOP2 (O35047)	Homologous-pairing protein 2 homolog	*Mus musculus*	Eukarya	217	Nucleus/DNA binding, meiotic recombination, double-strand break repair, positive regulation of transcription by RNA pol II [65,66]
DsvD (Q46582)	DsvD	*Desulfovibrio vulgaris*	Bacteria	78	Role in dissimilatory sulfite reduction, Possible Interaction with B- and Z-DNA by Its Winged-Helix Motif [67]
D2PEW5	Uncharacterised DNA binding protein	*Sulfolobus islandicus*	Archaea	59	DNA binding
feoC (B5XTS6)	Probable [Fe-S]-dependent transcriptional repressor	*Klebsiella pneumoniae*	Bacteria	79	DNA binding may function as a transcriptional regulator that controls feoABC expression [68]

Table 2. *Cont.*

Protein Symbol/ID	Protein Name	Organism	Domain	Protein Length	Cellular Localisation/Known Function
pefI (Q04822)	FaeA-like protein	*Salmonella typhimurium*	Bacteria	70	Regulation of transcription [69]
RPA2 (P15927)	Replication protein A 32 kDa subunit	*Homo sapiens*	Eukarya	270	Nucleus/DNA binding, multifunctional protein (DNA repairs, DNA replication, telomere maintenance, preventing G-quadruplex formation) [70–73]
CDC53 (Q12018)	Cell division control protein 53	*Saccharomyces cerevisiae*	Eukarya	815	Nucleus & Cytoplasm/DNA replication origin binding, cell division, protein ubiquitination [74]
CUL1 (Q13616)	Cullin-1	*Homo sapiens*	Eukarya	776	Nucleus & Cytoplasm/Protein ubiquitination, cell division, transcription regulation [75]
ANC2 (Q9UJX6)	Anaphase-promoting complex subunit 2	*Homo sapiens*	Eukarya	822	Nucleus & Cytoplasm/Component of the anaphase promoting complex/cyclosome (APC/C) [76]
SCC1 (Q12158)	Sister chromatid cohesion protein 1	*Saccharomyces cerevisiae*	Eukarya	566	Nucleus/Mitotic sister chromatid cohesion, double-strand break repair [77]
APC2 (Q12440)	Anaphase-promoting complex subunit 2	*Saccharomyces cerevisiae*	Eukarya	853	Nucleus & cytoplasm/Component of the anaphase promoting complex/cyclosome (APC/C) [78]
Rpc34 (Q921X6)	DNA-directed RNA polymerase III subunit RPC6	*Mus musculus*	Eukarya	316	Nucleus/Nuclear and cytosolic DNA sensor involved in innate immune response, defence response to the virus [79]
PBP2 (A0A0E3GTJ4)	Archaeal DNA polymerase holoenzyme (PBP2 subunit)	*Saccharolobus solfataricus*	Archaea	76	Enhances DNA synthesis [80]
Reut_B4095 (Q46TT3)	Putative DNA-binding protein	*Cupriavidus pinatubonensis*	Bacteria	95	DNA binding

The resulting hits from Table 2 are visualised in Figure 2, together with the "reference" structure of a Zα domain (PDB: 3f21), which was used as the query protein for the structural similarity searching. All 14 proteins show noticeable structural similarity to the functional Zα domain, as each of these structures contains three alpha-helices and two antiparallel beta-strands, in order, typical for the Zα domain.

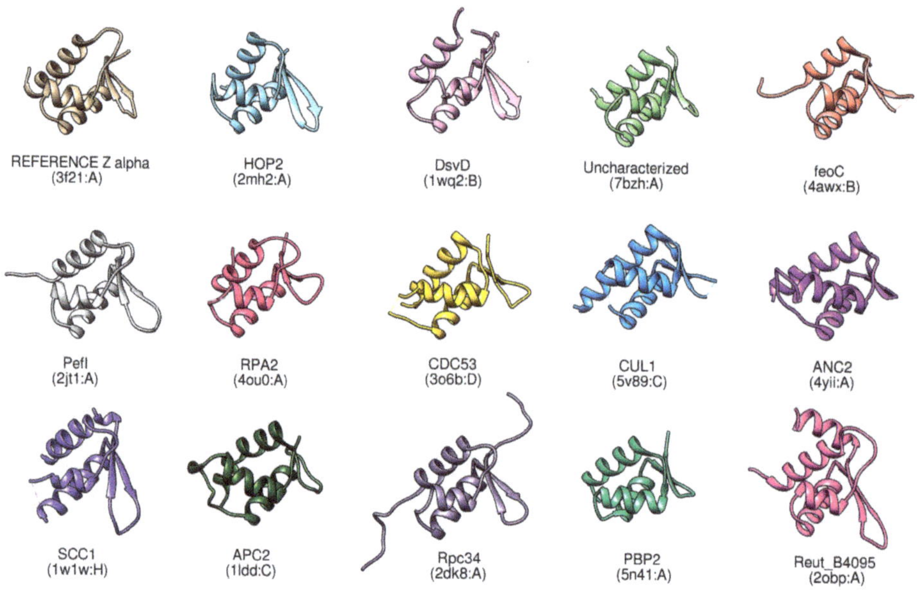

Figure 2. Comparison of the reference Zα domain (PDB: 3f21) (upper left corner) with the experimentally solved proteins (or their corresponding domains) having significant structural similarity (structures are ordered according to their similarity score to the reference structure (HOP2 best, DsvD second best, etc.).

The best new possible Z-DNA/Z-RNA binding protein found (based on the highest Q-score of its Zα domain), homologous-pairing protein 2 (HOP2), is widely conserved across the whole Eukarya domain. HOP2 proteins play an important role in meiotic recombination, particularly that of stimulating DMC1-mediated strand exchange that is necessary for homologous chromosome pairing during meiosis [81]. HOP2 forms a heterodimeric complex together with Meiotic nuclear division protein 1 homolog (MND1), and this HOP2/MND1 complex also promotes DMC1 mediated D-loop formation from double-strand DNA. Interestingly, a short 3bp deletion in the gene encoding HOP2 protein (leading to a deletion of a glutamic acid residue in the highly conserved C-terminal acidic domain) in humans causes "XX female gonadal dysgenesis" (XX-GD), which is a rare genetic disorder characterised for example by primary amenorrhea, uterine hypoplasia, or hypergonadotropic hypogonadism [82]. Another four proteins share a Cullin domain, particularly CDC53, CUL1, ANC2, and APC2. Proteins CDC53 (from *Saccharomyces cerevisiae*) and CUL1 (from *Homo sapiens*) are very distant functional homologs, and the same for ANC2 (from *Homo sapiens*) and APC2 (from *Saccharomyces cerevisiae*). Regarding Cullin domains and related ubiquitination processes, there are interesting links to viral diseases, see e.g., Rudnicka et al. [83]. Considering the current SARS-CoV-2 pandemic, it would be interesting to validate the potential of the viral RNA to form Z-RNA structures during replication, as was described for the influenza virus (H1N1 strain Puerto Rico/8/1934) virus in 2020 [40]. In this article, Zhang et al. found that replicating influenza A virus produces Z-RNAs and these are sensed by host ZBP1 in the nucleus of the host cell. This process led to the activation of specific protein kinases, resulting in nuclear rupture and unwanted necroptosis. From our newly described Z-DNA/Z-RNA binding proteins, protein Rpc34, which is subunit 6 of human RNA polymerase III, seems to have a direct association with a viral infection. For example, identical twins having a mutation in *POLR3F* (gene encoding Rpc34) had different susceptibility to the varicella-zoster virus in the CNS and lungs –

the patient with the POLR3F mutation exhibited impaired antiviral and inflammatory responses and increased viral replication [84].

Figure 3 shows a sequence alignment derived from the structural superposition of the predicted Zα domains from the analysed proteins to the Zα domain of the human protein ADAR1. All three alpha-helices are structurally conserved in the 14 possible Z-DNA/Z-RNA binding proteins. Similarly, beta-sheets of two or three strands are mostly preserved, except for in protein APC2. Interestingly, some amino acids in the predicted Zα domains were found to be repeatedly enriched in the exact positions of alignment—mainly in alpha helix 3, which is believed to be critical for Z-DNA/Z-RNA binding [52,60,85].

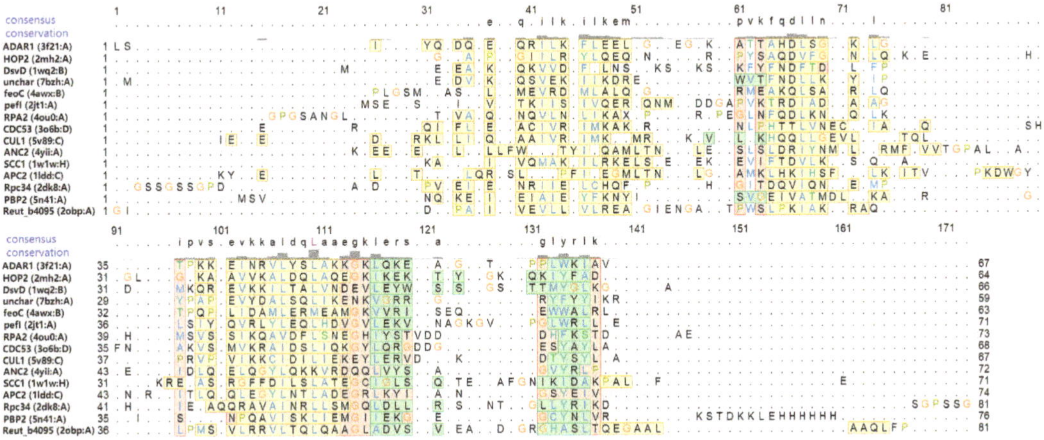

Figure 3. Sequence alignment is constructed from the structural superposition of the Zα domain of human ADAR1 protein (PDB: 3f21) and the 14 possible Z-DNA/Z-RNA binding proteins. The default colour of fully populated columns is light red, in addition, helices are coloured in yellow and strands in green. Letter colours correspond to the ClustalX colouring scheme.

Most of these 14 proteins identified (except for proteins CDC53 and CUL1, and proteins ANC2 and APC2) do not likely share a common evolutionary ancestor. Instead, the similar global fold of Zα 'domain' could be a result of convergent evolution [86,87] leading to preferential Z-DNA/Z-RNA structures binding. Currently known Z-DNA/Z-RNA binding proteins (ADAR, ZBP1, PKZ, E3L) are also not homologous, but rather analogous in their Z-DNA/Z-RNA binding function. This phenomenon is common in the case of other proteins which preferentially bind noncanonical forms of nucleic acids, such as G-quadruplex binding proteins [88] or cruciform binding proteins [89] (most of them don't have a common ancestor, but are analogous in their preferential interaction with G-quadruplexes, cruciforms, or another nucleic acid structures). In addition, it was found that some of the three-dimensional protein structures are widely conserved in non-homologous or unrelated DNA-binding proteins [90]. Then, the question arises we to whether the Zα domain is correctly annotated as a protein family (pfam ID: PF02295) as protein families are usually defined as groups of evolutionarily (not necessary functionally) related proteins. According to information deposited in the Pfam database, the HMM profile of this protein family was defined using only 5 seeds (regions 135–201 and 295–359 of human protein ADAR, region 137–203 of ADAR protein from *Rattus norvegicus*, region 7–71 of protein E3L from *Vaccinia virus*, and region 1–64 of protein ORF020 dsRNA-binding PKR inhibitor from *Orf virus* (Q6TVV0_ORFSA). This selection is problematic, as 3 of the 5 seed regions come from human and rat protein ADAR. The average length of the Zα domain is then 64.20 aa, with only 32% alignment identity. Therefore, we are sceptical about the current definition of the Zα domain on the level of the primary amino acid sequence. Nonetheless, further demystifying this issue is one motivation behind the scope of this paper, so we will continue with using the term 'Zα domain', in the sensu lato meaning, as the protein domain which preferentially interacts with Z-DNA/Z-RNA.

As the AlphaFold2 database [59] has provided putative structural models for thousands of proteins in several model organisms that have not yet been experimentally resolved, we sought to better understand which of these proteins may be involved in Z-DNA/Z-RNA binding. The ADAR1 Zα domain (PDB: 3f21) was chosen as a query structure for structural similarity searches using the RUPEE web server, which allows for the structural comparison with all AlphaFold2 models. RUPEE uses the TM-score to rank and quantify the structural similarity between protein alignments. On a scale from 0 to 1, a TM-score of over 0.5 is predicted to imply a similar fold. In a similar manner to the high Q-score threshold value used with PDBeFold, a TM-score of over 0.6 was chosen as a basis for the selection of hits from the structural alignment screen with RUPEE [91]. Since many of the proteins in the AlphaFold2 database do not yet have functional annotations, structural comparisons may further delineate their roles in cell survival.

Using the ADAR1 Zα domain (PDB: 3f21) as the query protein for the RUPEE web server, a total of 308 proteins were returned. Subsequent manual inspection of the alignments was performed to ensure that the putative Zα domains were structurally accessible and consisted primarily of basic residues that may be important for DNA-binding. A total of 185 unique proteins were selected after inspection, among which 59 proteins currently do not have complete functional annotation. Taking into consideration the previously annotated proteins that were predicted to contain one or more Zα domains, most have been assigned as putative transcriptional regulators—which further supports their potential to bind Z-DNA/Z-RNA. The probable [Fe-S]-dependent transcriptional repressor from *Escherichia coli* detected using RUPEE reflects the identification of the feoC protein from *Klebsiella* pneumoniae, detected using PDBeFold, that has been assigned the same function, which further validates the use of both structural comparison tools. In addition to feoC, additional similar proteins to Rpc34 and SCC1 were found, particularly DNA-directed RNA polymerase III subunit RPC3 (RNA polymerase III subunit C3) from *Leishmania infantum* and Rad21_Rec8 domain-containing protein from *Glycine max*. Interestingly, the uncharacterised proteins predicted to contain Zα domains were primarily found in the *Drosophila melanogaster*, *Methanocaldococcus jannaschii*, *Staphylococcus aureus*, and *Mycobacterium tuberculosis* proteomes (covering all three domains of life—Bacteria, Archaea, and Eukarya) The presence of proteins likely interacting with Z-DNA/Z-RNA in all domains of life further highlights the widespread occurrence of Z-DNA/Z-RNA and biological significance of such nucleic acid structures. The most numerous groups were uncharacterised proteins (59), transcriptional factors (56), and proteins related to ribosome biogenesis (49)—for further details see Supplementary Material S1. Both transcriptional factors and ribosomal proteins identified are in direct contact with DNA or RNA respectively, therefore their putative Z-DNA/Z-RNA binding ability is supported. The relatively large number of detected proteins, especially previously uncharacterised proteins, suggests that Z-DNA/Z-RNA binding domains may be more common than previously assumed. Further structural investigations may reveal the ability or extent of these proteins to bind Z-DNA/Z-RNA. Nonetheless, as the reliability of AlphaFold2 structural predictions still have some shortcomings [92], we have further proceeded only with 14 possible Z-DNA/Z-RNA binding proteins obtained from PDBeFold searches (experimentally solved structures).

2.2. *Domain Composition and Nuclear Localisation Signals within the Most Promising Z-DNA/Z-RNA Binding Proteins*

Figure 4 shows the position of regions that are structurally similar to the Zα domain of ADAR1 and the 14 possible Z-DNA/RNA binding proteins inferred in the PDBeFold search (Table 2 and Figure 2). Interestingly, these regions are exclusively located in the N′ (HOP2, Rpc34) or C′ terminal ends (RPA2, CDC53, CUL1, ANC2, SCC1, APC2) of proteins longer than 100 aa. These data are in congruence with a previous observation by Chiang et al. [43], where they depicted the position of Zα domains in six proteins with known Z-DNA/RNA function (Zα domains were always located at the N terminal end of longer proteins). These results potentially highlight the need for maximal exposure of the Zα domain to be able to interact with this type of non-canonical nucleic acid structure. AlphaFold structures of predicted Z-DNA/Z-RNA binding proteins from *Homo sapiens* are enclosed in Supplementary Material S2, together with highlighted domains with structural similarity to Zα. In addition, in protein HOP2, there is an isoform lacking the N-terminal region (ΔN) spanning the Zα domain structural homolog. In the study conducted by Uanschou et al. they found that the N′ terminal domain of the protein HOP2 is crucial for its DNA-binding function in *Arabidopsis thaliana* [93]. Nevertheless, HOP2 protein seems to be highly conserved across Eukaryotic organisms (typical N-terminal wHTH was predicted also in the mouse, rat, human, *Saccharomyces cerevisiae* and *Dictyostelium discoideum* proteomes according to models obtained from AlphaFold2 database—https://alphafold.ebi.ac.uk/search/text/hop2, (accessed on 25 October 2021)) [59]. The above-mentioned ΔN isoform is also present in the human proteome according to UniProt Sequence annotation (Isoform 3: Q9P2W1-3, aa residues 1–125 are missing). Finally, there are also two previously known examples of human proteins ADAR1 and DAI, where, in both cases, ΔN isoforms exist (which result in missing Zα domain). Regarding protein ADAR1, its short isoform ADAR1p110 is constitutively expressed and located in the nucleus, whereas the long isoform ADAR1p150 is interferon-inducible and undergoes shuffling between the cytoplasm and nucleus [94,95]. Both of these isoforms share a Zβ domain (which may not have Z-DNA-binding ability [60] and its function is still unknown [96]), A-to-I deaminase domain, three double-stranded RNA-binding domains, but the long P150 isoform has an extra Z-DNA/RNA-binding domain at its N-terminus [97].

All eukaryotic proteins found have at least theoretical possibility to be localised both in the cytoplasm and cell nucleus, as was checked in a literature search and using nuclear localisation signal prediction within primary amino acid sequences of these proteins (cNLS Mapper webserver, accessed from http://nls-mapper.iab.keio.ac.jp/cgi-bin/NLS_Mapper_form.cgi, (accessed on 11 November 2021)) [98] (Supplementary Material S3). It is worth mentioning that the overall amino acid composition of these fourteen proteins identified shows similar significant enrichments (isoleucine, lysine, aspartic acid) and depletion (cysteine) as observed previously by us [99].

2.3. Representative Molecular Docking of RPA2 Region Structurally Similar to Zα Domain and Z-DNA/Z-RNA

We carried out representative molecular docking (using theHDOCK web server [100], further details in Materials and Methods section) of the human RPA2 putative Z-DNA/Z-RNA binding domain to Z-DNA (Figure 5A) and Z-RNA (Figure 5B). RPA2 was selected for its important molecular function in DNA replication and the cellular response to DNA damage. Results of this analysis revealed key amino acid residues involved in Z-DNA and/or Z-RNA binding. In both cases, tyrosine at position 256 (considering the whole RPA2 protein) was involved, suggesting its critical role in interaction with left-handed nucleic acids. In both cases, alpha-helix 3 and two subsequent beta-sheets seem to play pivotal roles in Z-DNA/Z-RNA recognition. These results are in congruence with previous experimental models of known Zα domains interacting with Z-DNA/Z-RNA, where the tyrosine, lysine, asparagine and serine amino acid residues played key roles in interaction [21,52,101,102]. The dockings of the remaining 13 possible Z-DNA/Z-RNA binding proteins are enclosed in Supplementary Material S4 (10 best docking poses for all protein/nucleic acid combinations). The inspection of the best docking poses revealed that it in general follows the rules described above. Carrying out a detailed molecular dynamic study would be beneficial in subsequent research to shed more light on the stability of these complexes.

Figure 4. Position of Zα domain in ADAR1 and within 14 newly described possible Z-DNA/Z-RNA binding proteins. The exact position of Zα and its structural homologs is always highlighted in yellow.

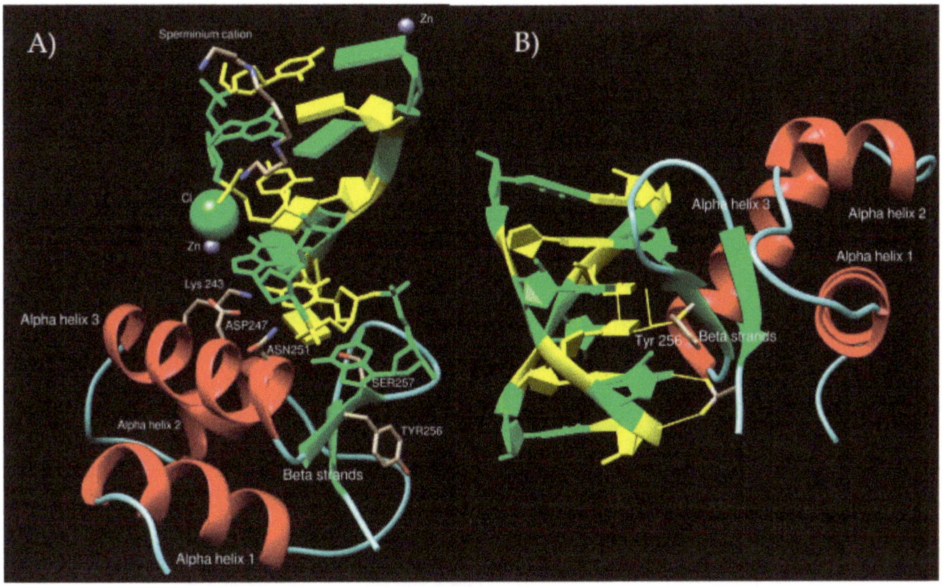

Figure 5. Representative molecular docking of human RPA2 Zα structural homolog to Z-DNA (**A**) and Z-RNA (**B**). Protein alpha helices are in red, beta-strands in green, coiled-coil regions in azure. Highlighting of Z-DNA/Z-RNA follows classic NDB colouring (guanines in green, cytosines in yellow).

2.4. Functional Enrichment and Interaction Network of Human Z-DNA/Z-RNA Binding Proteins

Finally, we aimed to better illustrate the possible functional interconnection between previously known human proteins ADAR and ZBP1, together with newly predicted human Z-DNA/Z-RNA binding proteins. We have constructed a STRING interaction network [103] made from two previously known human Z-DNA/Z-RNA binding proteins and five newly identified possible human Z-DNA/Z-RNA binding proteins containing structural similarity to the Zα domain. Additionally, the 50 closest interacting proteins were added via STRING (first shell of interactors) to better show possible pathways involving Z-DNA/Z-RNA binding and vice versa (Figure 6). This analysis has shown that newly identified possible Z-DNA/Z-RNA proteins (in humans) are quite distinct from two previously known human Z-DNA/Z-RNA interacting proteins ADAR and ZBP1 (blue cluster). Specifically, proteins RPA2 and HOP2 (syn. PSMC3IP) are both important members of the Meiotic Strand Invasion curated pathway [104] (azure cluster). POLR3F, the human homolog of mouse Rpc34, is interacting mainly with other subunits of RNA polymerase III complex, which is composed of 17 subunits and its structure was solved last year [105]. Interestingly, causative polymerase III mutations have been described in patients with hypersensitivity to viral infection [106,107]. The cluster containing human Cullin 1 protein (yellow) and a cluster containing ANAPC2 protein (red) are very tightly interconnected through functional interactions and involved in various cell cycle processes, including the proteasome-mediated ubiquitin-dependent protein catabolic process, the anaphase-promoting complex-dependent catabolic process, or activation of the innate immune response [108]. These results (Figure 6) reflect the current state of knowledge and do not consider the putative Z-DNA/Z-RNA binding function of proteins POLR3F, RPA2, HOP2/PSMC3IP, CUL1 and ANAPC2, which was first proposed in this manuscript. Once these proteins are validated as *bona fide* Z-DNA/Z-RNA binding in vitro (and their annotations are actualised within the STRING database), they will probably form a strong functional network by themselves (based on their Z-DNA/Z-RNA annotations).

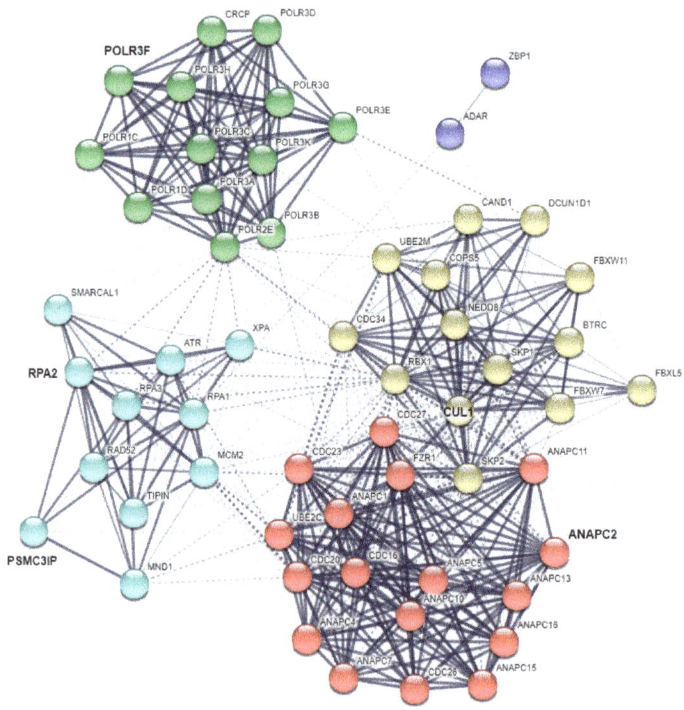

Figure 6. STRING interaction network of newly identified human possible Z-DNA/Z-RNA binding proteins (in bold and higher letter size), together with 2 previously known human Z-DNA/Z-RNA binding proteins (ZBP1 and ADAR), and also with 50 first shell interactors. Clustering was made using MCL inflation parameter (3), the resulting five clusters are highlighted in distinct colours. Line thickness indicates the strength of data support and edges between different clusters are dotted.

3. Materials and Methods

3.1. Collection of Experimentally-Validated Z-DNA/RNA Binding Protein Structures

A systematic review of existing literature sources deposited in the Web of Science (https://clarivate.com/webofsciencegroup/solutions/web-of-science/, (accessed on 18 August 2021)), NCBI PubMed (https://pubmed.ncbi.nlm.nih.gov/, (accessed on 18 August 2021)), or Google Scholar (https://scholar.google.com/, (accessed on 18 August 2021)) databases was done to identify all up-to-date known Z-DNA/RNA binding proteins containing at least one Zα or Zβ domain. The resulting list of these proteins can be found in Table 1. Where available, the information about experimentally solved 3D structures was gathered as well.

3.2. Structure-Based Similarity Searches

Structure-based similarity searches were performed using the PDBeFold and RUPEE web servers [64], accessed from https://www.ebi.ac.uk/msd-srv/ssm/cgi-bin/ssmserver, (accessed on 10 September 2021), and from https://ayoubresearch.com/, (accessed on 21 October 2021). As a query, the experimentally-resolved structure of the Zα domain was used (PDB: 3f21, chain:A). PDBeFold was used to structurally compare the query Zα domain to all known experimentally-resolved structures in PDB, and RUPEE was used to query against all AlphaFold2 models. Parameters were left to be Default using PDBeFold, except for the "precision", which was changed from "normal" to "high". Three settings were used for the RUPEE search: "Full-Length" (finding exact length matches of the query protein in the database protein), "Contains" (finding query protein inside database protein), and "Contained-In" options (small protein motif detection in query protein). The hits resulting from the "Full-Length", "Contained-In", and "Contains" modes using RUPEE were combined to identify the total list of putative unique proteins.

3.3. Structure Visualisation and Contacts/Clashes Depicting

All protein structures were visualised and graphically pre-processed in a standalone version of the UCSF Chimera Tool [109]. Prediction of contact amino acid residues was carried out using the Chimera function "Find clashes/contacts" with the following parameters: "VDW overlap" $\geq$ 0.4 angstroms; "subtractions of 0.4 from overlap for potentially H-bonding pairs"; "Ignoring contacts of pairs 2 or fewer bonds apart".

3.4. Structural Alignment Construction

Structural alignments of newly described Z-DNA/RNA binding proteins were done using Chimera structural analyses toolbox [110], particularly MatchMaker program was used with the following parameters: "Reference structure": 3f21; "Structures to match": 14 newly predicted proteins; "Chain pairing": Best aligning pair of chains between reference and match structures; "Alignment algorithm": Needleman-Wunsch; "Matrix": BLOSUM-62; "Gap opening penalty": 12; "Gap extension penalty": 1; "Include secondary structure score": 50%; "Compute secondary structure assignments": yes; "Iterate by pruning long atom pairs until no pair exceeds": 2.0 angstroms; "After superposition, compute structure-based multiple sequence alignment": yes; "Create alignment from superposition": choose all 15 protein structures; "Residue-residue distance cutoff": 5.0 angstroms; "Residue aligned in column if within cutoff of": at least one other; "Allow for circular permutation": no; "Iterate superposition/alignment": no.

3.5. Docking to Z-DNA/RNA

Docking of the putative RPA2 Zα domain (PDB: 4ou0:A) to Z-DNA (PDB: 4HIF) [111] and Z-RNA (PDB: 1T4X) [112] was done using HDOCK webserver (http://hdock.phys.hust.edu.cn/, (accessed on 30 December 2021)) [100] with default parameters. Protein structures were always submitted as a "receptor", and Z-DNA structure as a "ligand". The same procedure was repeated for the rest of the 14 possible Z-DNA/Z-RNA binding proteins. The resulting docking poses (best 10) are enclosed in Supplementary Material S4. The resulting models are sorted according to their HDOCK docking energy scores ("model 1" has the best energy score). Finally, the docking results were manually validated with respect to the existing literature, where main contact residues were determined (see Section 2.3 in Results and Discussion section).

3.6. Functional Enrichment Analysis

Functional enrichment analysis of 14 predicted Z-DNA/RNA binding proteins was done as follows: at first, homologous proteins were found in *Homo sapiens*, where available, and structural conservation of desired "Zα-like" fold was visually checked using AlphaFold prediction [59]. Secondly, five human proteins with conserved "Zα-like" fold (identified in this study) were uploaded to STRING webserver together with previously known Z-DNA/RNA binding proteins (https://string-db.org/cgi/input?sessionId=bVBUeCTKWYuE&input_page_show_search=on, (accessed on 12 December 2021)) [103] and 50 closest interacting proteins were automatically added via STRING (first shell of interactors).

4. Conclusions

Our analysis detected the Zα domain structural homologs in fourteen proteins that have not yet been described as Z-DNA/Z-RNA recognising proteins. These suggest that Z-DNA/Z-RNA recognition is more common and important in living systems than previously thought. Functional pathways interactions of the newly characterised proteins with a Zα domain indicate their involvement in innate immunity and other important molecular and biological pathways. These results also highlight the utility of structure-based similarity searches to elucidate the structure-function relationship of uncharacterised proteins or protein domains. Further experimental validation is required to determine the extent to which these proteins may bind to Z-DNA/Z-RNA.

Supplementary Materials: The following supporting information can be downloaded at: https://www.mdpi.com/article/10.3390/ijms23020768/s1, S1: RUPEE search through the whole AlphaFold database for proteins containing structural similarity to Zα domain; S2: AlphaFold structures of predicted human Z-DNA/Z-RNA proteins with highlighted Zα domains; S3: Nuclear localisation signals in newly described Z-DNA/Z-RNA binding proteins; S4: Top 10 docking poses for all possible Z-DNA/Z-RNA proteins and selected Z-DNA (PDB: 4HIF) and Z-RNA (PDB: 4HIF) structures.

Author Contributions: Conceptualisation, M.B. and P.P.; data curation, K.S.; formal analysis, M.B. and C.A.B.; methodology, M.B. and C.A.B.; resources, K.S. and M.B.; writing—original draft preparation, M.B., V.B. and K.S.; writing—review and editing, V.B., C.A.B. and J.Č.; visualisation, M.B.; supervision, P.P. and T.L.B. All authors have read and agreed to the published version of the manuscript.

Funding: This research was funded by the Czech Science Foundation, grant number 22-21903S. T.L.B. thanks the Wellcome Trust for support through an Investigator Award (200814/Z/16/Z; 2016–2021). C.A.B. was supported by Antibiotic Research UK (PHZJ/687).

Institutional Review Board Statement: Not applicable.

Informed Consent Statement: Not applicable.

Data Availability Statement: All data is contained within this article or Supplementary Material.

Acknowledgments: We would like to show our gratitude to Adriana Volná (University of Ostrava) for sharing their pearls of wisdom with us during the course of this research.

Conflicts of Interest: The authors declare no conflict of interest. The funders had no role in the design of the study; in the collection, analyses, or interpretation of data; in the writing of the manuscript, or in the decision to publish the results.

References

1. Guiblet, W.M.; Cremona, M.A.; Harris, R.S.; Chen, D.; Eckert, K.A.; Chiaromonte, F.; Huang, Y.-F.; Makova, K.D. Non-B DNA: A Major Contributor to Small- and Large-Scale Variation in Nucleotide Substitution Frequencies across the Genome. *Nucleic Acids Res.* **2021**, *49*, 1497–1516. [CrossRef]
2. Brázda, V.; Bartas, M.; Bowater, R.P. Evolution of Diverse Strategies for Promoter Regulation. *Trends Genet.* **2021**, *37*, 730–744. [CrossRef]
3. Lyons, S.M.; Kharel, P.; Akiyama, Y.; Ojha, S.; Dave, D.; Tsvetkov, V.; Merrick, W.; Ivanov, P.; Anderson, P. EIF4G Has Intrinsic G-Quadruplex Binding Activity That Is Required for TiRNA Function. *Nucleic Acids Res.* **2020**, *48*, 6223–6233. [CrossRef] [PubMed]
4. Mukherjee, A.K.; Sharma, S.; Chowdhury, S. Non-Duplex G-Quadruplex Structures Emerge as Mediators of Epigenetic Modifications. *Trends Genet.* **2019**, *35*, 129–144. [CrossRef]
5. Hayward, B.E.; Usdin, K. Mechanisms of Genome Instability in the Fragile X-Related Disorders. *Genes* **2021**, *12*, 1633. [CrossRef]
6. Schaich, M.A.; Van Houten, B. Searching for DNA Damage: Insights from Single Molecule Analysis. *Front. Mol. Biosci.* **2021**, *8*, 772877. [CrossRef] [PubMed]
7. Brázda, V.; Laister, R.C.; Jagelská, E.B.; Arrowsmith, C. Cruciform Structures Are a Common DNA Feature Important for Regulating Biological Processes. *BMC Mol. Biol.* **2011**, *12*, 33. [CrossRef]
8. Zhao, J.; Bacolla, A.; Wang, G.; Vasquez, K.M. Non-B DNA Structure-Induced Genetic Instability and Evolution. *Cell. Mol. Life Sci.* **2010**, *67*, 43–62. [CrossRef]
9. Herbert, A. Z-DNA and Z-RNA in Human Disease. *Commun. Biol.* **2019**, *2*, 7. [CrossRef]
10. Roy, R.; Chakraborty, P.; Chatterjee, A.; Sarkar, J. Comparative Review on Left-Handed Z-DNA. *Front. Biosci.* **2021**, *26*, 29–35.
11. Rajeswari, M.R. DNA Triplex Structures in Neurodegenerative Disorder, Friedreich's Ataxia. *J. Biosci.* **2012**, *37*, 519–532. [CrossRef]
12. Varshney, D.; Spiegel, J.; Zyner, K.; Tannahill, D.; Balasubramanian, S. The Regulation and Functions of DNA and RNA G-Quadruplexes. *Nat. Rev. Mol. Cell Biol.* **2020**, *21*, 459–474. [CrossRef]
13. Brázda, V.; Hároníková, L.; Liao, J.C.; Fojta, M. DNA and RNA Quadruplex-Binding Proteins. *Int. J. Mol. Sci.* **2014**, *15*, 17493–17517. [CrossRef] [PubMed]
14. Brázda, V.; Coufal, J.; Liao, J.C.C.; Arrowsmith, C.H. Preferential Binding of IFI16 Protein to Cruciform Structure and Superhelical DNA. *Biochem. Biophys. Res. Commun.* **2012**, *422*, 716–720. [CrossRef]

15. Schwartz, T.; Rould, M.A.; Lowenhaupt, K.; Herbert, A.; Rich, A. Crystal Structure of the Zalpha Domain of the Human Editing Enzyme ADAR1 Bound to Left-Handed Z-DNA. *Science* **1999**, *284*, 1841–1845. [CrossRef]
16. Placido, D.; Brown, B.A.; Lowenhaupt, K.; Rich, A.; Athanasiadis, A. A Left-Handed RNA Double Helix Bound by the Z α Domain of the RNA-Editing Enzyme ADAR1. *Structure* **2007**, *15*, 395–404. [CrossRef] [PubMed]
17. Ha, S.C.; Kim, D.; Hwang, H.-Y.; Rich, A.; Kim, Y.-G.; Kim, K.K. The Crystal Structure of the Second Z-DNA Binding Domain of Human DAI (ZBP1) in Complex with Z-DNA Reveals an Unusual Binding Mode to Z-DNA. *Proc. Natl. Acad. Sci. USA* **2008**, *105*, 20671–20676. [CrossRef] [PubMed]
18. Schwartz, T.; Behlke, J.; Lowenhaupt, K.; Heinemann, U.; Rich, A. Structure of the DLM-1–Z-DNA Complex Reveals a Conserved Family of Z-DNA-Binding Proteins. *Nat. Struct. Mol. Biol.* **2001**, *8*, 761–765. [CrossRef]
19. Kim, D.; Hur, J.; Park, K.; Bae, S.; Shin, D.; Ha, S.C.; Hwang, H.-Y.; Hohng, S.; Lee, J.-H.; Lee, S.; et al. Distinct Z-DNA Binding Mode of a PKR-like Protein Kinase Containing a Z-DNA Binding Domain (PKZ). *Nucleic Acids Res.* **2014**, *42*, 5937–5948. [CrossRef]
20. de Rosa, M.; Zacarias, S.; Athanasiadis, A. Structural Basis for Z-DNA Binding and Stabilization by the Zebrafish Z-DNA Dependent Protein Kinase PKZ. *Nucleic Acids Res.* **2013**, *41*, 9924–9933. [CrossRef]
21. Kuś, K.; Rakus, K.; Boutier, M.; Tsigkri, T.; Gabriel, L.; Vanderplasschen, A.; Athanasiadis, A. The Structure of the Cyprinid Herpesvirus 3 ORF112-Zα·Z-DNA Complex Reveals a Mechanism of Nucleic Acids Recognition Conserved with E3L, a Poxvirus Inhibitor of Interferon Response. *J. Biol. Chem.* **2015**, *290*, 30713–30725. [CrossRef]
22. Ha, S.C.; Lokanath, N.K.; Van Quyen, D.; Wu, C.A.; Lowenhaupt, K.; Rich, A.; Kim, Y.-G.; Kim, K.K. A Poxvirus Protein Forms a Complex with Left-Handed Z-DNA: Crystal Structure of a Yatapoxvirus Zα Bound to DNA. *Proc. Natl. Acad. Sci. USA* **2004**, *101*, 14367–14372. [CrossRef]
23. Nikpour, N.; Salavati, R. The RNA Binding Activity of the First Identified Trypanosome Protein with Z-DNA-Binding Domains. *Sci. Rep.* **2019**, *9*, 5904. [CrossRef] [PubMed]
24. Mitsui, Y.; Langridge, R.; Shortle, B.E.; Cantor, C.R.; Grant, R.C.; Kodama, M.; Wells, R.D. Physical and Enzymatic Studies on Poly d(I–C).Poly d(I–C), an Unusual Double-Helical DNA. *Nature* **1970**, *228*, 1166–1169. [CrossRef]
25. Wang, A.H.-J.; Quigley, G.J.; Kolpak, F.J.; Crawford, J.L.; van Boom, J.H.; van der Marel, G.; Rich, A. Molecular Structure of a Left-Handed Double Helical DNA Fragment at Atomic Resolution. *Nature* **1979**, *282*, 680–686. [CrossRef]
26. Ha, S.C.; Lowenhaupt, K.; Rich, A.; Kim, Y.-G.; Kim, K.K. Crystal Structure of a Junction between B-DNA and Z-DNA Reveals Two Extruded Bases. *Nature* **2005**, *437*, 1183–1186. [CrossRef] [PubMed]
27. Zhang, Y.; Cui, Y.; An, R.; Liang, X.; Li, Q.; Wang, H.; Wang, H.; Fan, Y.; Dong, P.; Li, J.; et al. Topologically Constrained Formation of Stable Z-DNA from Normal Sequence under Physiological Conditions. *J. Am. Chem. Soc.* **2019**, *141*, 7758–7764. [CrossRef] [PubMed]
28. Renčiuk, D.; Kypr, J.; Vorlíčková, M. CGG Repeats Associated with Fragile X Chromosome Form Left-Handed Z-DNA Structure. *Biopolymers* **2011**, *95*, 174–181. [CrossRef]
29. Bae, S.; Kim, D.; Kim, K.K.; Kim, Y.-G.; Hohng, S. Intrinsic Z-DNA Is Stabilized by the Conformational Selection Mechanism of Z-DNA-Binding Proteins. *J. Am. Chem. Soc.* **2011**, *133*, 668–671. [CrossRef]
30. Dumat, B.; Larsen, A.F.; Wilhelmsson, L.M. Studying Z-DNA and B- to Z-DNA Transitions Using a Cytosine Analogue FRET-Pair. *Nucleic Acids Res.* **2016**, *44*, e101. [CrossRef]
31. Beknazarov, N.; Jin, S.; Poptsova, M. Deep Learning Approach for Predicting Functional Z-DNA Regions Using Omics Data. *Sci. Rep.* **2020**, *10*, 19134. [CrossRef]
32. Champ, P.C.; Maurice, S.; Vargason, J.M.; Camp, T.; Ho, P.S. Distributions of Z-DNA and Nuclear Factor I in Human Chromosome 22: A Model for Coupled Transcriptional Regulation. *Nucleic Acids Res.* **2004**, *32*, 6501–6510. [CrossRef]
33. Ho, P.S.; Mooers, B.H. Z-DNA Crystallography. *Biopolymers* **1997**, *14*, 65–90. [CrossRef]
34. Herbert, A. ALU Non-B-DNA Conformations, Flipons, Binary Codes and Evolution. *R. Soc. Open Sci.* **2020**, *7*, 200222. [CrossRef] [PubMed]
35. Fleming, A.M.; Zhu, J.; Ding, Y.; Esders, S.; Burrows, C.J. Oxidative Modification of Guanine in a Potential Z-DNA-Forming Sequence of a Gene Promoter Impacts Gene Expression. *Chem. Res. Toxicol.* **2019**, *32*, 899–909. [CrossRef] [PubMed]
36. Li, H.; Xiao, J.; Li, J.; Lu, L.; Feng, S.; Dröge, P. Human Genomic Z-DNA Segments Probed by the Zα Domain of ADAR1. *Nucleic Acids Res.* **2009**, *37*, 2737–2746. [CrossRef]
37. Shin, S.-I.; Ham, S.; Park, J.; Seo, S.H.; Lim, C.H.; Jeon, H.; Huh, J.; Roh, T.-Y. Z-DNA-Forming Sites Identified by ChIP-Seq Are Associated with Actively Transcribed Regions in the Human Genome. *DNA Res.* **2016**, *23*, 477–486. [CrossRef] [PubMed]
38. Fogg, J.M.; Randall, G.L.; Pettitt, B.M.; Sumners, D.W.L.; Harris, S.A.; Zechiedrich, L. Bullied No More: When and How DNA Shoves Proteins Around. *Q. Rev. Biophys.* **2012**, *45*, 257–299. [CrossRef]
39. Hall, K.; Cruz, P.; Tinoco, I.; Jovin, T.M.; van de Sande, J.H. 'Z-RNA'—A Left-Handed RNA Double Helix. *Nature* **1984**, *311*, 584–586. [CrossRef]
40. Zhang, T.; Yin, C.; Boyd, D.F.; Quarato, G.; Ingram, J.P.; Shubina, M.; Ragan, K.B.; Ishizuka, T.; Crawford, J.C.; Tummers, B.; et al. Influenza Virus Z-RNAs Induce ZBP1-Mediated Necroptosis. *Cell* **2020**, *180*, 1115–1129.e13. [CrossRef]
41. Cer, R.Z.; Donohue, D.E.; Mudunuri, U.S.; Temiz, N.A.; Loss, M.A.; Starner, N.J.; Halusa, G.N.; Volfovsky, N.; Yi, M.; Luke, B.T.; et al. Non-B DB v2.0: A Database of Predicted Non-B DNA-Forming Motifs and Its Associated Tools. *Nucleic Acids Res.* **2013**, *41*, D94–D100. [CrossRef]

42. Herbert, A.; Lowenhaupt, K.; Spitzner, J.; Rich, A. Double-Stranded RNA Adenosine Deaminase Binds Z-DNA in Vitro. *Nucleic Acids Symp. Ser.* **1995**, *33*, 16–19.
43. Chiang, D.C.; Li, Y.; Ng, S.K. The Role of the Z-DNA Binding Domain in Innate Immunity and Stress Granules. *Front. Immunol.* **2021**, *11*, 3779. [CrossRef]
44. Lee, A.-R.; Kim, N.-H.; Seo, Y.-J.; Choi, S.-R.; Lee, J.-H. Thermodynamic Model for B-Z Transition of DNA Induced by Z-DNA Binding Proteins. *Molecules* **2018**, *23*, 2748. [CrossRef]
45. Wang, H.; Wang, G.; Zhang, L.; Zhang, J.; Zhang, J.; Wang, Q.; Billiar, T.R. ADAR1 Suppresses the Activation of Cytosolic RNA-Sensing Signaling Pathways to Protect the Liver from Ischemia/Reperfusion Injury. *Sci. Rep.* **2016**, *6*, 20248. [CrossRef]
46. Takaoka, A.; Wang, Z.; Choi, M.K.; Yanai, H.; Negishi, H.; Ban, T.; Lu, Y.; Miyagishi, M.; Kodama, T.; Honda, K.; et al. DAI (DLM-1/ZBP1) Is a Cytosolic DNA Sensor and an Activator of Innate Immune Response. *Nature* **2007**, *448*, 501–505. [CrossRef]
47. Kuriakose, T.; Kanneganti, T.-D. ZBP1: Innate Sensor Regulating Cell Death and Inflammation. *Trends Immunol.* **2018**, *39*, 123–134. [CrossRef] [PubMed]
48. Fischer, S.E.J.; Ruvkun, G. Caenorhabditis Elegans ADAR Editing and the ERI-6/7/MOV10 RNAi Pathway Silence Endogenous Viral Elements and LTR Retrotransposons. *Proc. Natl. Acad. Sci. USA* **2020**, *117*, 5987–5996. [CrossRef] [PubMed]
49. Kahmann, J.D.; Wecking, D.A.; Putter, V.; Lowenhaupt, K.; Kim, Y.-G.; Schmieder, P.; Oschkinat, H.; Rich, A.; Schade, M. The Solution Structure of the N-Terminal Domain of E3L Shows a Tyrosine Conformation That May Explain Its Reduced Affinity to Z-DNA in Vitro. *Proc. Natl. Acad. Sci. USA* **2004**, *101*, 2712–2717. [CrossRef] [PubMed]
50. Thakur, M.; Seo, E.J.; Dever, T.E. Variola Virus E3L Zα Domain, but Not Its Z-DNA Binding Activity, Is Required for PKR Inhibition. *RNA* **2014**, *20*, 214–227. [CrossRef]
51. Kim, Y.-G.; Muralinath, M.; Brandt, T.; Pearcy, M.; Hauns, K.; Lowenhaupt, K.; Jacobs, B.L.; Rich, A. A Role for Z-DNA Binding in Vaccinia Virus Pathogenesis. *Proc. Natl. Acad. Sci. USA* **2003**, *100*, 6974–6979. [CrossRef]
52. Deigendesch, N.; Koch-Nolte, F.; Rothenburg, S. ZBP1 Subcellular Localization and Association with Stress Granules Is Controlled by Its Z-DNA Binding Domains. *Nucleic Acids Res.* **2006**, *34*, 5007–5020. [CrossRef] [PubMed]
53. Ng, S.K.; Weissbach, R.; Ronson, G.E.; Scadden, A.D.J. Proteins That Contain a Functional Z-DNA-Binding Domain Localize to Cytoplasmic Stress Granules. *Nucleic Acids Res.* **2013**, *41*, 9786–9799. [CrossRef] [PubMed]
54. Taghavi, N.; Samuel, C.E. RNA-Dependent Protein Kinase PKR and the Z-DNA Binding Orthologue PKZ Differ in Their Capacity to Mediate Initiation Factor EIF2α-Dependent Inhibition of Protein Synthesis and Virus-Induced Stress Granule Formation. *Virology* **2013**, *443*, 48–58. [CrossRef]
55. Licht, K.; Jantsch, M.F. The Other Face of an Editor: ADAR1 Functions in Editing-Independent Ways. *Bioessays* **2017**, *39*, 1700129. [CrossRef]
56. Herbert, A.; Schade, M.; Lowenhaupt, K.; Alfken, J.; Schwartz, T.; Shlyakhtenko, L.S.; Lyubchenko, Y.L.; Rich, A. The Z α Domain from Human ADAR1 Binds to the Z-DNA Conformer of Many Different Sequences. *Nucleic Acids Res.* **1998**, *26*, 3486–3493. [CrossRef]
57. Kim, U.; Wang, Y.; Sanford, T.; Zeng, Y.; Nishikura, K. Molecular Cloning of CDNA for Double-Stranded RNA Adenosine Deaminase, a Candidate Enzyme for Nuclear RNA Editing. *Proc. Natl. Acad. Sci. USA* **1994**, *91*, 11457–11461. [CrossRef] [PubMed]
58. Ha, S.C.; Choi, J.; Hwang, H.-Y.; Rich, A.; Kim, Y.-G.; Kim, K.K. The Structures of Non-CG-Repeat Z-DNAs Co-Crystallized with the Z-DNA-Binding Domain, HZα ADAR1. *Nucleic Acids Res.* **2009**, *37*, 629–637. [CrossRef]
59. Jumper, J.; Evans, R.; Pritzel, A.; Green, T.; Figurnov, M.; Ronneberger, O.; Tunyasuvunakool, K.; Bates, R.; Žídek, A.; Potapenko, A.; et al. Highly Accurate Protein Structure Prediction with AlphaFold. *Nature* **2021**, *596*, 583–589. [CrossRef]
60. Athanasiadis, A.; Placido, D.; Maas, S.; Brown, B.A.; Lowenhaupt, K.; Rich, A. The Crystal Structure of the Z β Domain of the RNA-Editing Enzyme ADAR1 Reveals Distinct Conserved Surfaces among Z-Domains. *J. Mol. Biol.* **2005**, *351*, 496–507. [CrossRef]
61. Schade, M.; Turner, C.J.; Kühne, R.; Schmieder, P.; Lowenhaupt, K.; Herbert, A.; Rich, A.; Oschkinat, H. The Solution Structure of the Zα Domain of the Human RNA Editing Enzyme ADAR1 Reveals a Prepositioned Binding Surface for Z-DNA. *Proc. Natl. Acad. Sci. USA* **1999**, *96*, 12465–12470. [CrossRef] [PubMed]
62. Kim, K.; Khayrutdinov, B.I.; Lee, C.-K.; Cheong, H.-K.; Kang, S.W.; Park, H.; Lee, S.; Kim, Y.-G.; Jee, J.; Rich, A.; et al. Solution Structure of the Z β Domain of Human DNA-Dependent Activator of IFN-Regulatory Factors and Its Binding Modes to B- and Z-DNAs. *Proc. Natl. Acad. Sci. USA* **2011**, *108*, 6921–6926. [CrossRef]
63. Park, C.; Zheng, X.; Park, C.Y.; Kim, J.; Lee, S.K.; Won, H.; Choi, J.; Kim, Y.-G.; Choi, H.-J. Dual Conformational Recognition by Z-DNA Binding Protein Is Important for the B–Z Transition Process. *Nucleic Acids Res.* **2020**, *48*, 12957–12971. [CrossRef] [PubMed]
64. Krissinel, E.; Henrick, K. Secondary-Structure Matching (SSM), a New Tool for Fast Protein Structure Alignment in Three Dimensions. *Acta Cryst. D Biol. Cryst.* **2004**, *60*, 2256–2268. [CrossRef]
65. Leu, J.Y.; Chua, P.R.; Roeder, G.S. The Meiosis-Specific Hop2 Protein of S. Cerevisiae Ensures Synapsis between Homologous Chromosomes. *Cell* **1998**, *94*, 375–386. [CrossRef]
66. Chan, Y.-L.; Brown, M.S.; Qin, D.; Handa, N.; Bishop, D.K. The Third Exon of the Budding Yeast Meiotic Recombination Gene HOP2 Is Required for Calcium-Dependent and Recombinase Dmc1-Specific Stimulation of Homologous Strand Assimilation. *J. Biol. Chem.* **2014**, *289*, 18076–18086. [CrossRef]

67. Mizuno, N.; Voordouw, G.; Miki, K.; Sarai, A.; Higuchi, Y. Crystal Structure of Dissimilatory Sulfite Reductase D (DsrD) Protein—Possible Interaction with B- and Z-DNA by Its Winged-Helix Motif. *Structure* **2003**, *11*, 1133–1140. [CrossRef]
68. Hung, K.-W.; Tsai, J.-Y.; Juan, T.-H.; Hsu, Y.-L.; Hsiao, C.-D.; Huang, T.-H. Crystal Structure of the Klebsiella Pneumoniae NFeoB/FeoC Complex and Roles of FeoC in Regulation of Fe^{2+} Transport by the Bacterial Feo System. *J. Bacteriol.* **2012**, *194*, 6518–6526. [CrossRef]
69. Aramini, J.M.; Rossi, P.; Cort, J.R.; Ma, L.-C.; Xiao, R.; Acton, T.B.; Montelione, G.T. Solution NMR Structure of the Plasmid-Encoded Fimbriae Regulatory Protein PefI from Salmonella Enterica Serovar Typhimurium. *Proteins Struct. Funct. Bioinform.* **2011**, *79*, 335–339. [CrossRef]
70. Sleeth, K.M.; Sørensen, C.S.; Issaeva, N.; Dziegielewski, J.; Bartek, J.; Helleday, T. RPA Mediates Recombination Repair during Replication Stress and Is Displaced from DNA by Checkpoint Signalling in Human Cells. *J. Mol. Biol.* **2007**, *373*, 38–47. [CrossRef] [PubMed]
71. Grudic, A.; Jul-Larsen, A.; Haring, S.J.; Wold, M.S.; Lønning, P.E.; Bjerkvig, R.; Bøe, S.O. Replication Protein A Prevents Accumulation of Single-Stranded Telomeric DNA in Cells That Use Alternative Lengthening of Telomeres. *Nucleic Acids Res.* **2007**, *35*, 7267–7278. [CrossRef]
72. Erdile, L.F.; Wold, M.S.; Kelly, T.J. The Primary Structure of the 32-KDa Subunit of Human Replication Protein A. *J. Biol. Chem.* **1990**, *265*, 3177–3182. [CrossRef]
73. Maestroni, L.; Audry, J.; Luciano, P.; Coulon, S.; Géli, V.; Corda, Y. RPA and Pif1 Cooperate to Remove G-Rich Structures at Both Leading and Lagging Strand. *Cell Stress* **2020**, *4*, 48–63. [CrossRef] [PubMed]
74. Seol, J.H.; Feldman, R.M.; Zachariae, W.; Shevchenko, A.; Correll, C.C.; Lyapina, S.; Chi, Y.; Galova, M.; Claypool, J.; Sandmeyer, S.; et al. Cdc53/Cullin and the Essential Hrt1 RING-H2 Subunit of SCF Define a Ubiquitin Ligase Module That Activates the E2 Enzyme Cdc34. *Genes Dev.* **1999**, *13*, 1614–1626. [CrossRef]
75. Sweeney, M.A.; Iakova, P.; Maneix, L.; Shih, F.-Y.; Cho, H.E.; Sahin, E.; Catic, A. The Ubiquitin Ligase Cullin-1 Associates with Chromatin and Regulates Transcription of Specific c-MYC Target Genes. *Sci. Rep.* **2020**, *10*, 13942. [CrossRef] [PubMed]
76. Barford, D. Structural Interconversions of the Anaphase-Promoting Complex/Cyclosome (APC/C) Regulate Cell Cycle Transitions. *Curr. Opin. Struct. Biol.* **2020**, *61*, 86–97. [CrossRef] [PubMed]
77. Skibbens, R.V. Buck the Establishment: Reinventing Sister Chromatid Cohesion. *Trends Cell Biol.* **2010**, *20*, 507–513. [CrossRef]
78. Zachariae, W.; Shevchenko, A.; Andrews, P.D.; Ciosk, R.; Galova, M.; Stark, M.J.; Mann, M.; Nasmyth, K. Mass Spectrometric Analysis of the Anaphase-Promoting Complex from Yeast: Identification of a Subunit Related to Cullins. *Science* **1998**, *279*, 1216–1219. [CrossRef] [PubMed]
79. Chiu, Y.-H.; MacMillan, J.B.; Chen, Z.J. RNA Polymerase III Detects Cytosolic DNA and Induces Type-I Interferons Through the RIG-I Pathway. *Cell* **2009**, *138*, 576–591. [CrossRef]
80. Yan, J.; Beattie, T.R.; Rojas, A.L.; Schermerhorn, K.; Gristwood, T.; Trinidad, J.C.; Albers, S.V.; Roversi, P.; Gardner, A.F.; Abrescia, N.G.A.; et al. Identification and Characterization of a Heterotrimeric Archaeal DNA Polymerase Holoenzyme. *Nat. Commun.* **2017**, *8*, 15075. [CrossRef] [PubMed]
81. Enomoto, R.; Kinebuchi, T.; Sato, M.; Yagi, H.; Kurumizaka, H.; Yokoyama, S. Stimulation of DNA Strand Exchange by the Human TBPIP/Hop2-Mnd1 Complex. *J. Biol. Chem.* **2006**, *281*, 5575–5581. [CrossRef]
82. Zangen, D.; Kaufman, Y.; Zeligson, S.; Perlberg, S.; Fridman, H.; Kanaan, M.; Abdulhadi-Atwan, M.; Abu Libdeh, A.; Gussow, A.; Kisslov, I.; et al. XX Ovarian Dysgenesis Is Caused by a PSMC3IP/HOP2 Mutation That Abolishes Coactivation of Estrogen-Driven Transcription. *Am. J. Hum. Genet.* **2011**, *89*, 572–579. [CrossRef] [PubMed]
83. Rudnicka, A.; Yamauchi, Y. Ubiquitin in Influenza Virus Entry and Innate Immunity. *Viruses* **2016**, *8*, 293. [CrossRef]
84. Carter-Timofte, M.E.; Hansen, A.F.; Mardahl, M.; Fribourg, S.; Rapaport, F.; Zhang, S.-Y.; Casanova, J.-L.; Paludan, S.R.; Christiansen, M.; Larsen, C.S.; et al. Varicella-Zoster Virus CNS Vasculitis and RNA Polymerase III Gene Mutation in Identical Twins. *Neurol.-Neuroimmunol. Neuroinflamm.* **2018**, *5*, e500. [CrossRef]
85. Van Quyen, D.; Ha, S.C.; Lowenhaupt, K.; Rich, A.; Kim, K.K.; Kim, Y.-G. Characterization of DNA-Binding Activity of Zα Domains from Poxviruses and the Importance of the β-Wing Regions in Converting B-DNA to Z-DNA. *Nucleic Acids Res.* **2007**, *35*, 7714–7720. [CrossRef] [PubMed]
86. Chemes, L.B.; de Prat-Gay, G.; Sánchez, I.E. Convergent Evolution and Mimicry of Protein Linear Motifs in Host-Pathogen Interactions. *Curr. Opin. Struct. Biol.* **2015**, *32*, 91–101. [CrossRef]
87. Tomii, K.; Sawada, Y.; Honda, S. Convergent Evolution in Structural Elements of Proteins Investigated Using Cross Profile Analysis. *BMC Bioinform.* **2012**, *13*, 11. [CrossRef] [PubMed]
88. Brázda, V.; Červeň, J.; Bartas, M.; Mikysková, N.; Coufal, J.; Pečinka, P. The Amino Acid Composition of Quadruplex Binding Proteins Reveals a Shared Motif and Predicts New Potential Quadruplex Interactors. *Molecules* **2018**, *23*, 2341. [CrossRef] [PubMed]
89. Bartas, M.; Bažantová, P.; Brázda, V.; Liao, J.C.; Červeň, J.; Pečinka, P. Identification of Distinct Amino Acid Composition of Human Cruciform Binding Proteins. *Mol. Biol.* **2019**, *53*, 97–106. [CrossRef]
90. Sousounis, K.; Haney, C.E.; Cao, J.; Sunchu, B.; Tsonis, P.A. Conservation of the Three-Dimensional Structure in Non-Homologous or Unrelated Proteins. *Hum. Genom.* **2012**, *6*, 10. [CrossRef]
91. Zhang, Y.; Skolnick, J. TM-Align: A Protein Structure Alignment Algorithm Based on the TM-Score. *Nucleic Acids Res.* **2005**, *33*, 2302–2309. [CrossRef]

92. Perrakis, A.; Sixma, T.K. AI Revolutions in Biology. *EMBO Rep.* **2021**, *22*, e54046. [CrossRef]
93. Uanschou, C.; Ronceret, A.; Von Harder, M.; De Muyt, A.; Vezon, D.; Pereira, L.; Chelysheva, L.; Kobayashi, W.; Kurumizaka, H.; Schlögelhofer, P.; et al. Sufficient Amounts of Functional HOP2/MND1 Complex Promote Interhomolog DNA Repair but Are Dispensable for Intersister DNA Repair during Meiosis in Arabidopsis. *Plant Cell* **2013**, *25*, 4924–4940. [CrossRef] [PubMed]
94. Poulsen, H.; Nilsson, J.; Damgaard, C.K.; Egebjerg, J.; Kjems, J. CRM1 Mediates the Export of ADAR1 through a Nuclear Export Signal within the Z-DNA Binding Domain. *Mol. Cell. Biol.* **2001**, *21*, 7862–7871. [CrossRef]
95. Strehblow, A.; Hallegger, M.; Jantsch, M.F. Nucleocytoplasmic Distribution of Human RNA-Editing Enzyme ADAR1 Is Modulated by Double-Stranded RNA-Binding Domains, a Leucine-Rich Export Signal, and a Putative Dimerization Domain. *Mol. Biol. Cell* **2002**, *13*, 3822–3835. [CrossRef] [PubMed]
96. Kim, C. How Z-DNA/RNA Binding Proteins Shape Homeostasis, Inflammation, and Immunity. *BMB Rep.* **2020**, *53*, 453–457. [CrossRef] [PubMed]
97. Gallo, A.; Vukic, D.; Michalík, D.; O'Connell, M.A.; Keegan, L.P. ADAR RNA Editing in Human Disease; More to It than Meets the I. *Hum. Genet.* **2017**, *136*, 1265–1278. [CrossRef]
98. Kosugi, S.; Hasebe, M.; Tomita, M.; Yanagawa, H. Systematic Identification of Cell Cycle-Dependent Yeast Nucleocytoplasmic Shuttling Proteins by Prediction of Composite Motifs. *Proc. Natl. Acad. Sci. USA* **2009**, *106*, 10171–10176. [CrossRef]
99. Bartas, M.; Červeň, J.; Guziurová, S.; Slychko, K.; Pečinka, P. Amino Acid Composition in Various Types of Nucleic Acid-Binding Proteins. *Int. J. Mol. Sci.* **2021**, *22*, 922. [CrossRef]
100. Yan, Y.; Tao, H.; He, J.; Huang, S.-Y. The HDOCK Server for Integrated Protein–Protein Docking. *Nat. Protoc.* **2020**, *15*, 1829–1852. [CrossRef]
101. Nichols, P.J.; Bevers, S.; Henen, M.; Kieft, J.S.; Vicens, Q.; Vögeli, B. Recognition of Non-CpG Repeats in Alu and Ribosomal RNAs by the Z-RNA Binding Domain of ADAR1 Induces A-Z Junctions. *Nat. Commun.* **2021**, *12*, 793. [CrossRef]
102. Kim, D.; Lee, Y.-H.; Hwang, H.-Y.; Kim, K.K.; Park, H.-J. Z-DNA Binding Proteins as Targets for Structure-Based Virtual Screening. *Curr. Drug Targets* **2010**, *11*, 335–344. [CrossRef]
103. Szklarczyk, D.; Gable, A.L.; Lyon, D.; Junge, A.; Wyder, S.; Huerta-Cepas, J.; Simonovic, M.; Doncheva, N.T.; Morris, J.H.; Bork, P.; et al. STRING V11: Protein–Protein Association Networks with Increased Coverage, Supporting Functional Discovery in Genome-Wide Experimental Datasets. *Nucleic Acids Res.* **2019**, *47*, D607–D613. [CrossRef]
104. Jassal, B.; Matthews, L.; Viteri, G.; Gong, C.; Lorente, P.; Fabregat, A.; Sidiropoulos, K.; Cook, J.; Gillespie, M.; Haw, R.; et al. The Reactome Pathway Knowledgebase. *Nucleic Acids Res.* **2020**, *48*, D498–D503. [CrossRef]
105. Ramsay, E.P.; Abascal-Palacios, G.; Daiß, J.L.; King, H.; Gouge, J.; Pilsl, M.; Beuron, F.; Morris, E.; Gunkel, P.; Engel, C.; et al. Structure of Human RNA Polymerase III. *Nat. Commun.* **2020**, *11*, 6409. [CrossRef]
106. Ogunjimi, B.; Zhang, S.-Y.; Sørensen, K.B.; Skipper, K.A.; Carter-Timofte, M.; Kerner, G.; Luecke, S.; Prabakaran, T.; Cai, Y.; Meester, J.; et al. Inborn Errors in RNA Polymerase III Underlie Severe Varicella Zoster Virus Infections. *J. Clin. Investig.* **2017**, *127*, 3543–3556. [CrossRef] [PubMed]
107. Carter-Timofte, M.E.; Hansen, A.F.; Christiansen, M.; Paludan, S.R.; Mogensen, T.H. Mutations in RNA Polymerase III Genes and Defective DNA Sensing in Adults with Varicella-Zoster Virus CNS Infection. *Genes Immun.* **2019**, *20*, 214–223. [CrossRef]
108. The Gene Ontology Consortium Gene Ontology Consortium: Going Forward. *Nucleic Acids Res.* **2015**, *43*, D1049–D1056. [CrossRef]
109. Pettersen, E.F.; Goddard, T.D.; Huang, C.C.; Couch, G.S.; Greenblatt, D.M.; Meng, E.C.; Ferrin, T.E. UCSF Chimera—A Visualization System for Exploratory Research and Analysis. *J. Comput. Chem.* **2004**, *25*, 1605–1612. [CrossRef] [PubMed]
110. Meng, E.C.; Pettersen, E.F.; Couch, G.S.; Huang, C.C.; Ferrin, T.E. Tools for Integrated Sequence-Structure Analysis with UCSF Chimera. *BMC Bioinform.* **2006**, *7*, 339. [CrossRef] [PubMed]
111. Drozdzal, P.; Gilski, M.; Kierzek, R.; Lomozik, L.; Jaskolski, M. Ultrahigh-Resolution Crystal Structures of Z-DNA in Complex with Mn^{2+} and Zn^{2+} Ions. *Acta Cryst. D* **2013**, *69*, 1180–1190. [CrossRef] [PubMed]
112. Popenda, M.; Milecki, J.; Adamiak, R.W. High Salt Solution Structure of a Left-Handed RNA Double Helix. *Nucleic Acids Res.* **2004**, *32*, 4044–4054. [CrossRef] [PubMed]

International Journal of *Molecular Sciences*

Review

Interaction of Proteins with Inverted Repeats and Cruciform Structures in Nucleic Acids

Richard P. Bowater [1,†], Natália Bohálová [2,3,†] and Václav Brázda [2,*]

1 School of Biological Sciences, University of East Anglia, Norwich Research Park, Norwich NR4 7TJ, UK; r.bowater@uea.ac.uk
2 Department of Biophysical Chemistry and Molecular Oncology, Institute of Biophysics of the Czech Academy of Sciences, 61265 Brno, Czech Republic; nataliabohalova@gmail.com
3 Department of Experimental Biology, Faculty of Science, Masaryk University, Kamenice 5, 62500 Brno, Czech Republic
* Correspondence: vaclav@ibp.cz
† These authors contributed equally to this work.

Abstract: Cruciforms occur when inverted repeat sequences in double-stranded DNA adopt intra-strand hairpins on opposing strands. Biophysical and molecular studies of these structures confirm their characterization as four-way junctions and have demonstrated that several factors influence their stability, including overall chromatin structure and DNA supercoiling. Here, we review our understanding of processes that influence the formation and stability of cruciforms in genomes, covering the range of sequences shown to have biological significance. It is challenging to accurately sequence repetitive DNA sequences, but recent advances in sequencing methods have deepened understanding about the amounts of inverted repeats in genomes from all forms of life. We highlight that, in the majority of genomes, inverted repeats are present in higher numbers than is expected from a random occurrence. It is, therefore, becoming clear that inverted repeats play important roles in regulating many aspects of DNA metabolism, including replication, gene expression, and recombination. Cruciforms are targets for many architectural and regulatory proteins, including topoisomerases, p53, Rif1, and others. Notably, some of these proteins can induce the formation of cruciform structures when they bind to DNA. Inverted repeat sequences also influence the evolution of genomes, and growing evidence highlights their significance in several human diseases, suggesting that the inverted repeat sequences and/or DNA cruciforms could be useful therapeutic targets in some cases.

Keywords: cruciform; DNA base sequence; DNA structure; DNA supercoiling; epigenetics; genome stability; inverted repeat; replication; transcription

Citation: Bowater, R.P.; Bohálová, N.; Brázda, V. Interaction of Proteins with Inverted Repeats and Cruciform Structures in Nucleic Acids. *Int. J. Mol. Sci.* **2022**, *23*, 6171. https://doi.org/10.3390/ijms23116171

Academic Editor: Mateus Webba Da Silva

Received: 29 April 2022
Accepted: 30 May 2022
Published: 31 May 2022

1. Introduction

The wealth of DNA sequence information provided by genome sequencing projects has brought new insights into the primary sequences of genomes and also about possible sequence-dependent local secondary structures [1]. The primary base sequence alone is insufficient to decipher all principles that support basic molecular processes and those that maintain genomic and cellular stability. Inevitably, in-depth knowledge of epigenetic modifications and the local and global DNA structure is crucial for a full understanding of these processes. DNA molecules typically form two-stranded, right-handed helical B-form structures, which maximize the thermodynamic stability of the molecule [2]. However, a range of alternative (non-B) structures can also occur in DNA, and these are usually characterized by the occurrence of single-stranded regions (loops) and/or sites of disrupted base pair stacking (junctions between continuous B-form DNA and the alternative structure) [3]. Any disruption of stacking interactions or hydrogen bonds in base pairs alters the thermodynamic stability of the molecule, but non-B DNA structures can be favourable for some sequences under some environmental (and cellular) conditions. Although they were initially considered as in vitro artefacts, several local secondary DNA structures are now well characterized and confirmed to form in living cells under physiologically relevant conditions [4–6]. These sequence-dependent conformational changes give rise to triplexes [7,8], G-quadruplexes [5,9], i-motifs [10], R-loops [8,11], four-way junctions [12], and cruciforms [13–15]. The latter is formed in DNA molecules containing inverted repeat sequences, either uninterrupted or interspaced with several additional bases forming loops. Thus, cruciform structures consist of branch-points, stems, and loops (Figure 1A) [15]. The thermodynamic stability of cruciforms is influenced by their size, with stable cruciforms usually requiring the inverted repeat to be at least six bases in length (for the stem, or one half of the repeat). Cruciforms can also arise from imperfect inverted repeats, meaning that unpaired bases occur within the stems of the cruciform, although this means such structures are energetically disfavoured compared to the fully base-paired structure [15,16]. In addition to inverted repeat unit size and unpaired bases, the length of the loop is also a critical factor influencing the stability of such structures (Figure 1B). Analyses of inverted repeats in various genomes have shown they have a non-random distribution and a functional association with regulatory sites, including promoters [17,18].

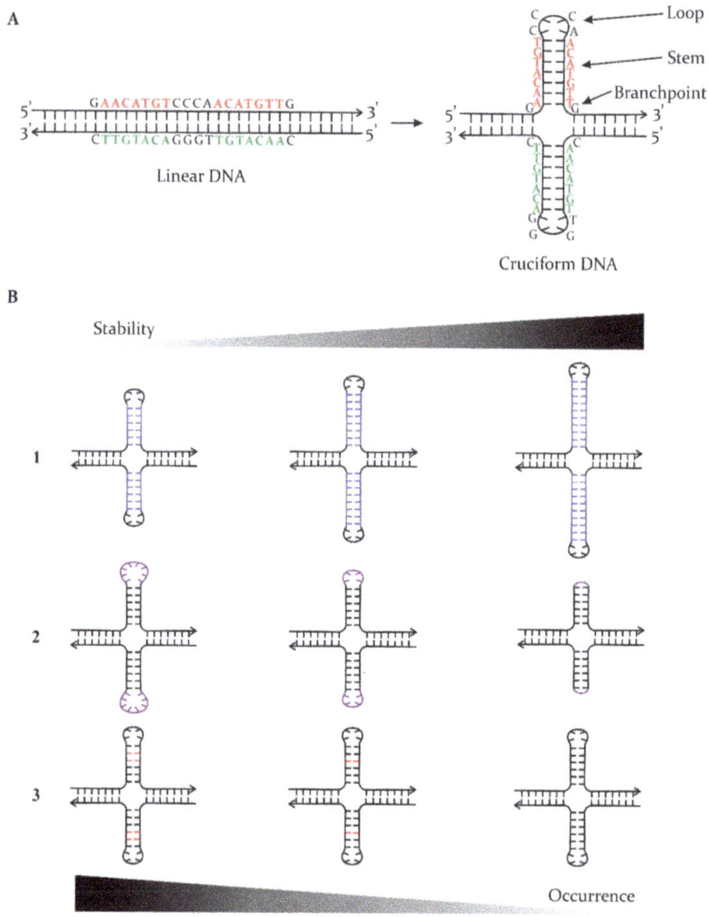

Figure 1. Inverted repeat sequences can form different types of double-stranded conformations.

(**A**) Transition of inverted repeat in a linear conformation to a double hairpin, cruciform state. For the sequence indicated, the cruciform structure consists of four branchpoints and two 7 bp-long stems, each with 4 nt loops. (**B**) Decisive factors for the resulting thermodynamic stability and genomic occurrence of cruciform structures are: (1) stem size indicated in blue; (2) loop length indicated in purple; (3) possible mismatches in base pairing indicated in red. The arrows at the top and bottom of part (B) highlight the relative stability and occurrence of the represented cruciforms, with the larger and darker part of the arrows indicating those that are most stable and are most likely to occur in genomes. For all schematic molecules, the arrow indicates the 3′-end of the DNA strand.

Inverted repeats and cruciforms have been found in all forms of life and appear to share similar functions and properties in many of them [3,6,17–21]. Inverted repeats are found in bacteria, eukaryotes, archaea, and viruses in higher amounts than would be expected from a random distribution of bases in both coding and non-coding regions, with a more pronounced frequency in non-coding regions. The frequency of inverted repeats in all organisms decreases with increasing length, but in most cases, the relative difference between expected and actual numbers tends to be higher for longer repeats [18]. As we describe in detail below, in all organisms, the presence of inverted repeats contributes to reduced genomic stability, primarily through the induction of inversions and the formation of hairpins and four-way junctions, which induce the stalling of polymerases and the generation of double-strand breaks. Cruciform conservation across all domains is, thus, a likely result of their involvement in essential molecular processes, such as opening of the DNA double helix during replication, transcription, and DNA damage repair [15].

2. Biophysical and Molecular Characterization of Cruciforms

The formation of cruciforms during the expression of genes was first postulated more than 50 years ago [22]. Their presence and function was subsequently studied both in vitro and in vivo, mostly for those located in plasmid DNAs from bacteria and yeasts [15]. The formation of cruciform structures requires the double-stranded helix of DNA to be opened, an energetically unfavourable process. A wide range of chemical and molecular probes have characterized properties that influence this process [6,23], with computer modelling methods helping to interpret experimental data [24]. Biophysical and molecular studies have clearly demonstrated that cruciforms are stable for some DNA molecules in vitro, but the situation has been less clear in vivo, mainly due to difficulties with studying the DNA structure inside cells. To assay for cruciform structures in cells, a range of probes of DNA structure have been used, including various factors that attack single-stranded regions of DNA, including psoralen and UV light cross-linking [6,25,26]. In some cases, the experiments cause the death of the cells, leading to studies being referred to as in situ to highlight that the cells are under physiological conditions, but may no longer be "living" [27]. Using *Escherichia coli* as a model, experiments have shown that large inverted repeats can be detected in cruciforms under some conditions, but sometimes at relatively low proportions of the total DNA [6]. Direct visualization of cruciforms in cells was attempted with a monoclonal antibody (2D3) shown to recognize cruciforms, but not heteroduplex slipped-stranded DNA containing a hairpin on one strand only [6,27]. Immunoprecipitation using this antibody revealed the presence of cruciform-containing DNA at a yeast replication origin, although it is unclear whether it specifically binds cruciforms or a panel of slipped-stranded DNA molecules [6,28]. Many methods continue to be used to study cruciform structures and their formation, from broad bioinformatic studies and electrophoretic in vitro assays to in vivo visualization by specific antibody interaction and single-molecule-level analyses [29–31]. Indeed, single-molecule manipulation of DNAs allowed cruciform formation, dynamics, and removal to be studied in real-time [32,33], as well as to reveal the mechanochemical properties of cruciform structure and cooperativity between opposing stem–loop structures [34].

In recent years, advances have been especially striking in high-resolution analyses of non-B DNA structures either as the nucleic acid alone or in combination with proteins.

In the context of this review, significant progress has been made in studies of four-way junctions, which are equivalent to the central part of cruciform structures—see Figure 1. Four-way junctions (often referred to as Holliday junctions) are critical intermediates in many DNA recombination and repair pathways [35], but it is important to recognize that such structures are usually formed by DNA molecules that do not contain inverted repeat sequences. A range of structural studies demonstrate that four-way junctions adopt different structures depending on the ionic environment and other factors [35,36]. X-ray crystallography and nuclear magnetic resonance (NMR) analyses of several DNA inverted repeat sequences confirm that they adopt the "stacked-X structure" in the absence of proteins, in which duplexes coaxially stack on each other. In thermodynamic terms, this type of structure has the most favourable energetics when monovalent or divalent cations are available to counteract the repelling interactions that occur between the negatively charged backbones, although cations are not an absolute requirement for the formation of stable cruciform structures. Figure 2 shows several views of a DNA inverted repeat structure determined at 2.10 Å for the sequence 5′-CCGGTACCGG-3′ [37], and similar structures have been observed for a variety of other inverted repeats [36]. The DNA forms a four-way junction in a "stacked-X" conformation (Figure 2). Two strands are "continuous" and are closest to a B-DNA conformation, while the other two strands make a tight U-turn and cross at the junction. The stacked-X structure is seen clearly in Figure 2A,B. For this complex, a Na^+ ion at its centre reduces electrostatic repulsion as the phosphodiester backbones come close to each other at the junction crossover (Figure 2C). Note that when the stacked-X structure is viewed from one face, Na^+ is relatively protected by the DNA backbones, but it is relatively accessible to the local environment from the opposite side. Molecular dynamics simulation of a decamer inverted repeat as a four-way junction confirms its twofold symmetry and that temperature and its structural integrity are preserved by a range of other parameters (i.e., the presence of ions, solvents, etc.) [38]. Epigenetic markers on DNA, such as hydroxymethyl and methyl substituents, can be accommodated without disrupting the structure or stability of the cruciform, although they open the structure to make the junction core more accessible [36]. The binding of proteins—usually enzymes—to four-way junctions can alter their conformation, although they can have dramatically different effects [36,39–41]. High-resolution structures that are currently available for these altered conformations of four-way DNA junctions with proteins bound are usually for sequences that are not inverted repeats. It is expected that DNA cruciforms formed by inverted repeats will have similar flexibility when proteins bind to them, but this still has to be verified by high-resolution structures.

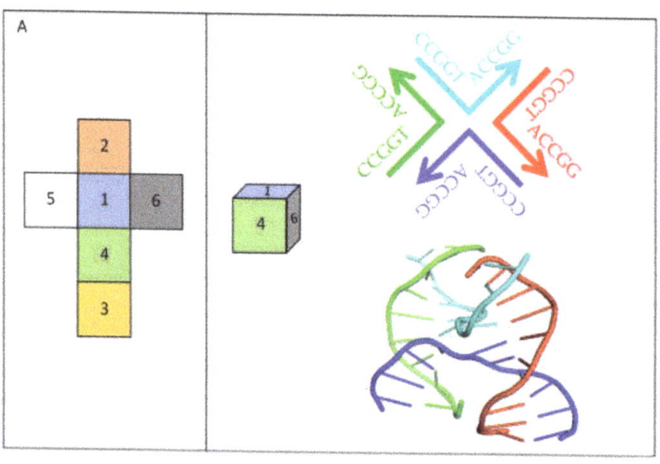

Figure 2. *Cont.*

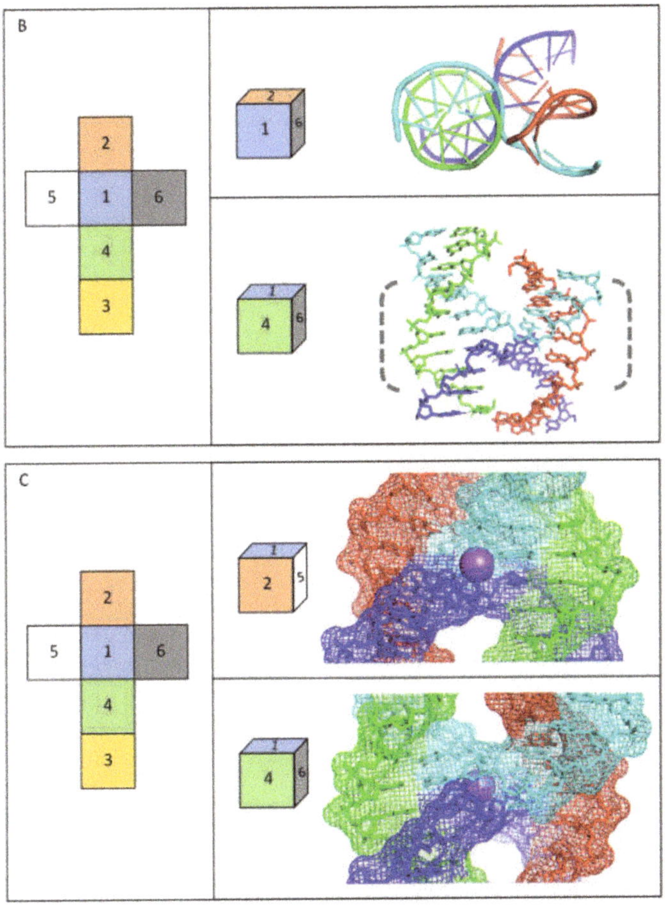

Figure 2. High-resolution structure of a cruciform (four-way junction) formed by an inverted repeat DNA sequence. Images show the X-ray crystallographic structure determined at 2.10 Å for DNA with the sequence 5′-CCGGTACCGG-3′ (1DCW) [37]. The DNA alone forms a four-way junction in a stacked-X conformation, in which duplexes coaxially stack, with each pair of stacked duplexes related by +30° to +60° (right-handed) rotation. The continuous (least distorted relative to B-DNA) strands are coloured as green and red, while those of the crossing strands (making a tight U-turn) are coloured blue and cyan. In each panel, the images show the structure visualised via different axes viewpoints as indicated by the coloured squares. (**A**) The upper image provides a schematic view of the molecule, the distinct strands (in different colours), and their sequences, with arrows indicating the 3′-ends of the DNA strands. The lower image presents the high-resolution structure of 1DCW, illustrating its arrangement of base pairs. (**B**) The upper image views the structure down the helix axis of one pair of stacked duplexes, while the lower image views it from a rotational shift of approximately 90°. (**C**) The images zoom in on the central part of the structure (dashed bracketed region in (**B**)) to highlight the electrostatic interactions, particularly close to the Na^{+} ion at its centre. The lower image views the same face of the dyad axis shown in (**B**), and the upper image shows the opposite face of the axis, viewed from a rotational shift of approximately 180°.

3. Presence of Inverted Repeats in Genomes

The various experimental methods referred to above have provided abundant evidence for the presence of inverted repeats in genomes across all forms of life [6,42]. Since the start of the 21st Century, the evidence has improved due to dramatic advances in sequencing technologies and bioinformatic analyses identifying genome sequences for many different organisms. Notably, it has been challenging to accurately sequence genomic regions that are rich in repeated bases for various reasons, but potentially including the presence of thermodynamically stable secondary structures [43]. Recent advances in sequencing methods mean that such problems can now usually be resolved, even for the human genome [44,45]. Here, we summarize the deepening understanding about the amounts of inverted repeats across all forms of life.

3.1. Viruses

Inverted repeats are found in higher numbers in many viral genomes than is expected from a random occurrence of bases [46]. This is true for many different types of viruses, but we illustrate this using Severe Acute Respiratory Syndrome Coronavirus 2 (SARS-CoV-2) and adeno-associated viruses (AAV), which have single-stranded RNA and DNA genomes, respectively. Using the SARS-CoV-2 virus genome as an example, a total of 1203 inverted repeats with stems of 6-13 bp in length were identified. The average frequency of their occurrence was 40.24 inverted repeats per 1000 nt, whereas it was 33.90 for the entire *Nidovirales* family to which SARS-CoV-2 belongs [42]. Recurrent mutations were shown to occur within inverted repeats with a higher frequency than would be expected from a random distribution of them [47,48]. Furthermore, an abundance of inverted repeats was found within 5′ untranslated regions of the *Nidovirales* family (Figure 3) [42]. In a different virus, AAV, terminal inverted repeats of 125 bases can form T-shaped hairpin structures by base-paring of two small internal inverted repeat sequences and large flanking inverted repeat sequences [49]. This terminal inverted repeat of AAV was determined as the binding site for several transcriptional transactivators and was shown to facilitate recombination of the viral genome with the cellular genome.

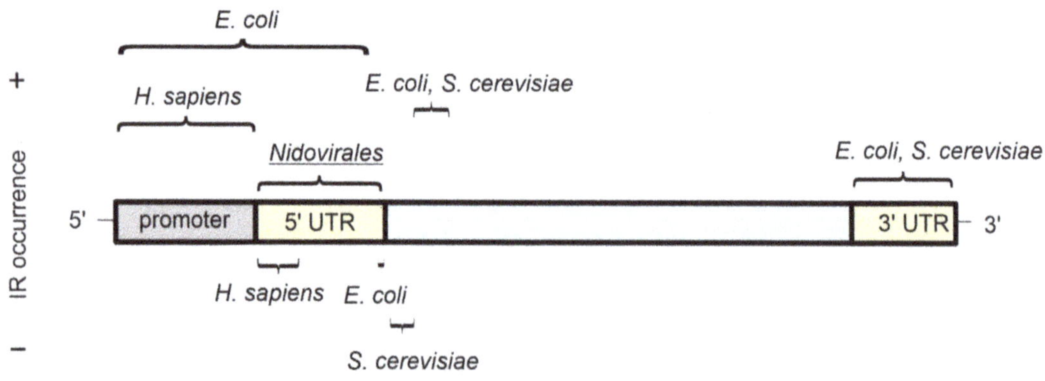

Figure 3. The occurrence of inverted repeat sequences in gene features as determined by bioinformatic analyses. An idealised gene and its regulatory sequences are shown, with UTR referring to "untranslated regions". A relative abundance (+) or depletion (−) of inverted repeats in the indicated genomes is highlighted above and below the idealised gene, respectively. For *E. coli* and *S. cerevisiae*, inverted repeats with a stem length from 5 bp and a spacer length up to 8 bp were considered [50,51], while for *H. sapiens* and viruses from the *Nidovirales* order, inverted repeats with the stem length from 6–30 bp and spacer length up to 10 bp were taken into account [18,42].

3.2. Prokaryotes

Early evidence for the presence of inverted repeats and cruciforms in genomes was obtained from studies across a range of bacteria, with a particular focus on *E. coli* [6]. Because bacterial DNA is often circular, it easily results in a negative supercoiled conformation [52], which can be an important factor in the formation of cruciforms. In the *E. coli* genome, short inverted repeats with arm lengths from 5 bp up to 20 bp are abundant in both coding and non-coding regions [19]. On average, there are nine inverted repeats per non-coding region, although a small proportion of regions contain the majority of the inverted repeats. The average arm length of the inverted repeats is approximately 6 bp, suggesting the sequences can form stable cruciforms. When comparing the genome with other proteobacteria, a significant number of identical inverted repeats are observed, providing evidence for evolutionary conservation [19]. Another study of genome sequences [53] performed similar analyses on 37 genomes of various prokaryotes, namely archaea, chlamydiales, firmicutes, proteobacteria, and others. For all bacteria, inverted repeats were found more frequently in non-coding regions. In almost all bacterial species examined, inverted repeats were found in genomes at a significantly higher frequency than the randomly generated sequences. Notably, only in two species, *Deinococcus radiodurans* and *Synechocystis* sp., were the observed number of inverted repeats statistically significantly lower than predicted by Markovian models of DNA sequences, although the reasons for the differences in these genomes are unclear. In archaea, the frequencies were higher than expected for five of eight species that were studied, but even in the five species that were higher, the difference was relatively small compared to that seen for bacteria. Mapping of the occurrence of inverted repeats in the *E. coli* genome [50] found that sequences with the potential to form cruciforms are enriched near stop codons and are part of terminators—and thus probably serve in the Rho-independent termination of transcription (Figure 3). Inverted repeats are also enriched within promoters, 5′-untranslated regions (UTRs), and in regions ~25–45 bp encompassing the start codon. It was also found that the small region ~5bp before the start codon has a statistically significant depletion of inverted repeats compared to 50 randomized genomes. Explanations for this observation could be that such a depletion prevents the formation of hairpin structures on the corresponding mRNA strands and also prevents disruption of the Shine Dalgarno sequence, both of which could negatively impact the initiation of translation.

For organisms that had complete genome sequences in 2020, about 36% of all bacteria and 75% of archaea have a prokaryotic immune system known as CRISPR/Cas [54]. CRISPR is an acronym for segments of clustered regularly interspaced short palindromic repeats, while Cas is the name of a group of proteins that associate with these regions. As the name implies, this system consists of sequences of inverted repeats, which are preceded by a leader sequence that is rich in adenine and thymine, and new spacers are integrated in its vicinity [55]. The nucleases Cas1 and Cas2 are the only Cas proteins that occur in all CRISPR/Cas systems, and both nucleases require a negatively supercoiled conformation to integrate new intervening sequences [56,57]. In vitro, the Cas1-Cas2 complex is able to integrate the new intervening sequence outside the CRISPR locus; however, the integration is non-random. In studying the specificity of integration of new intervening sequences, it was found that in the absence of the CRISPR locus, integration occurred preferentially in the vicinity of inverted repeats capable of forming cruciforms [56]. The CRISPR/Cas methodology is gaining widespread use across all organisms, but the potential impact of cruciforms on its implementation requires further analyses.

3.3. Eukaryotes

In eukaryotes, inverted repeats occur frequently in nuclear DNA and also in mitochondrial and plastid DNA, usually in even higher numbers than in nuclear DNA [20,21,58]. For example, in *Saccharomyces cerevisiae*, inverted repeats in mitochondrial DNA are 45-times more frequent than in its chromosomal DNA [17]. Correspondingly, inverted repeats have been demonstrated to impact evolution in mitochondria and in other genome contexts [59,60]. An overlay with annotated features revealed a statistically significant deficiency of inverted repeats in regions 20 bp downstream of the start codon [51]. In a similar way to examples already discussed for *E. coli* [50], inverted repeats in *S. cerevisiae* are enriched in the region ~ 30–60 bp downstream of the start codon and in close vicinity of positions corresponding to the ends of the mRNA (Figure 3) [51]. Whereas inverted repeats in *E. coli* are parts of intrinsic terminators and are GC-rich, inverted repeats in *S. cerevisiae* are parts of the polyA signal and are AT-rich. Therefore, inverted repeats in both organisms appear to play roles in transcription termination, although the sequences of the repeats are not preserved [50,51].

The effort to complete the sequence of the human genome is now successfully finished [61], with two chromosomes (8 and X) fully assembled already in 2021 [62,63]. Regions in chromosomes 8 and X that were uncharacterized in the current reference human genome assembly GRCh38 are now resolved and reveal a previous strong underestimation of the frequency of repeat tracts [64,65]. The difference of inverted repeat frequency between the two assemblies of chromosome 8 increases with the length of the inverted repeat, with up to twice as many for inverted repeats with an arm length of 30 bp [64]. When examining inverted repeats in promoters of the human genome [18], it was found that their frequency depends on the length of the repeat and its distance from the transcription start site. Shorter inverted repeats (6–11 bases for the size of the stem) are found primarily near the transcription start site, while longer repeats (14 bases and above for the size of the stem) are more frequent in regions that are at least 500 bp upstream from the transcription start site. In general, inverted repeats in the human genome are abundant upstream from the transcription start site, while downstream (in the direction of transcription), their presence is rarer (Figure 3). Some evidence suggests DNA is negatively supercoiled upstream of RNA polymerase [66], which will facilitate DNA strand separation and increase the likelihood that inverted repeats could form cruciforms [67]. The increased incidence of inverted repeats upstream of the transcription site would be consistent with these repeat sequences being involved in organizing and controlling promoter activities whether or not they form cruciforms [18,68]. It is also likely that the inverted repeats or potential cruciforms may impact differently on different transcription factors, as evidenced by promoters of genes involved in inflammatory, tumour, and developmental processes containing relatively high levels of inverted repeats, whereas promoters of metabolic-related genes contain lower levels of inverted repeats [18].

4. A Range of Proteins Interact with Cruciforms

Inverted repeats and cruciforms are targets for binding by many architectural and regulatory proteins. While many proteins have only weak sequence specificity, they are able to bind strongly to non-B-DNA structures, such as cruciforms [15]. Additionally, some proteins induce or stabilize cruciforms after binding to the nucleic acid. Cruciform binding proteins have been shown to have roles in chromatin remodelling, replication, and transcription regulation. Table 1 highlights the names and sources of proteins confirmed to interact with cruciforms, and details about the impact of some of these interactions have been discussed previously [15]. More recent findings in relation to the involvement of these interactions across the full range of cellular processes are described below.

Table 1. Proteins involved in interactions with cruciform structures. TF = transcription factor, chromatin AP = chromatin-associated protein. Adapted from [15]. * If no reference is listed for an entry, see [15] for further details.

Protein	Source	Function	Reference *
14-3-3	Eukaryotes	Replication, DNA repair, TF	[69]
A22	Coccinia virus	Junction-resolving enzyme	
AF10	*H. sapiens*	TF	
Bmh1, homolog of 14-3-3	*S. cerevisiae*	Replication, DNA repair, TF	
BRCA1	Mammals	Chromatin AP, DNA repair, TF	[70]
Cas1, Cas2	Archaea, Bacteria	Endonuclease, defence response to virus	[56,57]
Cce1	Yeast	Junction-resolving enzyme	[71]
Crp-1	*S. cerevisiae*	DNA repair	[72]
DEK	Mammals	Chromatin AP, replication, DNA repair	[73,74]
DNA-PK	Eukaryotes	DNA repair	

Table 1. *Cont.*

Protein	Source	Function	Reference *
Dps	*E. coli*	DNA repair, stress response	[75–77]
Endonuclease I	Phage T7	Junction-resolving enzyme	[78]
Endonuclease VII	Phage T4	Junction-resolving enzyme	
Estrogen receptor	Mammals	TF	
GEN1	Vertebrates	Junction-resolving enzyme	[79]
GF14, homolog of 14-3-3	Plants	Replication, stress response	
Helicases	all	Replication	[80,81]
Hjc, Hje	Archaea	Junction-resolving enzymes	
HMG protein family	all	Chromatin AP, DNA repair, TF	
Hop1	*S. cerevisiae*	DNA Repair	
HU	*E. coli*	Replication	[82]
IFI16	*H. sapiens*	Viral DNA recognition	[83,84]
Integrases	all	Junction-resolving enzyme	
MLH1-MLH3	Vertebrates	Junction-resolving enzyme	[85]
MLL (leukaemia)	*H. sapiens*	Replication	
MSH2	Mammals	Junction-resolving enzyme	[86]
Mus81-Eme1	Eukaryotes	Junction-resolving enzyme	
Mus81-Mms4	*S. cerevisiae*	Junction-resolving enzyme	[72,87]
MutH	Eukaryotes	Junction-resolving enzyme	
p53	*H. sapiens* and others	DNA repair, TF	[88]
p73	*H. sapiens* and others	DNA repair, TF	[89]
PARP-1	*H. sapiens* and others	DNA repair, TF	[90]
Rad51	Eukaryotes	Chromatin AP	[91]
Rad52-Rad59	Eukaryotes	Chromatin AP	[91]
Rad54	Eukaryotes	Chromatin AP	[91]
RecU	G+ bacteria	Junction-resolving enzyme	
RepC	Bacteria	Replication	[92]
Rif1	Mammals	DNA repair, TF	[93,94]
Rmi-1	Yeast	DNA repair, TF	
RusA	*E. coli*	Junction-resolving enzyme	
RuvC	*E. coli*	Junction-resolving enzyme	
S16	*E. coli*	Replication	
SbcCD	*E. coli*	Junction-resolving enzyme	[95]
Smc	*S. cerevisiae*	DNA repair, TF	
Topoisomerase I	Eukaryotes	Chromatin AP	
Topoisomerase II	Eukaryotes	Chromatin AP	[96]
TRF2	*H. sapiens*	Junction-resolving enzyme	
Vlf-1	Baculoviruses	Replication	
WRN(Werner syndrome)	*H. sapiens*	Replication	
XPF, XPG protein families	Eukaryotes	Junction-resolving enzyme	[97]
Ydc2	*S. pombe*	Junction-resolving enzyme	
Yen1, homolog of GEN1	*S. cerevisiae*	Junction-resolving enzyme	[98]

Cruciform formation is enabled by DNA negative supercoiling, which is unevenly spread through genomes and is tightly regulated, mainly by topoisomerases (TOPs) [15]. In eukaryotes, TOP1 relaxes DNA supercoiling generated by transcription, replication, and chromatin remodelling through the introduction of a single-strand break, and it binds to Holliday junctions, whereas TOP2 changes the DNA topology and is capable of generating transient DNA double-strand breaks [99]. TOP2 has been shown to recognize and cleave cruciform structures [15]. TOP2 and a member of the HMG family, chromatin-stabilizing protein Hmo1, preserve negative supercoiling at gene boundaries and are suggested to instigate the formation of cruciforms, thus directing TOP1 and RNA polymerase II to coding regions [96].

Inverted repeats located in the promoter regions of genes are preferentially bound by many transcription factors (Table 1), such as PARP-1, BRCA1, ER, and p53 [15,70,90]. The tumour suppressor protein p53 is critical for protection against many human cancers.

Most tumorigenic p53 mutations occur in its central domain, which binds to specific DNA sequences, referred to as response elements. Such response elements with a propensity to form cruciforms are favoured for binding by p53 both in vitro and in vivo [14,100]. The protein p73 is a member of the p53 family and has essential functions in several signaling pathways involved in development, differentiation, DNA damage responses, and cancer. Like its p53 homolog, p73 shows a preference for binding to its target sequence in cruciform structures [89]. Yeast-based assays revealed that p73-mediated transactivation correlated with the relative propensity of a response element to form a cruciform [89].

Another protein showing a preference for binding to DNA cruciforms is interferon-inducible protein 16 (IFI16), a sensor of foreign DNA in human cells. Upon DNA recognition, the protein oligomerizes, forms a filament, and triggers an innate immune response [101]. Besides its role in the immune response, IFI16 represses the transcription of viral genes [102]. IFI16 showed a preference for binding to negatively supercoiled plasmid over linear DNA in vitro, stabilizing local DNA structures such as cruciforms and quadruplexes [83]. Importantly, the binding pattern varies dependent on secondary structures in the DNA: with linear DNA, the protein interacts cooperatively, leading to non-specific filamentous aggregates of a higher molecular weight being formed, but in the presence of cruciforms, the protein binds to DNA selectively, forming more compact globular complexes [83,84]. The functional role of the different binding patterns remains unclear, but provides a possible explanation for the distinct roles of IFI16 in antiviral defence.

Cruciforms have also been demonstrated to influence various aspects of DNA replication. A range of studies confirmed cruciform formation in the origins of replication in bacteria, yeast, and mammalian cells [15,103]. Furthermore, several proteins involved in replication bind to cruciform structures, such as S16, MLL, WRN, and 14-3-3 (Table 1). Replication initiator protein C (RepC), which is encoded by the pT181 plasmid of *Staphylococcus aureus*, binds to a specific DNA sequence, which is able to form a cruciform and creates a nick that allows replication to begin [104]. It is proposed that cruciforms are formed passively due to the natural supercoiling of DNA, but their formation is necessary for RepC cleavage of DNA [92]. Rap1-interacting factor 1 (Rif1) is a mammalian protein involved in regulating the timing of DNA replication, mediating the repair of double-stranded DNA breaks, and replication fork restart [93]. The C-terminal region CII of RIF1 is critical for replication fork protection, and recent structural analyses identified that it preferentially binds cruciform structures [93,94]. Rif1 accumulates on stalled replication forks and possibly protects reversed forks, which could involve cruciform structures in vivo.

Cruciforms also influence other aspects of replication. Cruciforms formed ahead of a replication fork could stop their movement, which would temporarily stop replication. Such problems can be resolved by the formation of reversed replication forks at the four-way junctions, followed by homologous recombination and branch migration in order to restart replication [105]. Since cruciforms share structural similarity with Holliday junctions, cruciform-binding proteins are likely to be involved in these (or related) processes. For example, AT-rich cruciform cleavage is mediated by the Holliday junction resolvase GEN-1 in human cells [79,106], with GEN1 splitting the cruciform diagonally, creating two hairpins healed by DNA ligases [79]. The tips of these hairpins are then cleaved by Artemis proteins and joined by non-homologous end joining. The resulting heteroduplexes are repaired by proteins associated with mismatch repair (MMR), for which the template would normally be selected according to the strand where the nick is not ligated. Since, however, both strands are fully ligated, the template is chosen randomly and may result in translocation between two palindromic AT-rich repeats at different chromosomal locations that do not share a complete sequence homology. The involvement of other resolvases in this type of process, such as Mus81 in human cells, was rejected [79]. However, in *S. cerevisiae*, Mus81-Mms4 was able to process recombination intermediates that arose during the repair of stalled replication forks and double-stranded breaks after being stimulated by Crp1, a protein that specifically binds to DNA four-way junctions [72,107].

Notably, long inverted repeats with an arm length of more than 150–200 nucleotides and with a spacer between the repeats being shorter than 50–60 nucleotides are almost impossible to clone into *E. coli*, mainly due to the action of SbcCD endonuclease/exonuclease, which can cleave hairpin structures, leading to DNA double-strand breaks [95,108]. It was confirmed that such long inverted repeats are converted to cruciform DNA before they encounter the replication fork, creating SbcCD-sensitive hairpin structures on both leading and lagging strands that transiently impede replication fork movement [109].

Another example of a protein able to bind to cruciforms is DNA-binding protein from starved cells (Dps), which is produced in stationary-phase *E. coli* cells on a large scale, reaching 85,000–180,000 molecules per cell. The main role of Dps is to protect cells from oxidative stress, UV- and γ-radiation, and metal ion toxicity, which it does via its ferroxidase activity [75]. Dps also regulates transcription by competing for binding sites with other transcription factors [76]. Dps protein binding to DNA does not depend on sequence, but a non-random distribution of Dps binding sites was observed with significant correlation with inverted repeats, suggesting the protein may interact with specific structures in DNA [76,77].

5. Inverted Repeats and Cruciforms as Potential Therapeutics in Human Disease

Evidence presented so far clearly demonstrates that cruciforms can form within DNA molecules in cells and that proteins bind to them, but the physiological significance of these observations remains unclear, particularly for human cells. However, a recent analysis of 1000 human genomes estimated that the probability of occurrence of pathology-associated single-nucleotide polymorphism variants is 14-times higher in inverted repeats than in other genome sites [110], and their role has been shown in germline mutagenesis with implications for evolution and genetic diseases [111]. Single-nucleotide polymorphism variants in inverted repeats have been linked with many human neuronal disorders, mental retardation, and various cancers. Moreover, when amplified genomic regions are determined for various cancer types [112], short palindromes are observed to facilitate these processes and lead to cancer progression [113]. Due to the presence of inverted repeats in multiple parts of genomes that are associated with regulatory functions, cruciforms are likely to be involved in several basic biological processes with physiological and pathological importance (Figure 4).

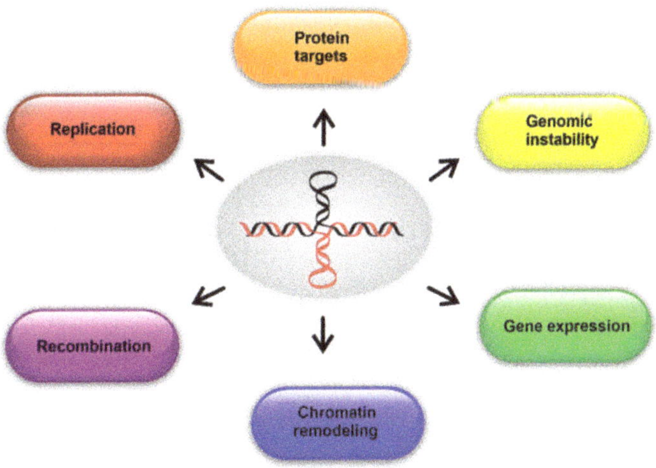

Figure 4. Cellular processes influenced by cruciform structures.

A range of local DNA structures are suggested as good therapeutic targets for human disease [9,114]. Considering that cruciforms formed by inverted repeats are hotspots of DNA breakpoints and for mutations with various pathologies [27,48], the detailed knowledge presented within this review provides an important background for their use as therapeutic targets. Incomplete assemblies of genomes present significant problems in that sequences with good potential to form local DNA structures are often not characterized properly, and until recently, many repeat tracts have not been identified because sequencing technologies have not been able to cope with them [64]. Fortunately, contemporary sequencing technologies allow the complete assembly of even very complex genomes, including the human genome [63]. As described above, recent data of complete human chromosomes identified inverted repeats in the human genome that had previously not been seen [64]. The improved understanding of the widespread nature of these regulatory sequences will make it possible to judge more accurately whether their targeting is feasible for specific human diseases.

The range of structures that can be adopted within DNA have important impacts on genome integrity and genome plasticity. Thus, it is not surprising that cruciforms (and four-way junctions) play critical roles in the maintenance of genomic stability, with a concomitant impact on essential cellular processes [15,115]. For example, this is observed directly through their identification as hotspots of genomic rearrangements [115]. Molecular mechanisms have been inferred for how these types of structures mediate such rearrangements in the human genome [116], such as by Holliday junction resolvases mediating chromosomal translocations, as discussed above. Inverted repeats are frequently found at fragile sites in the genome that are prone to chromosome breakage, as shown for the fragile site FRA16D, where a variable-length AT repeat forms a cruciform that stalls replication [117]. The relative position and size of inverted repeats is also important in relation to their effects on genome stability. These parameters impact translocation frequency, with an inverted repeat arm size of up to 100 bp correlating with translocation breakpoints in human cancer genomes [97]. The involvement of structure-specific nucleases on the fragility of inverted repeats also depends on the distance between them and their transcriptional status [87]. The association of several human diseases with mutations of DNA helicases has also suggested possible roles for cruciforms in the diseases [118]. Although cruciforms may be important for basic biological processes, if they are not resolved by helicases, their presence could lead to transcription stop or delay and to chromosome breakage during replication. Dysfunction of these helicases can lead to various diseases, for example Werner's syndrome, which is associated with mutations in the WRN helicase [119]. Inverted repeats also play a key role in the transposition and reorganization of transposable elements as demonstrated in several disease models, for example in Williams–Beuren syndrome, where insertions and deletions are associated with genomic regions that have an abundant number of inverted repeats [120].

Cruciforms are already used for various applications in medicine. For example, a cruciform DNA nanostructure is used for targeted delivery of doxorubicin to cancer cells [121] and was used to treat colon cancer [122]. It has also been demonstrated that cruciforms in gene promoters impact transcription upon oxidative modification of 2′-Deoxyguanosine [123]. The association of cruciforms with the regulation of transcription [90], as discussed above, opens other therapeutic windows where the specific levels of gene expression are influenced by the their presence and stability in promoter regions. An important tool allowing such approaches is the monoclonal antibody with specificity to the cruciform structure, although up to now, this has only been used for research purposes [28,69,124]. Currently, there are no small molecules that specifically recognize cruciforms, but it is likely that compounds will soon be designed that impact cruciform–protein interactions.

6. Conclusions

DNA molecules that contain inverted repeat sequences are able to adopt fully base-paired "linear" conformations and cruciforms that contain several unpaired regions. The structures of cruciforms (and four-way junctions) have been best characterized in vitro, including in complexes with proteins from prokaryotes and eukaryotes that bind to hairpins and four-way junctions. The structure of the cruciform influences the thermodynamic stability of the DNA, and paired regions of at least 6 bp are usually required to offset the energetically unfavourable folding of the junction and loop regions. In recent years, significant advances have been made in identifying high-resolution analyses of unusual DNA structures, either as the nucleic acid alone or in combination with proteins. A range of structural studies has demonstrated that cruciforms and four-way junctions adopt different structures depending on the ionic environment and other factors, including whether or not proteins are interacting with them. High-resolution structures that are currently available for four-way DNA junctions are usually for sequences that are not inverted repeats, but it is expected that structures formed by inverted repeats will have similar flexibility, although this still has to be verified by high-resolution structures. It will be useful to confirm at high resolution whether proteins bind to the junction, stem, or loops, or whether this is protein-dependent.

Detailed studies of many organisms have identified that inverted repeats are widespread in natural genomes. Indeed, in most cases, they are found at higher levels than expected if these were present at just random frequencies. This suggests that these types of sequences and/or their structures have functions in cells. In most eukaryotes, inverted repeats occur in higher amounts near promoters and transcriptional terminators, whereas in prokaryotes, they occur more frequently close to terminators. Both observations suggest these sequences and/or their cruciform structures have roles in regulating transcription. A similar increase in the amount of inverted repeats occurs near the origins of replication in eukaryotes, suggesting that the proteins involved in the initiation of replication may bind to these sequences and/or the structures within them.

The presence of inverted repeats can have negative effects on genome stability, and they have been shown to promote mutations and are, thus, an important driver of evolution. When examined in relation to human diseases, such as a range of cancers, genetic rearrangements are often abundant and complex, meaning it can be difficult to unravel the events that start and then lead to a certain genotype. Clearly, amplifications of inverted repeats have important impacts on the mechanisms involved in carcinogenesis, but their exact roles in diseases remain unclear; those that exist in the human genome could have a much greater role in initiating recombination events than is currently appreciated.

Although inverted repeats have been the subject of many studies over the last 50 years, their distribution has recently become an increased focus of research due to developments in sequencing and computer software. It is now clear that inverted repeats are conserved and not randomly distributed in genomes, suggesting that they play important roles in nucleic acid metabolism. In the future, advances with in vitro and in vivo methods will allow experimental examination of the predictions from bioinformatics analyses, facilitating thorough investigations into the effects of cruciforms on cellular processes, providing a deeper understanding of the resulting effects on human disease.

Author Contributions: Conceptualization, R.P.B. and V.B.; formal data and literature analysis, N.B. and V.B.; writing—original draft preparation, review and editing, R.P.B., N.B. and V.B. All authors have read and agreed to the published version of the manuscript.

Funding: This research was funded by The Czech Science Foundation (No. 22-21903S) to V.B.

Institutional Review Board Statement: Not applicable.

Informed Consent Statement: Not applicable.

Conflicts of Interest: The authors declare no conflict of interest.

References

1. Sato, M.P.; Ogura, Y.; Nakamura, K.; Nishida, R.; Gotoh, Y.; Hayashi, M.; Hisatsune, J.; Sugai, M.; Takehiko, I.; Hayashi, T. Comparison of the Sequencing Bias of Currently Available Library Preparation Kits for Illumina Sequencing of Bacterial Genomes and Metagenomes. *DNA Res.* **2019**, *26*, 391–398. [CrossRef] [PubMed]
2. Oprzeska-Zingrebe, E.A.; Meyer, S.; Roloff, A.; Kunte, H.-J.; Smiatek, J. Influence of Compatible Solute Ectoine on Distinct DNA Structures: Thermodynamic Insights into Molecular Binding Mechanisms and Destabilization Effects. *Phys. Chem. Chem. Phys.* **2018**, *20*, 25861–25874. [CrossRef] [PubMed]
3. Brazda, V.; Fojta, M.; Bowater, R.P. Structures and Stability of Simple DNA Repeats from Bacteria. *Biochem. J.* **2020**, *477*, 325–339. [CrossRef] [PubMed]
4. Summers, P.A.; Lewis, B.W.; Gonzalez-Garcia, J.; Porreca, R.M.; Lim, A.H.M.; Cadinu, P.; Martin-Pintado, N.; Mann, D.J.; Edel, J.B.; Vannier, J.B.; et al. Visualising G-Quadruplex DNA Dynamics in Live Cells by Fluorescence Lifetime Imaging Microscopy. *Nat. Commun.* **2021**, *12*, 162. [CrossRef]
5. Di Antonio, M.; Ponjavic, A.; Radzevičius, A.; Ranasinghe, R.T.; Catalano, M.; Zhang, X.; Shen, J.; Needham, L.-M.; Lee, S.F.; Klenerman, D.; et al. Single-Molecule Visualization of DNA G-Quadruplex Formation in Live Cells. *Nat. Chem.* **2020**, *12*, 832–837. [CrossRef]
6. Poggi, L.; Richard, G.-F. Alternative DNA Structures In Vivo: Molecular Evidence and Remaining Questions. *Microbiol. Mol. Biol. Rev.* **2021**, *85*, e00110-20. [CrossRef]
7. Brown, J.A. Unraveling the Structure and Biological Functions of RNA Triple Helices. *Wiley Interdiscip. Rev. RNA* **2020**, *11*, e1598. [CrossRef]
8. Neil, A.J.; Liang, M.U.; Khristich, A.N.; Shah, K.A.; Mirkin, S.M. RNA–DNA Hybrids Promote the Expansion of Friedreich's Ataxia (GAA)n Repeats via Break-Induced Replication. *Nucleic Acids Res.* **2018**, *46*, 3487–3497. [CrossRef]
9. Kosiol, N.; Juranek, S.; Brossart, P.; Heine, A.; Paeschke, K. G-Quadruplexes: A Promising Target for Cancer Therapy. *Mol. Cancer* **2021**, *20*, 40. [CrossRef]
10. Martella, M.; Pichiorri, F.; Chikhale, R.V.; Abdelhamid, M.A.S.; Waller, Z.A.E.; Smith, S.S. I-Motif Formation and Spontaneous Deletions in Human Cells. *Nucleic Acids Res.* **2022**, *50*, gkac158. [CrossRef]
11. Niehrs, C.; Luke, B. Regulatory R-Loops as Facilitators of Gene Expression and Genome Stability. *Nat. Rev. Mol. Cell Biol.* **2020**, *21*, 167–178. [CrossRef]
12. Tye, S.; Ronson, G.E.; Morris, J.R. A Fork in the Road: Where Homologous Recombination and Stalled Replication Fork Protection Part Ways. *Semin. Cell Dev. Biol.* **2021**, *113*, 14–26. [CrossRef] [PubMed]
13. Palecek, E. Local Supercoil-Stabilized DNA Structures. *Crit. Rev. Biochem. Mol. Biol.* **1991**, *26*, 151–226. [CrossRef]
14. Brázda, V.; Fojta, M. The Rich World of P53 DNA Binding Targets: The Role of DNA Structure. *Int. J. Mol. Sci.* **2019**, *20*, 5605. [CrossRef] [PubMed]
15. Brázda, V.; Laister, R.C.; Jagelská, E.B.; Arrowsmith, C. Cruciform Structures Are a Common DNA Feature Important for Regulating Biological Processes. *BMC Mol. Biol.* **2011**, *12*, 33. [CrossRef] [PubMed]
16. Benham, C.J.; Savitt, A.G.; Bauer, W.R. Extrusion of an Imperfect Palindrome to a Cruciform in Superhelical DNA: Complete Determination of Energetics Using a Statistical Mechanical Model. *J. Mol. Biol.* **2002**, *316*, 563–581. [CrossRef] [PubMed]
17. Čutová, M.; Manta, J.; Porubiaková, O.; Kaura, P.; Šťastný, J.; Jagelská, E.B.; Goswami, P.; Bartas, M.; Brázda, V. Divergent Distributions of Inverted Repeats and G-Quadruplex Forming Sequences in Saccharomyces Cerevisiae. *Genomics* **2020**, *112*, 1897–1901. [CrossRef]
18. Brázda, V.; Bartas, M.; Lýsek, J.; Coufal, J.; Fojta, M. Global Analysis of Inverted Repeat Sequences in Human Gene Promoters Reveals Their Non-Random Distribution and Association with Specific Biological Pathways. *Genomics* **2020**, *112*, 2772–2777. [CrossRef]
19. Lavi, B.; Karin, E.L.; Pupko, T.; Hazkani-Covo, E. The Prevalence and Evolutionary Conservation of Inverted Repeats in Proteobacteria. *Genome Biol. Evol.* **2018**, *10*, 918–927. [CrossRef]
20. Brázda, V.; Lýsek, J.; Bartas, M.; Fojta, M. Complex Analyses of Short Inverted Repeats in All Sequenced Chloroplast DNAs. *BioMed Res. Int.* **2018**, *2018*, 1097018. [CrossRef]
21. Čechová, J.; Lýsek, J.; Bartas, M.; Brázda, V. Complex Analyses of Inverted Repeats in Mitochondrial Genomes Revealed Their Importance and Variability. *Bioinformatics* **2018**, *34*, 1081–1085. [CrossRef] [PubMed]
22. Gierer, A. Model for DNA and Protein Interactions and the Function of the Operator. *Nature* **1966**, *212*, 1480–1481. [CrossRef] [PubMed]
23. Murchie, A.I.; Bowater, R.; Aboul-ela, F.; Lilley, D.M. Helix Opening Transitions in Supercoiled DNA. *Biochim. Biophys. Acta BBA Gene Struct. Expr.* **1992**, *1131*, 1–15. [CrossRef]
24. Zhabinskaya, D.; Benham, C.J. Competitive Superhelical Transitions Involving Cruciform Extrusion. *Nucleic Acids Res.* **2013**, *41*, 9610–9621. [CrossRef] [PubMed]
25. Neelsen, K.J.; Chaudhuri, A.R.; Follonier, C.; Herrador, R.; Lopes, M. Visualization and Interpretation of Eukaryotic DNA Replication Intermediates In Vivo by Electron Microscopy. In *Functional Analysis of DNA and Chromatin*; Methods in Molecular Biology; Stockert, J.C., Espada, J., Blázquez-Castro, A., Eds.; Humana Press: Totowa, NJ, USA, 2014; pp. 177–208. ISBN 978-1-62703-706-8.

26. Torregrosa-Muñumer, R.; Goffart, S.; Haikonen, J.A.; Pohjoismäki, J.L.O. Low Doses of Ultraviolet Radiation and Oxidative Damage Induce Dramatic Accumulation of Mitochondrial DNA Replication Intermediates, Fork Regression, and Replication Initiation Shift. *Mol. Biol. Cell* **2015**, *26*, 4197–4208. [CrossRef] [PubMed]
27. Correll-Tash, S.; Lilley, B.; Iv, H.S.; Mlynarski, E.; Franconi, C.P.; McNamara, M.; Woodbury, C.; Easley, C.A.; Emanuel, B.S. Double Strand Breaks (DSBs) as Indicators of Genomic Instability in PATRR-Mediated Translocations. *Hum. Mol. Genet.* **2021**, *29*, 3872–3881. [CrossRef]
28. Rekvig, O.P. The Anti-DNA Antibodies: Their Specificities for Unique DNA Structures and Their Unresolved Clinical Impact—A System Criticism and a Hypothesis. *Front. Immunol.* **2022**, *12*, 808008. [CrossRef]
29. Brázda, V.; Kolomazník, J.; Lýsek, J.; Hároníková, L.; Coufal, J.; Št'astný, J. Palindrome Analyser-A New Web-Based Server for Predicting and Evaluating Inverted Repeats in Nucleotide Sequences. *Biochem. Biophys. Res. Commun.* **2016**, *478*, 1739–1745. [CrossRef]
30. Gibbs, D.R.; Dhakal, S. Homologous Recombination under the Single-Molecule Fluorescence Microscope. *Int. J. Mol. Sci.* **2019**, *20*, 6102. [CrossRef]
31. Stefanovsky, V.Y.; Moss, T. The Cruciform DNA Mobility Shift Assay: A Tool to Study Proteins That Recognize Bent DNA. *Methods Mol. Biol. Clifton NJ* **2015**, *1334*, 195–203. [CrossRef]
32. Ramreddy, T.; Sachidanandam, R.; Strick, T.R. Real-Time Detection of Cruciform Extrusion by Single-Molecule DNA Nanomanipulation. *Nucleic Acids Res.* **2011**, *39*, 4275–4283. [CrossRef] [PubMed]
33. Shaheen, C.; Hastie, C.; Metera, K.; Scott, S.; Zhang, Z.; Chen, S.; Gu, G.; Weber, L.; Munsky, B.; Kouzine, F.; et al. Non-Equilibrium Structural Dynamics of Supercoiled DNA Plasmids Exhibits Asymmetrical Relaxation. *Nucleic Acids Res.* **2022**, *50*, 2754–2764. [CrossRef] [PubMed]
34. Mandal, S.; Selvam, S.; Cui, Y.; Hoque, M.E.; Mao, H. Mechanical Cooperativity in DNA Cruciform Structures. *ChemPhysChem* **2018**, *19*, 2627–2634. [CrossRef] [PubMed]
35. Lilley, D.M.J. Holliday Junction-Resolving Enzymes-Structures and Mechanisms. *FEBS Lett.* **2017**, *591*, 1073–1082. [CrossRef]
36. Ho, P.S. Structure of the Holliday Junction: Applications beyond Recombination. *Biochem. Soc. Trans.* **2017**, *45*, 1149–1158. [CrossRef]
37. Eichman, B.F.; Vargason, J.M.; Mooers, B.H.M.; Ho, P.S. The Holliday Junction in an Inverted Repeat DNA Sequence: Sequence Effects on the Structure of Four-Way Junctions. *Proc. Natl. Acad. Sci. USA* **2000**, *97*, 3971–3976. [CrossRef]
38. Yadav, R.K.; Yadava, U. Molecular Dynamics Simulation of Hydrated d(CGGGTACCCG)4 as a Four-Way DNA Holliday Junction and Comparison with the Crystallographic Structure. *Mol. Simul.* **2016**, *42*, 25–30. [CrossRef]
39. Kulkarni, D.S.; Owens, S.N.; Honda, M.; Ito, M.; Yang, Y.; Corrigan, M.W.; Chen, L.; Quan, A.L.; Hunter, N. PCNA Activates the MutLγ Endonuclease to Promote Meiotic Crossing Over. *Nature* **2020**, *586*, 623–627. [CrossRef]
40. Yan, J.; Hong, S.; Guan, Z.; He, W.; Zhang, D.; Yin, P. Structural Insights into Sequence-Dependent Holliday Junction Resolution by the Chloroplast Resolvase MOC1. *Nat. Commun.* **2020**, *11*, 1417. [CrossRef]
41. Wendorff, T.J.; Berger, J.M. Topoisomerase VI Senses and Exploits Both DNA Crossings and Bends to Facilitate Strand Passage. *eLife* **2018**, *7*, e31724. [CrossRef]
42. Bartas, M.; Brázda, V.; Bohálová, N.; Cantara, A.; Volná, A.; Stachurová, T.; Malachová, K.; Jagelská, E.B.; Porubiaková, O.; Červeň, J.; et al. In-Depth Bioinformatic Analyses of Nidovirales Including Human SARS-CoV-2, SARS-CoV, MERS-CoV Viruses Suggest Important Roles of Non-Canonical Nucleic Acid Structures in Their Lifecycles. *Front. Microbiol.* **2020**, *11*, 1583. [CrossRef] [PubMed]
43. Treangen, T.J.; Salzberg, S.L. Repetitive DNA and Next-Generation Sequencing: Computational Challenges and Solutions. *Nat. Rev. Genet.* **2012**, *13*, 36–46. [CrossRef]
44. Altemose, N.; Logsdon, G.A.; Bzikadze, A.V.; Sidhwani, P.; Langley, S.A.; Caldas, G.V.; Hoyt, S.J.; Uralsky, L.; Ryabov, F.D.; Shew, C.J.; et al. Complete Genomic and Epigenetic Maps of Human Centromeres. *Science* **2022**, *376*, eabl4178. [CrossRef] [PubMed]
45. Hoyt, S.J.; Storer, J.M.; Hartley, G.A.; Grady, P.G.S.; Gershman, A.; de Lima, L.G.; Limouse, C.; Halabian, R.; Wojenski, L.; Rodriguez, M.; et al. From Telomere to Telomere: The Transcriptional and Epigenetic State of Human Repeat Elements. *Science* **2022**, *376*, eabk3112. [CrossRef]
46. Spanò, M.; Lillo, F.; Miccichè, S.; Mantegna, R.N. Inverted Repeats in Viral Genomes. *Fluct. Noise Lett.* **2005**, *5*, L193–L200. [CrossRef]
47. Bartas, M.; Goswami, P.; Lexa, M.; Červeň, J.; Volná, A.; Fojta, M.; Brázda, V.; Pečinka, P. Letter to the Editor: Significant Mutation Enrichment in Inverted Repeat Sites of New SARS-CoV-2 Strains. *Brief. Bioinform.* **2021**, *22*, bbab129. [CrossRef]
48. Goswami, P.; Bartas, M.; Lexa, M.; Bohálová, N.; Volná, A.; Červeň, J.; Červeňová, V.; Pečinka, P.; Špunda, V.; Fojta, M.; et al. SARS-CoV-2 Hot-Spot Mutations Are Significantly Enriched within Inverted Repeats and CpG Island Loci. *Brief. Bioinform.* **2021**, *22*, 1338–1345. [CrossRef]
49. Berns, K.I. The Unusual Properties of the AAV Inverted Terminal Repeat. *Hum. Gene Ther.* **2020**, *31*, 518–523. [CrossRef]
50. Miura, O.; Ogake, T.; Ohyama, T. Requirement or Exclusion of Inverted Repeat Sequences with Cruciform-Forming Potential in Escherichia Coli Revealed by Genome-Wide Analyses. *Curr. Genet.* **2018**, *64*, 945–958. [CrossRef]
51. Miura, O.; Ogake, T.; Yoneyama, H.; Kikuchi, Y.; Ohyama, T. A Strong Structural Correlation between Short Inverted Repeat Sequences and the Polyadenylation Signal in Yeast and Nucleosome Exclusion by These Inverted Repeats. *Curr. Genet.* **2019**, *65*, 575–590. [CrossRef]

52. Lal, A.; Dhar, A.; Trostel, A.; Kouzine, F.; Seshasayee, A.S.N.; Adhya, S. Genome Scale Patterns of Supercoiling in a Bacterial Chromosome. *Nat. Commun.* **2016**, *7*, 11055. [CrossRef] [PubMed]
53. Lillo, F.; Basile, S.; Mantegna, R.N. Comparative Genomics Study of Inverted Repeats in Bacteria. *Bioinformatics* **2002**, *18*, 971–979. [CrossRef] [PubMed]
54. Pourcel, C.; Touchon, M.; Villeriot, N.; Vernadet, J.-P.; Couvin, D.; Toffano-Nioche, C.; Vergnaud, G. CRISPRCasdb a Successor of CRISPRdb Containing CRISPR Arrays and Cas Genes from Complete Genome Sequences, and Tools to Download and Query Lists of Repeats and Spacers. *Nucleic Acids Res.* **2020**, *48*, D535–D544. [CrossRef] [PubMed]
55. Makarova, K.S.; Grishin, N.V.; Shabalina, S.A.; Wolf, Y.I.; Koonin, E.V. A Putative RNA-Interference-Based Immune System in Prokaryotes: Computational Analysis of the Predicted Enzymatic Machinery, Functional Analogies with Eukaryotic RNAi, and Hypothetical Mechanisms of Action. *Biol. Direct* **2006**, *1*, 7. [CrossRef] [PubMed]
56. Nuñez, J.K.; Lee, A.S.Y.; Engelman, A.; Doudna, J.A. Integrase-Mediated Spacer Acquisition during CRISPR-Cas Adaptive Immunity. *Nature* **2015**, *519*, 193–198. [CrossRef] [PubMed]
57. Moch, C.; Fromant, M.; Blanquet, S.; Plateau, P. DNA Binding Specificities of Escherichia Coli Cas1-Cas2 Integrase Drive Its Recruitment at the CRISPR Locus. *Nucleic Acids Res.* **2017**, *45*, 2714–2723. [CrossRef]
58. Zhang, R.; Ge, F.; Li, H.; Chen, Y.; Zhao, Y.; Gao, Y.; Liu, Z.; Yang, L. PCIR: A Database of Plant Chloroplast Inverted Repeats. *Database J. Biol. Databases Curation* **2019**, *2019*, baz127. [CrossRef]
59. Liu, X.; Wu, X.; Tan, H.; Xie, B.; Deng, Y. Large Inverted Repeats Identified by Intra-Specific Comparison of Mitochondrial Genomes Provide Insights into the Evolution of Agrocybe Aegerita. *Comput. Struct. Biotechnol. J.* **2020**, *18*, 2424–2437. [CrossRef]
60. Damas, J.; Carneiro, J.; Gonçalves, J.; Stewart, J.B.; Samuels, D.C.; Amorim, A.; Pereira, F. Mitochondrial DNA Deletions Are Associated with Non-B DNA Conformations. *Nucleic Acids Res.* **2012**, *40*, 7606–7621. [CrossRef]
61. Nurk, S.; Koren, S.; Rhie, A.; Rautiainen, M.; Bzikadze, A.V.; Mikheenko, A.; Vollger, M.R.; Altemose, N.; Uralsky, L.; Gershman, A.; et al. The Complete Sequence of a Human Genome. *Science* **2022**, *376*, 44–53. [CrossRef]
62. Miga, K.H.; Koren, S.; Rhie, A.; Vollger, M.R.; Gershman, A.; Bzikadze, A.; Brooks, S.; Howe, E.; Porubsky, D.; Logsdon, G.A.; et al. Telomere-to-Telomere Assembly of a Complete Human X Chromosome. *Nature* **2020**, *585*, 79–84. [CrossRef] [PubMed]
63. Logsdon, G.A.; Vollger, M.R.; Hsieh, P.; Mao, Y.; Liskovykh, M.A.; Koren, S.; Nurk, S.; Mercuri, L.; Dishuck, P.C.; Rhie, A.; et al. The Structure, Function and Evolution of a Complete Human Chromosome 8. *Nature* **2021**, *593*, 101–107. [CrossRef] [PubMed]
64. Brazda, V.; Bohalova, N.; Bowater, R.P. New Telomere to Telomere Assembly of Human Chromosome 8 Reveals a Previous Underestimation of G-Quadruplex Forming Sequences and Inverted Repeats. *Gene* **2021**, *810*, 146058. [CrossRef] [PubMed]
65. Bohálová, N.; Mergny, J.-L.; Brázda, V. Novel G-Quadruplex Prone Sequences Emerge in the Complete Assembly of the Human X Chromosome. *Biochimie* **2021**, *191*, 87–90. [CrossRef]
66. Forth, S.; Sheinin, M.Y.; Inman, J.; Wang, M.D. Torque Measurement at the Single Molecule Level. *Annu. Rev. Biophys.* **2013**, *42*, 583–604. [CrossRef]
67. Ma, J.; Wang, M.D. DNA Supercoiling during Transcription. *Biophys. Rev.* **2016**, *8*, 75–87. [CrossRef]
68. Yamamoto, Y.; Miura, O.; Ohyama, T. Cruciform Formable Sequences within Pou5f1 Enhancer Are Indispensable for Mouse ES Cell Integrity. *Int. J. Mol. Sci.* **2021**, *22*, 3399. [CrossRef]
69. Brázda, V.; Cechová, J.; Coufal, J.; Rumpel, S.; Jagelská, E.B. Superhelical DNA as a Preferential Binding Target of 14-3-3γ Protein. *J. Biomol. Struct. Dyn.* **2012**, *30*, 371–378. [CrossRef]
70. Brázda, V.; Hároníková, L.; Liao, J.C.C.; Fridrichová, H.; Jagelská, E.B. Strong Preference of BRCA1 Protein to Topologically Constrained Non-B DNA Structures. *BMC Mol. Biol.* **2016**, *17*, 14. [CrossRef]
71. Samoilova, E.O.; Krasheninnikov, I.A.; Levitskii, S.A. Interaction between Saccharomyces Cerevisiae Mitochondrial DNA-Binding Protein Abf2p and Cce1p Resolvase. *Biochemistry* **2016**, *81*, 1111–1117. [CrossRef]
72. Phung, H.T.T.; Tran, D.H.; Nguyen, T.X. The Cruciform DNA-Binding Protein Crp1 Stimulates the Endonuclease Activity of Mus81-Mms4 in Saccharomyces Cerevisiae. *FEBS Lett.* **2020**, *594*, 4320–4337. [CrossRef] [PubMed]
73. Deutzmann, A.; Ganz, M.; Schönenberger, F.; Vervoorts, J.; Kappes, F.; Ferrando-May, E. The Human Oncoprotein and Chromatin Architectural Factor DEK Counteracts DNA Replication Stress. *Oncogene* **2015**, *34*, 4270–4277. [CrossRef] [PubMed]
74. Martinez-Useros, J.; Rodriguez-Remirez, M.; Borrero-Palacios, A.; Moreno, I.; Cebrian, A.; del Pulgar, T.G.; del Puerto-Nevado, L.; Vega-Bravo, R.; Puime-Otin, A.; Perez, N.; et al. DEK Is a Potential Marker for Aggressive Phenotype and Irinotecan-Based Therapy Response in Metastatic Colorectal Cancer. *BMC Cancer* **2014**, *14*, 965. [CrossRef] [PubMed]
75. Calhoun, L.N.; Kwon, Y.M. Structure, Function and Regulation of the DNA-Binding Protein Dps and Its Role in Acid and Oxidative Stress Resistance in Escherichia Coli: A Review. *J. Appl. Microbiol.* **2011**, *110*, 375–386. [CrossRef]
76. Antipov, S.S.; Tutukina, M.N.; Preobrazhenskaya, E.V.; Kondrashov, F.A.; Patrushev, M.V.; Toshchakov, S.V.; Dominova, I.; Shvyreva, U.S.; Vrublevskaya, V.V.; Morenkov, O.S.; et al. The Nucleoid Protein Dps Binds Genomic DNA of Escherichia Coli in a Non-Random Manner. *PLoS ONE* **2017**, *12*, e0182800. [CrossRef]
77. Melekhov, V.V.; Shvyreva, U.S.; Timchenko, A.A.; Tutukina, M.N.; Preobrazhenskaya, E.V.; Burkova, D.V.; Artiukhov, V.G.; Ozoline, O.N.; Antipov, S.S. Modes of Escherichia Coli Dps Interaction with DNA as Revealed by Atomic Force Microscopy. *PLoS ONE* **2015**, *10*, e0126504. [CrossRef]
78. Freeman, A.D.J.; Déclais, A.-C.; Lilley, D.M.J. The Importance of the N-Terminus of T7 Endonuclease I in the Interaction with DNA Junctions. *J. Mol. Biol.* **2013**, *425*, 395–410. [CrossRef]

79. Inagaki, H.; Ohye, T.; Kogo, H.; Tsutsumi, M.; Kato, T.; Tong, M.; Emanuel, B.S.; Kurahashi, H. Two Sequential Cleavage Reactions on Cruciform DNA Structures Cause Palindrome-Mediated Chromosomal Translocations. *Nat. Commun.* **2013**, *4*, 1592. [CrossRef]
80. Li, D.; Lv, B.; Zhang, H.; Lee, J.Y.; Li, T. Disintegration of Cruciform and G-Quadruplex Structures during the Course of Helicase-Dependent Amplification (HDA). *Bioorg. Med. Chem. Lett.* **2015**, *25*, 1709–1714. [CrossRef]
81. Boyer, A.-S.; Grgurevic, S.; Cazaux, C.; Hoffmann, J.-S. The Human Specialized DNA Polymerases and Non-B DNA: Vital Relationships to Preserve Genome Integrity. *J. Mol. Biol.* **2013**, *425*, 4767–4781. [CrossRef]
82. Bettridge, K.; Verma, S.; Weng, X.; Adhya, S.; Xiao, J. Single-Molecule Tracking Reveals That the Nucleoid-Associated Protein HU Plays a Dual Role in Maintaining Proper Nucleoid Volume through Differential Interactions with Chromosomal DNA. *Mol. Microbiol.* **2021**, *115*, 12–27. [CrossRef] [PubMed]
83. Brázda, V.; Coufal, J.; Liao, J.; Arrowsmith, C. Preferential Binding of IFI16 Protein to Cruciform Structure and Superhelical DNA. *Biochem. Biophys. Res. Commun.* **2012**, *422*, 716–720. [CrossRef] [PubMed]
84. Hároníková, L.; Coufal, J.; Kejnovská, I.; Jagelská, E.B.; Fojta, M.; Dvořáková, P.; Muller, P.; Vojtesek, B.; Brázda, V. IFI16 Preferentially Binds to DNA with Quadruplex Structure and Enhances DNA Quadruplex Formation. *PLoS ONE* **2016**, *11*, e0157156. [CrossRef]
85. Cannavo, E.; Sanchez, A.; Anand, R.; Ranjha, L.; Hugener, J.; Adam, C.; Acharya, A.; Weyland, N.; Aran-Guiu, X.; Charbonnier, J.-B.; et al. Regulation of the MLH1–MLH3 Endonuclease in Meiosis. *Nature* **2020**, *586*, 618–622. [CrossRef] [PubMed]
86. Rogacheva, M.V.; Manhart, C.M.; Chen, C.; Guarne, A.; Surtees, J.; Alani, E. Mlh1-Mlh3, a Meiotic Crossover and DNA Mismatch Repair Factor, Is a Msh2-Msh3-Stimulated Endonuclease. *J. Biol. Chem.* **2014**, *289*, 5664–5673. [CrossRef]
87. Saada, A.A.; Costa, A.B.; Sheng, Z.; Guo, W.; Haber, J.E.; Lobachev, K.S. Structural Parameters of Palindromic Repeats Determine the Specificity of Nuclease Attack of Secondary Structures. *Nucleic Acids Res.* **2021**, *49*, 3932–3947. [CrossRef]
88. Brázda, V.; Čechová, J.; Battistin, M.; Coufal, J.; Jagelská, E.B.; Raimondi, I.; Inga, A. The Structure Formed by Inverted Repeats in P53 Response Elements Determines the Transactivation Activity of P53 Protein. *Biochem. Biophys. Res. Commun.* **2017**, *483*, 516–521. [CrossRef]
89. Čechová, J.; Coufal, J.; Jagelská, E.B.; Fojta, M.; Brázda, V. P73, like Its P53 Homolog, Shows Preference for Inverted Repeats Forming Cruciforms. *PLoS ONE* **2018**, *13*, e0195835. [CrossRef]
90. Feng, X.; Xie, F.-Y.; Ou, X.-H.; Ma, J.-Y. Cruciform DNA in Mouse Growing Oocytes: Its Dynamics and Its Relationship with DNA Transcription. *PLoS ONE* **2020**, *15*, e0240844. [CrossRef]
91. Marie, L.; Symington, L.S. Mechanism for Inverted-Repeat Recombination Induced by a Replication Fork Barrier. *Nat. Commun.* **2022**, *13*, 32. [CrossRef]
92. Pastrana, C.L.; Carrasco, C.; Akhtar, P.; Leuba, S.H.; Khan, S.A.; Moreno-Herrero, F. Force and Twist Dependence of RepC Nicking Activity on Torsionally-Constrained DNA Molecules. *Nucleic Acids Res.* **2016**, *44*, 8885–8896. [CrossRef] [PubMed]
93. Sukackaite, R.; Jensen, M.R.; Mas, P.J.; Blackledge, M.; Buonomo, S.B.; Hart, D.J. Structural and Biophysical Characterization of Murine Rif1 C Terminus Reveals High Specificity for DNA Cruciform Structures. *J. Biol. Chem.* **2014**, *289*, 13903–13911. [CrossRef] [PubMed]
94. Mukherjee, C.; Tripathi, V.; Manolika, E.M.; Heijink, A.M.; Ricci, G.; Merzouk, S.; de Boer, H.R.; Demmers, J.; van Vugt, M.A.T.M.; Chaudhuri, A.R. RIF1 Promotes Replication Fork Protection and Efficient Restart to Maintain Genome Stability. *Nat. Commun.* **2019**, *10*, 3287. [CrossRef] [PubMed]
95. Eykelenboom, J.K.; Blackwood, J.K.; Okely, E.; Leach, D.R.F. SbcCD Causes a Double-Strand Break at a DNA Palindrome in the Escherichia Coli Chromosome. *Mol. Cell* **2000**, *30*, 644–651. [CrossRef]
96. Achar, Y.J.; Adhil, M.; Choudhary, R.; Gilbert, N.; Foiani, M. Negative Supercoil at Gene Boundaries Modulates Gene Topology. *Nature* **2020**, *577*, 701–705. [CrossRef]
97. Lu, S.; Wang, G.; Bacolla, A.; Zhao, J.; Spitser, S.; Vasquez, K.M. Short Inverted Repeats Are Hotspots for Genetic Instability: Relevance to Cancer Genomes. *Cell Rep.* **2015**, *10*, 1674–1680. [CrossRef]
98. Carreira, R.; Aguado, F.J.; Hurtado-Nieves, V.; Blanco, M.G. Canonical and Novel Non-Canonical Activities of the Holliday Junction Resolvase Yen1. *Nucleic Acids Res.* **2021**, *50*, 259–280. [CrossRef]
99. Vos, S.M.; Tretter, E.M.; Schmidt, B.H.; Berger, J.M. All Tangled up: How Cells Direct, Manage and Exploit Topoisomerase Function. *Nat. Rev. Mol. Cell Biol.* **2011**, *12*, 827–841. [CrossRef]
100. Coufal, J.; Jagelská, E.B.; Liao, J.C.C.; Brázda, V. Preferential Binding of P53 Tumor Suppressor to P21 Promoter Sites That Contain Inverted Repeats Capable of Forming Cruciform Structure. *Biochem. Biophys. Res. Commun.* **2013**, *441*, 83–88. [CrossRef]
101. Unterholzner, L.; Keating, S.E.; Baran, M.; Horan, K.A.; Jensen, S.B.; Sharma, S.; Sirois, C.M.; Jin, T.; Latz, E.; Xiao, T.S.; et al. IFI16 Is an Innate Immune Sensor for Intracellular DNA. *Nat. Immunol.* **2010**, *11*, 997–1004. [CrossRef]
102. Johnson, K.E.; Bottero, V.; Flaherty, S.; Dutta, S.; Singh, V.V.; Chandran, B. IFI16 Restricts HSV-1 Replication by Accumulating on the HSV-1 Genome, Repressing HSV-1 Gene Expression, and Directly or Indirectly Modulating Histone Modifications. *PLoS Pathog.* **2014**, *10*, e1004503. [CrossRef] [PubMed]
103. Toleikis, A.; Webb, M.R.; Molloy, J.E. OriD Structure Controls RepD Initiation during Rolling-Circle Replication. *Sci. Rep.* **2018**, *8*, 1206. [CrossRef] [PubMed]
104. Noirot, P.; Bargonetti, J.; Novick, R.P. Initiation of Rolling-Circle Replication in PT181 Plasmid: Initiator Protein Enhances Cruciform Extrusion at the Origin. *Proc. Natl. Acad. Sci. USA* **1990**, *87*, 8560–8564. [CrossRef]

105. Liao, H.; Ji, F.; Helleday, T.; Ying, S. Mechanisms for Stalled Replication Fork Stabilization: New Targets for Synthetic Lethality Strategies in Cancer Treatments. *EMBO Rep.* **2018**, *19*, e46263. [CrossRef]
106. Rass, U.; Compton, S.A.; Matos, J.; Singleton, M.R.; Ip, S.C.Y.; Blanco, M.G.; Griffith, J.D.; West, S.C. Mechanism of Holliday Junction Resolution by the Human GEN1 Protein. *Genes Dev.* **2010**, *24*, 1559–1569. [CrossRef] [PubMed]
107. Chen, S.; Geng, X.; Syeda, M.Z.; Huang, Z.; Zhang, C.; Ying, S. Human MUS81: A Fence-Sitter in Cancer. *Front. Cell Dev. Biol.* **2021**, *9*, 657305. [CrossRef] [PubMed]
108. Leach, D.R. Long DNA Palindromes, Cruciform Structures, Genetic Instability and Secondary Structure Repair. *Bioessays* **1994**, *16*, 893–900. [CrossRef] [PubMed]
109. Lai, P.J.; Lim, C.T.; Le, H.P.; Katayama, T.; Leach, D.R.F.; Furukohri, A.; Maki, H. Long Inverted Repeat Transiently Stalls DNA Replication by Forming Hairpin Structures on Both Leading and Lagging Strands. *Genes Cells* **2016**, *21*, 136–145. [CrossRef]
110. Ganapathiraju, M.K.; Subramanian, S.; Chaparala, S.; Karunakaran, K.B. A Reference Catalog of DNA Palindromes in the Human Genome and Their Variations in 1000 Genomes. *Hum. Genome Var.* **2020**, *7*, 40. [CrossRef]
111. Guiblet, W.M.; Cremona, M.A.; Harris, R.S.; Chen, D.; Eckert, K.A.; Chiaromonte, F.; Huang, Y.-F.; Makova, K.D. Non-B DNA: A Major Contributor to Small- and Large-Scale Variation in Nucleotide Substitution Frequencies across the Genome. *Nucleic Acids Res.* **2021**, *49*, 1497–1516. [CrossRef]
112. Tanaka, H.; Watanabe, T. Mechanisms Underlying Recurrent Genomic Amplification in Human Cancers. *Trends Cancer* **2020**, *6*, 462–477. [CrossRef] [PubMed]
113. Tanaka, H.; Tapscott, S.J.; Trask, B.J.; Yao, M.-C. Short Inverted Repeats Initiate Gene Amplification through the Formation of a Large DNA Palindrome in Mammalian Cells. *Proc. Natl. Acad. Sci. USA* **2002**, *99*, 8772–8777. [CrossRef] [PubMed]
114. Lopes-Nunes, J.; Oliveira, P.A.; Cruz, C. G-Quadruplex-Based Drug Delivery Systems for Cancer Therapy. *Pharmaceuticals* **2021**, *14*, 671. [CrossRef] [PubMed]
115. Miklenić, M.S.; Svetec, I.K. Palindromes in DNA—A Risk for Genome Stability and Implications in Cancer. *Int. J. Mol. Sci.* **2021**, *22*, 2840. [CrossRef]
116. Inagaki, H.; Kato, T.; Tsutsumi, M.; Ouchi, Y.; Ohye, T.; Kurahashi, H. Palindrome-Mediated Translocations in Humans: A New Mechanistic Model for Gross Chromosomal Rearrangements. *Front. Genet.* **2016**, *7*, 125. [CrossRef]
117. Kaushal, S.; Wollmuth, C.E.; Das, K.; Hile, S.E.; Regan, S.B.; Barnes, R.P.; Haouzi, A.; Lee, S.M.; House, N.C.M.; Guyumdzhyan, M.; et al. Sequence and Nuclease Requirements for Breakage and Healing of a Structure-Forming (AT)n Sequence within Fragile Site FRA16D. *Cell Rep.* **2019**, *27*, 1151–1164.e5. [CrossRef]
118. Brosh, R.M., Jr.; Matson, S.W. History of DNA Helicases. *Genes* **2020**, *11*, 255. [CrossRef]
119. Datta, A.; Brosh, R.M., Jr. New Insights into DNA Helicases as Druggable Targets for Cancer Therapy. *Front. Mol. Biosci.* **2018**, *5*, 59. [CrossRef]
120. Savvateeva-Popova, E.V.; Zhuravlev, A.V.; Brázda, V.; Zakharov, G.A.; Kaminskaya, A.N.; Medvedeva, A.V.; Nikitina, E.A.; Tokmatcheva, E.V.; Dolgaya, J.F.; Kulikova, D.A.; et al. Drosophila Model for the Analysis of Genesis of LIM-Kinase 1-Dependent Williams-Beuren Syndrome Cognitive Phenotypes: INDELs, Transposable Elements of the Tc1/Mariner Superfamily and MicroRNAs. *Front. Genet.* **2017**, *8*, 123. [CrossRef]
121. Abnous, K.; Danesh, N.M.; Ramezani, M.; Charbgoo, F.; Bahreyni, A.; Taghdisi, S.M. Targeted Delivery of Doxorubicin to Cancer Cells by a Cruciform DNA Nanostructure Composed of AS1411 and FOXM1 Aptamers. *Expert Opin. Drug Deliv.* **2018**, *15*, 1045–1052. [CrossRef]
122. Yao, F.; An, Y.; Li, X.; Li, Z.; Duan, J.; Yang, X.-D. Targeted Therapy of Colon Cancer by Aptamer-Guided Holliday Junctions Loaded with Doxorubicin. *Int. J. Nanomed.* **2020**, *15*, 2119–2129. [CrossRef] [PubMed]
123. Fleming, A.M.; Zhu, J.; Jara-Espejo, M.; Burrows, C.J. Cruciform DNA Sequences in Gene Promoters Can Impact Transcription upon Oxidative Modification of 2′-Deoxyguanosine. *Biochemistry* **2020**, *59*, 2616–2626. [CrossRef] [PubMed]
124. Kurahashi, H.; Inagaki, H.; Yamada, K.; Ohye, T.; Taniguchi, M.; Emanuel, B.S.; Toda, T. Cruciform DNA Structure Underlies the Etiology for Palindrome-Mediated Human Chromosomal Translocations. *J. Biol. Chem.* **2004**, *279*, 35377–35383. [CrossRef] [PubMed]

International Journal of *Molecular Sciences*

MDPI

Article

CNBP Binds and Unfolds In Vitro G-Quadruplexes Formed in the SARS-CoV-2 Positive and Negative Genome Strands

Georgina Bezzi [1], Ernesto J. Piga [1], Andrés Binolfi [1,2] and Pablo Armas [1,*]

1 Instituto de Biología Molecular y Celular de Rosario (IBR), Consejo Nacional de Investigaciones Científicas y Técnicas (CONICET), Universidad Nacional de Rosario (UNR), Ocampo y Esmeralda, Rosario S200EZP, Santa Fe, Argentina; bezzi@ibr-conicet.gov.ar (G.B.); piga@ibr-conicet.gov.ar (E.J.P.); binolfi@ibr-conicet.gov.ar (A.B.)

2 Plataforma Argentina de Biología Estructural y Metabolómica (PLABEM), Ocampo y Esmeralda, Rosario S200EZP, Santa Fe, Argentina

* Correspondence: armas@ibr-conicet.gov.ar; Tel.: +54-341-423-7070 (ext. 654)

Abstract: The Coronavirus Disease 2019 (COVID-19) pandemic has become a global health emergency with no effective medical treatment and with incipient vaccines. It is caused by a new positive-sense RNA virus called severe acute respiratory syndrome-related coronavirus 2 (SARS-CoV-2). G-quadruplexes (G4s) are nucleic acid secondary structures involved in the control of a variety of biological processes including viral replication. Using several G4 prediction tools, we identified highly putative G4 sequences (PQSs) within the positive-sense (+gRNA) and negative-sense (−gRNA) RNA strands of SARS-CoV-2 conserved in related betacoronaviruses. By using multiple biophysical techniques, we confirmed the formation of two G4s in the +gRNA and provide the first evidence of G4 formation by two PQSs in the −gRNA of SARS-CoV-2. Finally, biophysical and molecular approaches were used to demonstrate for the first time that CNBP, the main human cellular protein bound to SARS-CoV-2 RNA genome, binds and promotes the unfolding of G4s formed by both strands of SARS-CoV-2 RNA genome. Our results suggest that G4s found in SARS-CoV-2 RNA genome and its negative-sense replicative intermediates, as well as the cellular proteins that interact with them, are relevant factors for viral genes expression and replication cycle, and may constitute interesting targets for antiviral drugs development.

Keywords: COVID-19; SARS-CoV-2; coronavirus; G-quadruplex; CNBP

Citation: Bezzi, G.; Piga, E.J.; Binolfi, A.; Armas, P. CNBP Binds and Unfolds In Vitro G-Quadruplexes Formed in the SARS-CoV-2 Positive and Negative Genome Strands. *Int. J. Mol. Sci.* **2021**, *22*, 2614. https://doi.org/10.3390/ijms22052614

Academic Editor: Vaclav Brazda

Received: 15 January 2021
Accepted: 20 February 2021
Published: 5 March 2021

Publisher's Note: MDPI stays neutral with regard to jurisdictional claims in published maps and institutional affiliations.

1. Introduction

By the end of 2019, an unexpected outbreak of a new severe acute respiratory syndrome (SARS) termed by the World Health Organization (WHO) as Coronavirus Disease 2019 (COVID-19) emerged in Wuhan (China) [1]. The cause was infection by a highly contagious new SARS-related coronavirus named SARS-CoV-2 [2], which rapidly spread around the world. On 11 March 2020, WHO declared COVID-19 a "pandemic" condition. In one year, the global number of confirmed infections was nearly 90 million and the death toll over 1.9 millon, which is spread over more than 200 countries and with repetitive and constantly increasing waves of contagion (https://covid19.who.int/, accessed on 14 January 2021). Despite the enormous efforts of the medical and scientific community and the pharmaceutical industry, so far no clearly effective pharmacological treatments have been found for treating SARS-CoV-2 infection and disease, and the first immunizations with vaccines developed in record times are beginning. Therefore, there is urgency for the development of specific drugs and novel treatments.

SARS-CoV-2 is a betacoronavirus genus of the *Coronaviridae* family, and is one of the seven types of the *Coronaviridae* family of viruses which could infect humans. Four of them, two alphacoronaviruses (HCoV-229E and HCoV-NL63) and two betacoronaviruses (HCoV-HKU1 and HCoV-OC43), are common around the world and cause mild diseases [3]. The other three human betacoronaviruses are more recent, causing severe acute respiratory outbreaks. SARS-CoV emerged in 2002 and 2003 in Guangdong province (China) [4]. The Middle East Respiratory Syndrome Coronavirus (MERS-CoV) was identified in Saudi Arabia in 2012 [5], and in 2019 SARS-CoV-2 caused the COVID-19 pandemic. SARS-CoV and MERS-CoV originated from bats, and it appears to be so for SARS-CoV-2 as well, which shows fairly close relatedness with three bat-derived coronaviruses, bat-SL-CoVZC45 and bat-SL-CoVZXC21 [6], as well as RaTG13 [1]. SARS-CoV is the most genetically related to SARS-CoV-2 among human coronaviruses [6], sharing a high nucleotide sequence identity (79.7%). However, the replication rate of SARS-CoV-2 is higher than that of SARS-CoV [7]. All these viruses are enveloped viruses with positive-sense single-stranded RNA genomes of about 30 kb in length and have similar structures, genomic organizations, and replicative cycles [3,8]. Upon infection of a host cell, the positive-sense RNA genome (+gRNA) is released and ready to be translated by the protein synthesis machinery of the infected host cell to express a set of viral proteins crucial for viral replication [9]. Both, replication of the viral genome and transcription of positive-sense subgenomic RNAs (+sgRNAs) involve the synthesis of negative-strand RNA genome (−gRNA) and negative-strand subgenomic RNAs (−sgRNAs) intermediates [10]. As other RNA viruses, SARS-CoV-2 is dependent on effectively engaging host cell factors such as regulators of RNA stability, processing, localization, and translation to facilitate replication and production of new viral particles [11]. On the other hand, the infected host cell must detect the pathogen and activate appropriate innate immune response pathways to restrict virus infection [11]. Recent comprehensive studies have begun to identify expression changes or modifications in the host cell transcriptome [12,13] and proteome [14–16] as well as cellular proteins interacting with viral proteins [17,18] and with viral genome [19], as approaches to identify cellular pathways relevant for viral infection and replication. However, a more detailed understanding of the molecular mechanisms and interactions occurring during SARS-CoV-2 infection is required to design efficient therapeutic strategies.

Viral RNA genomes have some intrinsic characteristics that favor or obstruct the viral genome expression and replication. For instance, the folding of specific regions of the genomic RNA molecule into stable secondary structures may act as specific hallmarks for the attachment of cellular or viral RNA processing machinery, but may also be roadblocks for viral RNA metabolism [20]. Among these structures, G-quadruplexes (G4s) are stable four-stranded structures formed in G-rich DNA or RNA sequences that can be formed by the folding on itself of a single-stranded molecule [21]. The structure is characterized by the stacking of two or more planar arranges of four G nucleobases (called G-tetrads) stabilized by lateral Hoogsteen-type hydrogen bonds and by the coordination of monovalent cations, mainly K^+ (Figure 1a). These structures may occur in putative G-quadruplex sequences (PQSs) presenting at least four contiguous tracts of two or more guanine nucleotides interspersed with short nucleotide sequences forming the G4 loops. Depending on the relative orientation of the G-tracts, G4s may be parallel (with four G-tracts in the same relative orientation), antiparallel (with two G-tracts in opposite orientation in respect to the other two), or hybrid (with one G-tract in opposite orientation in respect to the other three). G4s have received extensive attention during the last two decades due to their involvement in the regulation of cellular processes such as transcription, replication, translation, and telomere maintenance and the development of specific G4 ligands with promising anticancer effects [21].

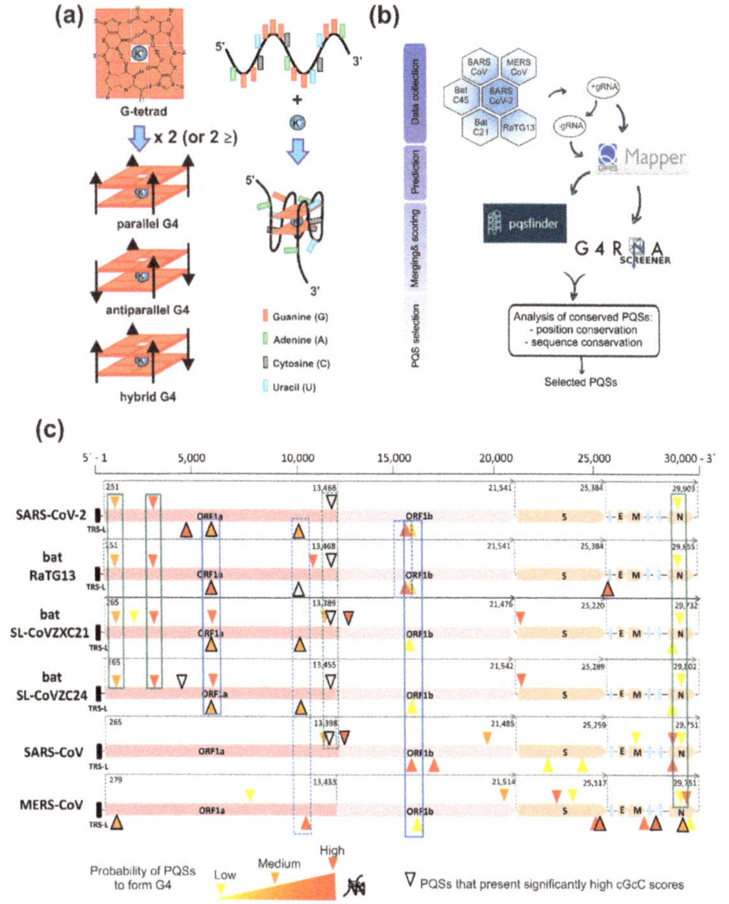

Figure 1. Identification and selection of PQSs in the genomes of SARS-CoV-2 and other members of the *Coronaviridae* family. (**a**) Cartoon representing the formation of G4s. Left: G-tetrads are formed by the planar arrange of four G nucleobases stabilized by lateral Hoogsteen-type hydrogen bonds. The stacking of two or more tetrads and the coordination of K^+ form the G4 structure. Depending on the relative orientation of the G-tracts, G4s may be parallel, antiparallel, or hybrid. Right: G4 formation by the folding on itself of a G-rich RNA strand with at least four contiguous G-tracts interspersed with short nucleotide loops. The formation of a parallel G4 is represented. (**b**) Schematic summary of the bioinformatic workflow conducted for PQSs identification and selection in the positive- and negative-sense RNA genomes of SARS-CoV-2, RaTG13, bat-SL-CoVZC45, bat-SL-CoVZXC21, SARS-CoV, and MERS-CoV. (**c**) Schematic representation of the location of the selected PQSs in the analyzed genomes. The organization of the coding regions for the main viral proteins are represented for each virus: ORF1a (open-reading frame 1a), ORF1b (open-reading frame 1b), S (spike), E (envelope), M (membrane) and N (nucleocapsid), black rectangle represents transcription-regulatory sequences (TRS) in the 5′ untranslated region (UTR). PQSs found in the positive-sense strand are represented above the genome while PQSs found on the negative-sense strand represented below the genome. PQSs are represented with the following color code: PQSs with high probability to form G4 (red), PQSs with medium probability to form G4 (orange) and PQSs with low probability to form G4 (yellow). PQSs that present significantly high cGcC scores (>150) are highlighted using thick edges. PQSs conserved in position and sequence are indicated with solid green line boxes (for positive-sense strand) and solid blue line boxes (for negative-sense strand). Dashed lines indicate PQSs conserved in position that were not selected due to lack of sequence conservation, to conservation in less than four genomes or to not overpassing the selection criterion.

Besides in human beings, G4s are widespread across nucleic acids of all the taxonomic phyla [22] including *Bacteria* [23], *Archaea* [24], and viruses [25,26]. Critical roles for viral G4s have been described in human viruses, including immunodeficiency virus (HIV), herpes simplex virus (HSV), Epstein-Barr virus (EBV), human papillomavirus (HPV), hepatitis B virus (HBV), Nipah virus, hepatitis C virus (HCV), Zika virus, and Ebola virus [27–29], and some G4-specific compounds have shown powerful antiviral activity by targeting G4 structures [28,29]. Very recent reports have initiated the search for G4s in the genomes of human coronaviruses, including SARS-CoV-2 [30–36]. Beyond the prediction of PQSs, some works have demonstrated the formation of a few G4s in vitro by PQSs found in the +gRNA [30,34–36] and one of them was demonstrated to be formed within cultured human cells and controls the translation efficiency of the nucleocapsid N protein [35,36].

Here, we have used a novel G4 prediction pipeline for the identification of the PQSs with high probability of G4 formation within the +gRNA and −gRNA of SARS-CoV-2 and related betacoronaviruses. We also performed a predictive analysis of the putative consequence of natural nucleotide variations reported for SARS-CoV-2 on the probability of G4 formation in the selected PQSs, showing that some synonymous mutations may alter G4 formation with putative consequences in their regulatory functions. Then, using multiple biophysical techniques, we confirmed the formation of two G4s in the +gRNA and provide the first evidence of G4 formation by two PQSs in the −gRNA of SARS-CoV-2. Finally, we performed biophysical and molecular approaches to demonstrate for the first time that cellular nucleic acid binding protein (CNBP), the main cellular protein bound to SARS-CoV-2 RNA genome in infected human cells [19], binds and promotes the unfolding of G4s formed by both +gRNA and −gRNA of SARS-CoV-2. Our results suggest that G4s found in SARS-CoV-2 +gRNA and −gRNA, and the cellular proteins that interact with them, are important elements for viral replication cycle and may be novel targets for developing antiviral drugs against COVID-19.

2. Results and Discussion

2.1. G4 Prediction and Selection in the SARS-CoV-2 Positive and Negative Genome

During the COVID-19 pandemic, mainly along the last half of 2020, several works analyzed the SARS-CoV-2 RNA genome to seek PQSs. All of them analyzed the +gRNA [30–36], while a few made a superficial overview of PQSs on the −gRNA [32,33,36]. Although −gRNA and −sgRNAs are minority in respect of their positive-sense RNA counterparts and represent only about 1% of viral RNA [9,10], negative-sense RNAs are key intermediates functioning as templates for +gRNA replication and +sgRNAs transcription. These processes are mediated by the replicase-transcriptase complex (RTC) formed by several non-structural proteins (nsps) [9,10]. Consequently, −gRNA and −sgRNAs may contain G4s with putative regulative functions on these processes. In addition, most of the previous PQSs analyses on SARS-CoV-2 genome have used different bioinformatics prediction tools, some of them designed for DNA-G4, and a variety of criteria for selecting the best PQS candidates, mainly including higher prediction scores [30–36] combined with lower potential thermodynamic stability of secondary structures competitive with G4s [32], uniqueness in SARS-CoV-2 and conservation among variants of SARS-CoV-2 [34], and conservation among human coronaviruses [36]. Although there is divergence in G4 prediction tools and selection criteria, none of the predictions have found PQSs with four tracts of three consecutive guanines (with the potential of forming three-tetrads G4s), and have only found PQSs with four tracts of two consecutive guanines (with the potential of forming two-tetrads G4s). Although two-tetrads G4s are less stable than the three-tetrads G4s, especially in vivo, it is known that the RNA G4s are more stable than their DNA counterparts [37] and several emerging studies have demonstrated the formation of two-tetrads G4s in viral sequences [38–44].

Here, we have performed a predictive analysis of PQSs on the +gRNA and −gRNA sequences from three coronaviruses that infect humans and cause the most severe health consequences: the SARS-CoV-2, SARS-CoV (or SARS-CoV-1), and MERS-CoV. We also included genomes from three bat coronaviruses probably related to SARS-CoV-2 origin: bat-SL-CoVZC45 and bat-SL-CoVZXC21 [6], as well as RaTG13, which presents a closer phylogenetic relationship with SARS-CoV-2 [1]. First, we downloaded the +gRNA sequences of the analyzed viruses and obtained for each one the respective reverse complement sequence for −gRNA analysis. Then, we performed an initial analysis of the PQSs found in the +gRNA and −gRNA obtained by using Quadruplex forming G-Rich Sequences (QGRS) Mapper online prediction software [45] for the identification of canonical two-tetrads PQSs (with four two-guanine tracts) and loops of extended lengths (from 1 to 15 nucleotides), i.e., $G_{2+}N_{1-15}G_{2+}N_{1-15}G_{2+}N_{1-15}G_{2+}$. The retrieved PQSs were then analyzed using two additional predictors: PQSfinder Web and G4RNA screener. PQSfinder Web [46] is an algorithm that supports DNA and RNA sequences but was validated primarily on DNA sequences and has been trained with G4-seq data. G4RNA screener [47] is a web algorithm that identifies regions in RNA sequences prone to fold into G4 based on three scoring systems: cGcC (Consecutive G over consecutive C ratio) [48], G4H (G4Hunter) [49], and G4NN (G4 Neural Network) [50]. Results from this analysis are detailed in Supplementary Table S1 (for +gRNA, each virus in a different tab) and Supplementary Table S2 (for −gRNA, each virus in a different tab). Based on the scores obtained using the five G4 predictors, we defined the following selection criterion: PQSs that were found with QGRS Mapper and which scores for the other four predictors were over the defined threshold for each predictor were classified with high probability to form G4 (highlighted in red), those PQSs that were found with QGRS Mapper and which scores for at least three of other four predictors were over the defined threshold for each predictor with at least one of those scores significantly high were classified with medium probability to form G4 (highlighted in orange), and those PQSs that were found with QGRS Mapper and which scores for at least three of other four predictors were over the defined threshold for each predictor with none of those scores significantly high were classified with low probability to form G4 (highlighted in yellow). In addition, we highlighted those PQSs that present significantly high cGcC scores (marked with thick edges), although some of them did not fulfill the selection criterion. A summary of the bioinformatic workflow is presented in Figure 1b, while a summary of the numbers of this analysis is represented in Table 1 and a schematic location of the selected PQSs on the explored viral genomes is shown in Figure 1c.

Our results show that PQSs predicted by QGRS Mapper are scattered along the genomes of the five analyzed viruses showing 29 to 49 PQSs in the +gRNA and 16 to 38 PQSs in the −gRNA (Table 1). SARS-CoV-2 and RaTG13 present an intermediate number of PQSs in both strands, with 37 PQSs in the +gRNA and 19 PQSs in the −gRNA of SARS-CoV-2 while 20 PQSs in the −gRNA of RaTG13. SARS-CoV is the one that displays the highest amount of initial PQSs predicted for the +gRNA followed by MERS-CoV and SARS-CoV-2, while bat-SL-CoVZC45 and bat-SL-CoVZXC21 show the lowest amount of initial PQSs predicted in the +gRNA. A similar order is observed for the number of initial PQSs predicted for the −gRNA, except for the fact that MERS-CoV presents the higher number of initial PQSs followed by SARS-CoV, SARS-CoV-2 and RaTG13, while bat-SL-CoVZC45 and bat-SL-CoVZXC21 again show the lowest amount of initial PQSs predicted in the −gRNA. The numbers of PQSs predicted by QGRS Mapper partially correlate with genomes G content, being the genomes of SARS-CoV and MERS-CoV which show the highest G% in both + and −gRNA and are the ones that present the highest numbers of PQSs, while the genomes of bat-SL-CoVZC45 and bat-SL-CoVZXC21 show the lowest G% in both + and −gRNA and are the ones that present the lowest numbers of PQSs. Curiously, SARS-CoV-2 and RaTG13 show the lowest G% in both + and −gRNA, but even so shows higher numbers of PQSs than the genomes of bat-SL-CoVZC45 and bat-SL-CoVZXC21, probably indicating that SARS-CoV-2 and RaTG13 may have gained PQSs during their evolution from the putative common ancestor shared with bat-SL-CoVZC45 and bat-SL-CoVZXC21. In agreement with this, the lower number of PQSs and lower G% found in SARS-CoV-2 compared to SARS-CoV may be related to the fact that SARS-CoV-2 replicates faster than SARS-CoV because G4 structures may represent an obstacle for viral proteins translation and RNA dependent RNA synthesis [36]. Noteworthy, the numbers of selected PQSs by our criterion using several predictors show a clear difference from the numbers of initial PQSs predicted by QGRS Mapper. SARS-CoV-2 presents the lowest number of selected PQSs in the +gRNA (only ≈8% of the PQSs originally predicted), probably indicating a negative selection of PQSs capable of forming stable G4s in this virus. This is in agreement with the previously reported data indicating that SARS-CoV-2 displays general PQSs poverty when compared to the virus realm, its PQS density being in the lower end of results from the *Coronaviridae* family, which itself is in the lower end of the (+) ssRNA Group IV [34] and the PQS frequency in SARS-CoV-2 is significantly lower than expected from its base composition [33]. On the contrary, the SARS-CoV-2 −gRNA presents the highest percentage of selected PQSs from the initially predicted PQSs by QGRS Mapper (≈26%), and a similar tendency is observed for bat-SL-CoVZC45, bat-SL-CoVZXC21, and MERS-CoV, while a lower percentage is observed for RaTG13 and SARS-CoV. This may indicate a positive selection of PQSs capable of forming stable G4s in the −gRNA of SARS-CoV-2 with potential regulatory functions in replication/transcription. This fact may be the consequence that −gRNA and −sgRNAs are not templates for translation and their evolution may not be constrained by the negative effect of the G4s on viral proteins translation.

Table 1. Numbers of found and selected PQSs for +gRNA and −gRNA of the analyzed viruses.

Betacoronavirus	Genome Accession Number	% G	Number of PQSs in QGRS Mapper	Selected PQSs				% of Selected over Predicted PQSs
				Low Probability to Form G4	Medium Probability to Form G4	High Probability to Form G4	Total	
+gRNA								
SARS-CoV-2	NC_045512.2	19.6	37	1	1	1	3	8.1
RaTG13	MN996532.2	19.6	37	1	2	1	4	10.8
Bat-SL-CoVZXC21	MG772934.1	20.1	32	2	2	4	8	25.0
Bat-SL-CoVZC45	MG772933.1	20.2	29	1	1	3	5	17.2
SARS-CoV	NC_004718.3	20.8	49	2	2	2	6	12.2
MERS-CoV	NC_019843.3	20.9	40	3	1	2	6	15.0
−gRNA								
SARS-CoV-2	NC_045512.2	18.4	19	-	3	2	5	26.3
RaTG13	MN996532.2	18.4	20	-	1	3	4	20.0
Bat-SL-CoVZXC21	MG772934.1	18.7	18	2	2	-	4	22.2
Bat-SL-CoVZC45	MG772933.1	18.7	16	2	2	-	4	25.0
SARS-CoV	NC_004718.3	20.0	29	2	-	3	5	17.2
MERS-CoV	NC_019843.3	20.3	38	2	2	5	9	23.7

Visual analysis of the location of the selected PQSs on the explored viral genomes (Figure 1c) shows that SARS-CoV-2 and bat coronaviruses display a higher PQS density in the genome region coding for the nsps (encoded by ORF1ab) and lower PQS density in the genome region coding for the structural proteins. Based on the location of the selected PQSs, we identified six of them (three from the +gRNA and three from the −gRNA) that are conserved in position in at least four viral genomes and analyzed their sequence conservation (Supplementary Figure S1). From the six selected PQSs conserved in position, five of them (three from the +gRNA and two from the −gRNA) show very high sequence conservation among SARS-CoV-2 and bat coronaviruses (>88% except for +28,880 PQS of RaTG13) and lower identity % with SARS-CoV and MERS-CoV (when they had PQSs conserved in position), while one selected PQS conserved in position from the −gRNA is neither conserved among SARS-CoV-2 and bat coronaviruses nor with SARS-CoV and MERS-CoV. It is noticeable that one of the PQSs that presented significantly high cGcC scores but did not fulfill our selection criterion (position +13,385 in SARS-CoV-2 +gRNA, see Supplementary Table S1) is conserved in position and sequence among the six viral genomes studied (Figure 1c and Supplementary Figure S1) and has been previously shown to fold in vitro as G4 [30,35]. Interestingly, this PQS is located very near (≈80 nucleotides upstream) the slippery sequence that causes the ribosomal frameshift that controls the transition from the translation of the ORF1a to the translation of the ORF1ab, making it an attractive PQS forming a G4 with putative function in the regulation of this process together with the pseudoknot structure already described [51].

In this work, we focused the following experimental studies on the SARS-CoV-2 PQSs highly conserved among the explored virus genomes. Conservation in PQSs candidates is a trace of maintenance through natural selection and indicates that selected PQSs may be relevant elements for the biological fitness of these viruses, beyond SARS-CoV-2 and extended to those viruses that share the conserved PQSs. Table 2 shows the main characteristics of the five selected PQSs with conserved positions and sequences among at least four of the explored coronaviruses. All of the selected PQSs had been previously predicted using different strategies and predictors, and for those in the +gRNA (+644, +3467 and +28,903) there are experimental evidences of G4 formation. However, until now, no experimental analysis of G4 formation has been performed for PQSs predicted in the −gRNA.

Table 2. Information of selected SARS-CoV-2 PQSs.

PQS Name	Genome	Length	Sequence	Prediction Scores					Reference of Previous Prediction	Reference of Experimental Evidence of G4 Formation
				G4RNA Screener			QGRS Mapper	PQS Finder		
				cGcC	G4H	G4NN				
+644	ORF1ab nsp1	20	GGUAAUAAAGGAGCUGGUGG C G G C	17	0.8	0.69	30	20	[30–34,36]	[36]
+3467	ORF1ab nsp3	17	GGAGGAGGUGUUGCAGGA A A G A A GAAG U U	18	1	0.87	30	24	[30–34,36]	[34]
+28,903	N	15	GGCUGGCAAUGGCGG UAU UU UU UUUAU AU ----- C	5.33	0.87	0.57	33	27	[30–34,36]	[34–36]
−13,963	ORF1ab nsp12	18	GGAUCUGGGUAAGGAAGG G G G AA	22	1.11	0.97	34	23	[32,33,36]	-
−23,877	ORF1ab nsp3	17	GGAUAUGGUUGGUUUGG AA C C A A	160	0.94	0.02	34	24	[32,33,36]	-

Note: nucleotide variations indicated below each PQS are highlighted according to their consequence to disrupt G-tracts and impede PQSs or reduce PQSs scores (red), do not affect G-tracts and may be neutral for PQSs (yellow) or produce G-tracts extension and increase PQSs scores (green). Dashes indicate single nucleotide deletions and are highlighted with the same color code as substitutions in respect to their effect on PQSs.

Finally, we performed an analysis of variations within these PQSs using GISAID database (https://www.gisaid.org/, accessed on 2 January 2021) [52]. Table 2 shows the reported variations and highlights those that may lead to impede PQSs and those that produce G tracts extension (and probably higher propensity to form stable G4s). Supplementary Table S3 contains further information about the GISAID mutations found in the selected PQSs, including frequency, codon, protein, amino acid change, and scores for the PQSs predictors used in this work. Of note, all the analyzed nucleotidic changes show very low frequencies (<1%) and most of them (28/48) disrupt G-tracts and may lead to impede PQSs or reduce PQSs scores, while only 8/46 produce G-tracts extension and may increase PQSs scores, the other 10/46 changes being those that do not disturb G-tracts and may be neutral for PQSs (although some of them present variations in scores which may lead to G4 stabilization or destabilizations). The PQS that showed the higher number of variations is +28,903, mainly of the G4-disruptive type (16/20) and none of the G4-stabilizing type. Interestingly, all the G4-stabilizing variations are synonymous or silent mutations with no consequence in the encoded amino acids, while most of the G4-disruptive variations are not synonymous (missense or frameshift) mutations producing changes in the encoded amino acids (27/28). Considering that single-nucleotide and short variations in PQSs may affect G4s formation or stability with consequences in transcriptional [53–55] and translational [56–58] control, it would be important to analyze the mutations occurring within PQSs not only for their effects on encoded proteins, but also for the putative effects in G4-regulated processes.

On the other hand, non-conserved PQSs that are unique for a particular virus may also play a central role in the ability of the virus to adapt to new environmental challenges and infect and replicate in novel hosts. This could be the case of SARS-CoV-2 PQSs of the −gRNA in positions −25,003 (which is unique for this virus) and −13,134 (which is only conserved in RaTG13 genome, Figure 1c and Supplementary Figure S1c) or the one in position −19,865, whose sequence is not fully conserved (except in RaTG13 genome, which shows a higher conservation but the PQS in this position did not fulfill our selection criterion, Figure 1c and Supplementary Figure S1b). None of these PQSs were selected for further study in this work, but they remain as interesting candidates to study SARS-CoV-2 specific G4s.

Although many of the identified PQSs were previously described by other approaches, our selection criterion has highlighted some new PQSs, mainly those in the −gRNA, as interesting candidates to perform experimental studies.

2.2. Confirmation That the Selected PQSs Fold In Vitro as G4

The five selected PQSs were further studied in their capability to form G4 structures in vitro using synthetic RNA oligoribonucleotides for four different spectroscopic approaches: Circular Dichroism (CD) Spectroscopy (Figure 2a and Supplementary Figure S2), 1D ^{1}H Nuclear Magnetic Resonance (NMR) (Figure 2b), Thermal Difference Spectroscopy (TDS) (Supplementary Figure S3), and Thioflavin T (ThT) fluorescence (Supplementary Figure S4).

For PQSs +3467 and −23,877, CD spectra have the typical pattern of peaks associated with parallel G4 structure, showing an increase of a positive peak around 263 nm and a negative peak around 240 nm in response to the presence of increasing K^+ concentrations (Figure 2a). K^+ is considered the main intracellular G4-stabilizing cation [59] and the CD positive peaks of these two PQSs easily reached the maximum intensity with K^+ concentrations above 10 mM. The characteristic G4 spectra were not observed in the presence of Li^+, which plays a neutral role in G4 folding and stability [59]. CD melting curves (Supplementary Figure S2b,e) showed that these G4s are stable structures with estimated Tm of 58.5 °C (for PQSs +3467) and 51.6 °C (for PQS −23,877) and high values of ΔG, indicative of high stabilities for two-tetrads G4s. In addition, 1D ^{1}H NMR showed defined signals around 11–12 ppm (Figure 2b), confirming the presence of Hoogsteen bonds and G4 structures. In agreement with the former results, TDS spectra showed the typical G4 signature with two positive peaks around 243 and 273 nm and a negative peak at 295 nm (Supplementary Figure S3), and ThT fluorescence assays showed that these folded PQSs notably enhance ThT fluorescence above 30-fold for +3467 and above 50-fold for −23,877 (Supplementary Figure S4). In coincidence, the prediction of the secondary structures of these PQSs by RNAfold predicts the G4 structures (at 20 °C) and NUPACK and RNAfold software do not predict stable secondary structures that may compete with G4 formation (Supplementary Table S4).

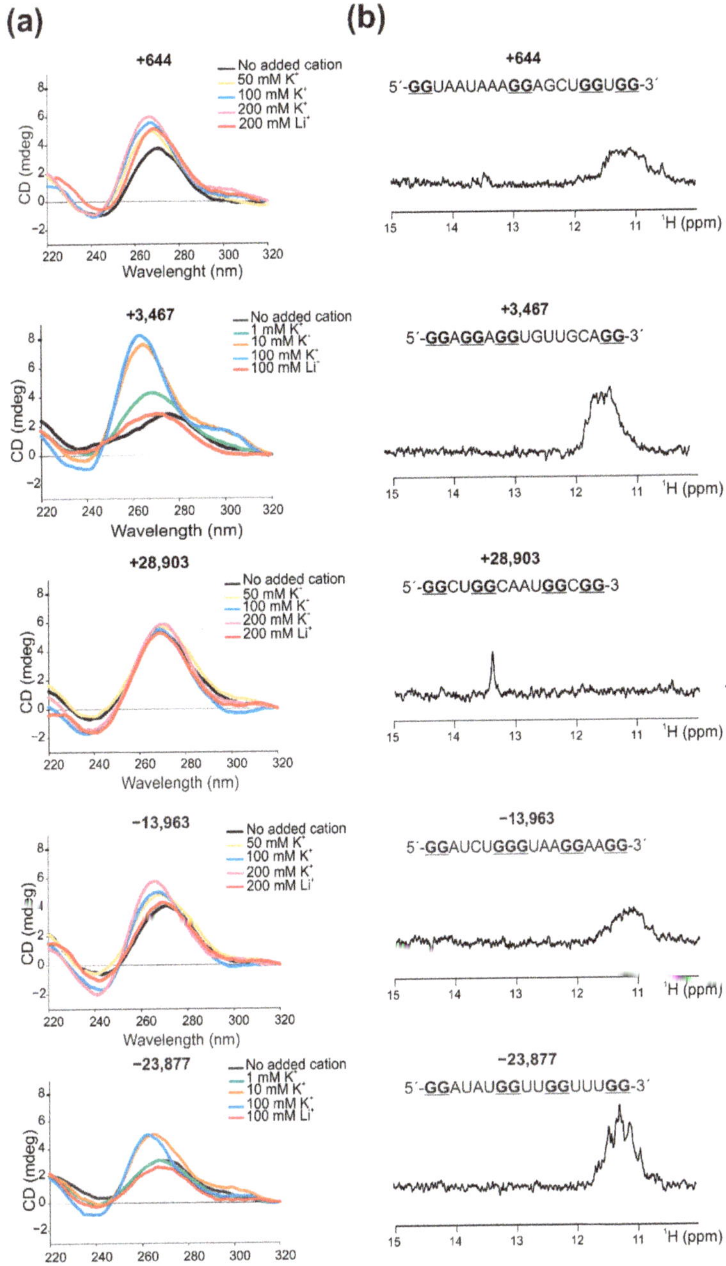

Figure 2. Evidence by CD and NMR spectroscopy of the in vitro G4 structures formed by the selected PQSs. (**a**) CD spectra were obtained for each RNA sequence (named by the PQS position) folded in the absence and in the presence of increasing K^+ concentrations, or in the presence of Li^+ at the highest concentration used for K^+. Concentrations are indicated for each plot. (**b**) 1D ^1H NMR spectra obtained for each RNA sequence (named by the PQS position) folded in the presence of K^+ at the highest concentration used for CD. RNA sequence for each PQS are represented above NMR spectra, and guanine nucleotides predicted to participate in the G4 formation are indicated in bold and underlined.

In the case of PQSs +644 and −13,963, CD spectra also showed the peaks associated with parallel G4s, but with a milder increase in the positive peak with the increase of K^+, which was not observed with Li^+ (Figure 2a). For these two PQSs, high K^+ concentrations (up to 200 mM) were needed so as to observe a clear G4 spectra, probably indicating that these G4s are less stable or less prone to fold. CD melting (Supplementary Figure S2a,d) showed that these G4s are less stable than PQSs +3467 and −23,877, showing estimated Tm of 52 °C (for PQS +644) and 48.5 °C (for PQS −13,963) and ΔG values slightly lower than those for PQSs +3467 and −23,877. 1D ^{1}H NMR also showed G4 signatures but signals were less intense than those for +3467 and −23,877 (Figure 2b), suggesting that there is a less amount of G4 probably due to lower stabilities and loose global conformation. TDS spectra showed the typical G4 signatures for the PQS +644 and a less defined spectrum for the PQS −13,963 (Supplementary Figure S3), while ThT fluorescence assay showed that both PQSs increase ThT fluorescence barely above 10-fold (Supplementary Figure S4). For these PQSs, RNAfold does not predict G4 structures and both, RNAfold and NUPACK, do not predict stable secondary structures for +644 and predict a weak stem-loop structure with three base pairs for −13,963 (Supplementary Table S4), indicating that the G4s might not compete with stable alternative structures.

PQS +28,903 showed CD spectra with a positive peak centered at 270 nm that remain unperturbed upon K^+ additions, even reaching K^+ concentration of 200 mM (Figure 2a). In addition, similar spectra were observed in the absence of added monovalent cation or in the presence of Li^+, suggesting that this sequence do not adopt a G4 structure. CD melting (Supplementary Figure S2c) showed that the observed structure is one of the less stable ones, with an estimated Tm = 49.3 °C and the lowest calculated ΔG value. 1D ^{1}H NMR did not show G4 signals in the 11–12 ppm region but showed a clear signal around 13.7 ppm (Figure 2b), indicating that this sequence does not form Hoogsteen bonds (i.e., G4 structures) but instead contains Watson-Crick base pairs at some extent [60]. TDS spectrum displayed a very low signal and poor defined spectrum (Supplementary Figure S3), which further supports the absence of G4 and is probably compatible with self-complementary duplex structure [61] and ThT fluorescence assay showed that this PQS does not significantly increase ThT fluorescence (only 3-fold) (Supplementary Figure S4). In coincidence, although RNAfold and NUPACK do not predict intermolecular duplex (self-dimers), they predict a relatively stable stem-loop structure with four base pairs for this PQS (Supplementary Table S4), which may compete with G4 formation and may account for the signatures observed in NMR and TDS spectra.

Overall, our data showed that, except for the PQS +28,903, the other selected PQSs form G4 structures, +3467 and −23,877 being the ones with higher stability and/or propensity to form, followed by +644 and −13,963. In agreement with our results, the PQS +644 has been reported to fold as G4 in vitro by using Thioflavin T (ThT) fluorescence assay and CD [36]. Similarly, the PQS +3467 has been reported to fold as G4 in vitro by using CD and 1D ^{1}H NMR [34], showing very similar spectra for both methods. Surprisingly, the PQS +28,903 was also reported to fold as G4 in vitro by several works [34–36], not only by N-methyl mesoporphyrin IX (NMM) and ThT fluorescence assays, CD and 1D ^{1}H NMR, but also by native PAGE mobility assays, fluorescence resonance energy transfer (FRET) combined with stopped flow, and PCR-stop assays in combination with PQS mutations and G4 stabilizing ligands. In addition, the PQS +28,903 was also informed to be formed within living cells, where it is capable of inhibiting the translation of a reporter gene (GFP) [36] and of the SARS-CoV-2 N protein [35] upon incubation with G4 stabilizing ligands, positioning this PQS as an interesting target for the design of SARS-CoV-2 antiviral strategies. In our experimental conditions, this PQS was not able to form a defined and stable G4, although it formed a stable secondary structure containing Watson-Crick bonds, as was evident in the ^{1}H NMR spectrum. Interestingly, in previously reported NMR spectra, similar Watson-Crick bonds peaks were observed around 13 ppm [34]. In summary, our results confirm the formation of G4 by two PQS found in SARS-CoV-2 +gRNA (i.e., +644 and +3467), and inform for the first time the formation of G4 by two PQSs found in SARS-CoV-2 −gRNA (i.e., −13,963 and −23,877), while they could not confirm the folding as G4 structure of the PQS +28,903 found in SARS-CoV-2 +gRNA. This suggests that other PQSs than the +28,903 may also be interesting targets for testing their biological role and antiviral strategies specific for G4.

2.3. Cellular Nucleic Acids Binding Protein (CNBP) Interacts with Some SARS-CoV-2 G4s and Promotes Their Unfolding

Viral reproduction depends at some points on host cellular machinery. The antiviral strategies that target viral proteins are usually effective only against specific viral strains and fails even for closely related viral species or mutant virus from the same species. However, targeting host proteins needed for viral replication cycle is a better strategy to achieve a wide range response toward viruses that make use of common cellular pathways. This is why, since the COVID-19 pandemic outbreak, several scientific groups around the world have made efforts to describe human cellular proteins interacting with SARS-CoV-2 viral components, not only to better understand the mechanism of viral infection and the host innate immune response, but also to discover new targets for antiviral therapy. The first studies on SARS-CoV-2-infected human cells have focused on characterizing changes in the host cell transcriptome [12,13] or proteome [14–16] and interactions between viral proteins and host proteins [17,18], revealing cellular pathways relevant to productive infection. However, these studies could not reveal how viral RNA is regulated during infection or how viral infection remodels host cell RNA metabolism to enable its replication. A bioinformatic approach has recently predicted human RNA-binding proteins sites in SARS-CoV-2 RNA proposing three highly promising candidates (SRSF7, HNRNPA1, and TRA2A) that are involved in cellular RNA metabolism and share multiple RGG-rich novel interesting quadruplex interaction (NIQI) motifs common to most G4 binding proteins [33]. A more recent work has identified 104 human proteins that directly and specifically bind to SARS-CoV-2 RNAs in infected human cells by using RNA antisense purification and quantitative mass spectrometry (RAP–MS) [19]. Among the identified cellular proteins, CNBP, also referred to as zinc finger protein 9 (ZNF9), was the human protein most significantly enriched in RAP–MS. CNBP was even more enriched than the 15 viral proteins found in the same study, which comprised of 5 structural proteins (included the N, S and M proteins) and 10 nsps known to bind viral RNA. Moreover, CNBP was strongly upregulated in a proteome analysis of human cells after viral infection [19], and, among all SARS-CoV-2 RNA interactome proteins, CNBP had the most significant effect on virus-induced cell death in a genome-wide CRISPR perturbation screen designed to identify host factors that affect cell survival after SARS-CoV-2 infection [62]. CNBP is a highly conserved nuclear-cytoplasmic protein with nucleic acid chaperone activity [63] that preferentially binds to G-enriched RNA or DNA single-stranded sequences [64,65]. CNBP contains an RGG-rich NIQI motif and has been recently described as a transcriptional regulator that unfolds G4 in the promoters of *c-MYC* and *KRAS* oncogenes and in the *NOG* developmental gene [66]. On the other hand, CNBP has been also reported to boost global translation by resolving G4 structures in the 5′ UTRs of mRNAs [65]. Considering this scenario, we decided to assay the binding and function of CNBP on the G4s identified in this work.

Electrophoretic mobility shift assays (EMSAs) were performed to evaluate CNBP capability to bind to the PQSs folded in the presence of K^+ or Li^+ (Figure 3a). PQSs +644, +3467 and −23,877 clearly interacted with CNBP, as evidenced by a shift of the PQSs mobilities. Binding of CNBP to PQS −13,963 was less evident, as only mild shifts were observed at high CNBP concentrations. Instead, the PQS +28,903 did not interact with CNBP in our experimental conditions. For the PQSs that showed the best interaction (i.e., +644, +3467 and −23,877), a slightly higher affinity was observed in the condition folded in presence of Li^+ than in the condition folded in presence of K^+, since in the Li^+ condition shift is observed (and/or free probes are consumed) at lower CNBP concentrations. This is in agreement with a previous report proposing that CNBP promotes G4s unfolding by shifting the equilibrium between G4 and unfolded (single-stranded) states towards the unfolded state through preferentially binding to the unfolded sequence, thus avoiding G4 re-folding [66]. To evaluate CNBP ability to unfold SARS-CoV-2 G4s studied in this work, we performed CD spectra of previously folded G4 incubated with CNBP or with BSA as an unspecific protein with no G4 unfolding activity (Figure 3b). CNBP caused a distortion and reduction of the characteristic G4 peaks of the spectra of the PQSs +644, +3467, −13,963 and −23,877, but it did not significantly affect the spectrum of the PQS +28,903. BSA did not affect any of the CD spectra. These results indicate that CNBP is capable of binding in vitro to the PQSs +644, +3467, −13,963 and −23,877 and unfolding the preexistent G4s that they form. Instead, the PQS +28,903 did not interact with CNBP and the structure detected by CD was not significantly affected by the protein.

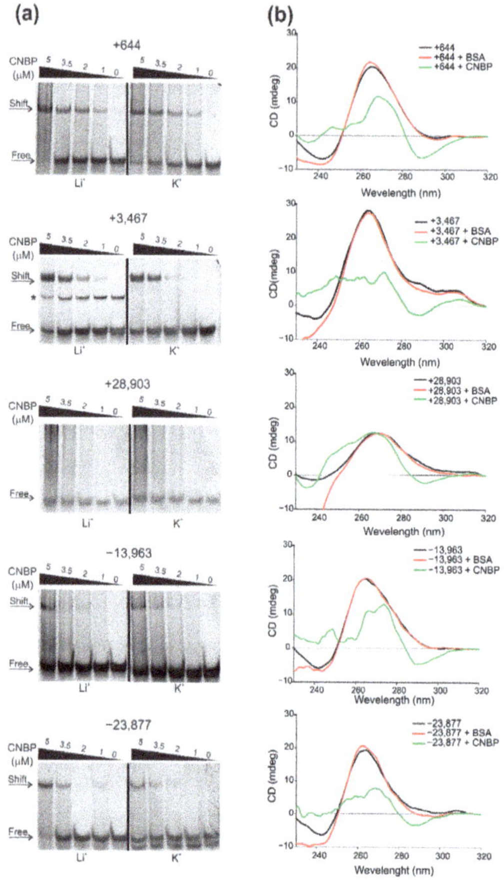

Figure 3. CNBP binds and unfolds the G4s formed by PQSs in the +gRNA and −gRNA of SARS-CoV-2. (**a**) Representative EMSAs performed using PQSs (named by the PQS position) as probes folded in the presence of Li^+ (left) or K^+ (right) and then incubated in the absence or presence of increasing concentrations of CNBP. Free and shifted probes are indicated by arrows at the left of the gels. The +3467 probe folded in the presence of Li^+ presents a minority band (marked with *) of lower mobility probably due to a self-assembled dimeric or multimeric complex. (**b**) CD spectra obtained for 8 μM oligonucleotides (named by the PQS position) folded as G4 in the presence of K^+ and incubated in the absence of protein or in the presence of CNBP (1:1 molar ratio) or BSA. For EMSAs and CD, K^+ and Li^+ concentrations used for G4 folding were 100 mM (+3467 and −23,877) or 200 mM (+644, +28,903 and −13,963).

Among several diverse functions assigned to CNBP, it was reported to induce the transcription of sustained pro-inflammatory cytokines by binding to specific short sequences in their promoters and activate its own transcription in a positive feedback mechanism of autoregulation in response to stimulation with lipopolysaccharide [67]. CNBP also induces IL-12β (Il12b) mRNA synthesis in response to diverse microbial pathogens that engage multiple pattern recognition receptors [68]. *Cnbp*-deficient mice fail to mount protective IL-12 and IFN-γ responses in vivo, resulting in a reduced Th1 cell immune response and an inability to control parasite replication [68]. Based on these data, CNBP has been suggested as a key transcriptional regulator required for activating and maintaining the immune response. These findings are consistent with CNBP-depleted cells being sensitized to virus-induced cell death [19], which suggests that CNBP may act as an antiviral regulator. Enhanced crosslinking and immunoprecipitation (eCLIP) in SARS-CoV-2-infected cells has shown CNBP binding along SARS-CoV-2 +gRNA with several strongly enriched binding sites [19]. However, it remains to be determined if CNBP role on SARS-CoV-2 replication is due to its action on viral RNA or to its action on cytokine and other cellular genes expression regulation. Here, we provide evidence that CNBP is capable of interacting and unfolding SARS-CoV-2 G4s in both the +gRNA and the −gRNA with putative regulative functions in SARS-CoV-2 gene expression and replication.

2.4. Integration of Results with Putative Functions of G4 and CNBP in SARS-CoV-2 Replicative Cycle

Replication cycle of the SARS-CoV-2 (Figure 4) initiates with the binding of the virus to the host cell by interaction of the S protein with its receptor, the angiotensin-converting enzyme 2 (ACE2) (Figure 4, step 1). Following receptor binding, the virus enters host cell by endocytosis and then to the cytosol via proteolytic cleavage of S protein, followed by fusion of the viral and cellular membranes (Figure 4, step 2). After the +gRNA enters the host cell, it is translated by cytosolic cellular ribosomes to synthesize the viral components of RTC. RTC is formed by some cellular proteins and up to 16 viral polypeptides, including the RNA dependent RNA polymerase (RdRp or nsp12), the RNA helicase (nsp13) and proteases derived from the proteolytic cleavage of the polyprotein encoded by the viral ORF1ab (Figure 4, step 3). Then, viral RNA replication (Figure 4, step 4) and transcription (Figure 4, step 5) take place attached to cytoplasmic membranes and involve coordinated processes of both continuous and discontinuous RNA synthesis complementary to the +gRNA that produces both −gRNAs and −sgRNAs. New copies of +gRNAs (replication) and +sgRNAs (transcription) are produced using the minority −gRNAs and −sgRNAs as intermediate templates. While +gRNA functions as mRNA for the synthesis of nsps encoded by the ORF1ab, +sgRNAs serve as mRNAs for the translation of structural and accessory genes encoded downstream of the replicase polyproteins. All +sgRNAs are 3′ co-terminal with the full-length +gRNA and thus form a set of nested RNAs. Structural proteins S, E, and M are translated from +sgRNAs (mRNAs) and inserted into the endoplasmic reticulum (ER) (Figure 4, step 6). These proteins move along the secretory pathway into the endoplasmic reticulum-Golgi intermediate compartment (ERGIC) where the viral +gRNAs that are encapsidated by the N protein bud into the membrane resulting in formation of the new mature virus particles (Figure 4, step 7). Following assembly, virions are transported to the cell surface in vesicles and released by exocytosis (Figure 4, step 8). This viral replication cycle consists of several steps involving different RNA molecules functioning as templates for translation and/or RNA dependent RNA synthesis. All of these steps may be sensitive to regulation by RNA secondary structures such as G4s, which may be modulated by viral and/or cellular G4 interacting proteins such as CNBP.

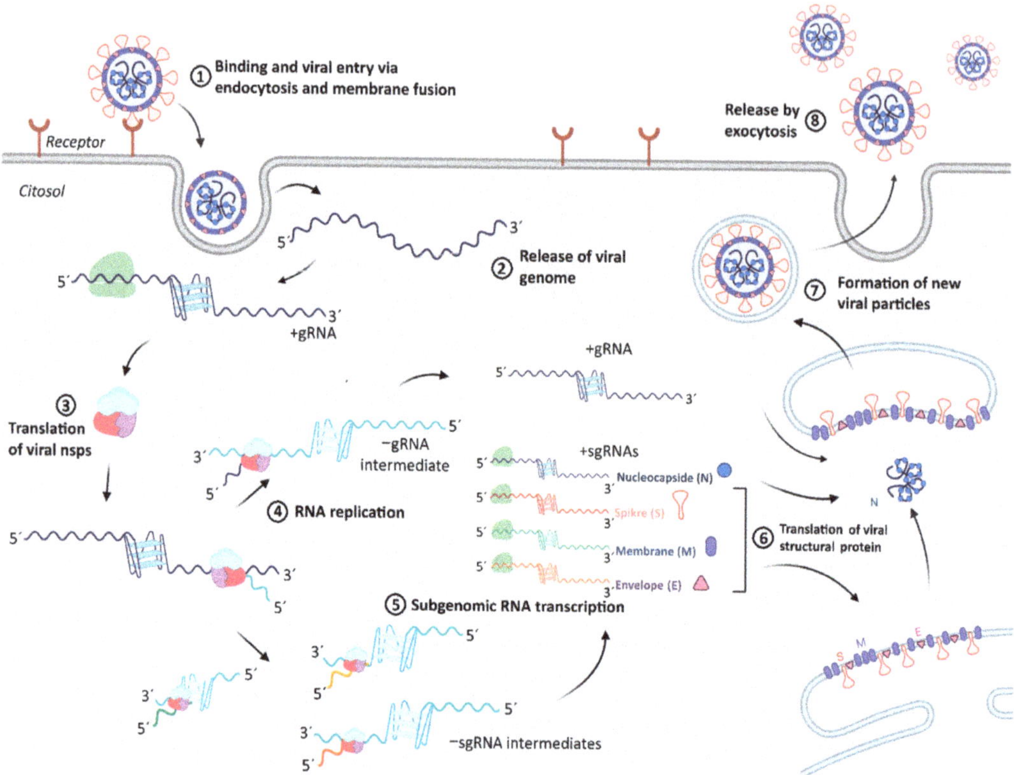

Figure 4. Possible role of G4s in the SARS-CoV-2 and coronaviruses replication cycle. Different steps of viral infection and replicative cycle are detailed: (1) binding to host cell receptor, (2) entry to host cell, (3) translation of viral nsps from ORF1ab, (4) viral RNA replication, (5) viral RNA transcription, (6) viral structural proteins translation, (7) encapsidation of viral +gRNA and formation of mature virus particles, (8) viral release. Viral G4s and the G4-interacting proteins, such as CNBP, may participate in the regulation of the efficiency of steps 3–7, which could be targets of drug candidates for antiviral therapies. Created with BioRender.com.

Our bioinformatic analysis revealed that several PQSs are found in both +gRNA and −gRNA, and some of them were selected based on the scores retrieved from several PQSs predictors, including some specially designed for the identification of RNA G4s. From the five selected PQSs, four of them, two from the +gRNA and two from the −gRNA, fold in vitro as stable G4s. Both selected PQSs from the +gRNA that fold as G4 in vitro overlap with the ORF1ab. The PQS +644 overlaps with the region coding the nsp1, which promotes cellular mRNA degradation and blocks host cell translation, resulting in innate immune response blockage. The PQS +3467 overlaps with the region coding the nsp3, a large transmembrane protein comprising several different domains whose precise functions have not been entirely clarified yet. Nsp3 contains the SARS Unique Domain (SUD) that deserves special attention since it is present only in SARS-type coronaviruses and has been associated with the increased pathogenicity of this viral family [69]. Interestingly, SUD has been shown to bind G4s. However, PQS +3467 does not overlap with the SUD coding region. G4s located in the ORF of mRNAs may reduce protein expression levels by acting as roadblocks to the ribosome [70]. Therefore, the ORF1ab translation by cellular ribosomes may be regulated by the +644 and +3467 G4s, especially if they are stabilized by G4 ligands with interesting potential as antiviral compounds. In fact, this translation blocking effect of viral G4s has been proposed for the PQS +28,903, which did not show G4 folding in our assay conditions but was demonstrated to fold as G4 and inhibit the translation of a reporter gene [36] and of the N protein [35] in living cells. Similar functions of RNA G4s were also reported in studies of some other viruses [43,71].

G4s not only act as roadblocks for ribosomes, but also disturb the progression of DNA polymerase [72], RNA polymerase [73], and reverse transcriptase [74], which show processive movement on template nucleotides and should unwind the G4s to continue their reactions. Similarly, RdRp activity could also be inhibited by G4s present in the template RNA. Therefore, the G4s formed in +gRNA (e.g., +644 and +3467 selected and characterized in this work) may act as regulator roadblocks for RdRp catalyzed synthesis of −gRNA. G4s formed in +gRNA could also interfere with RdRp catalyzed synthesis of −sgRNAs intermediates, probably for PQSs other than those selected in this work and overlapped with structural proteins coding region. Moreover, the G4s formed in the −gRNA (e.g., −23,877 and −13,963, both characterized in this work and overlapping the ORF1ab) may act as regulator roadblocks for RdRp catalyzed synthesis of new copies of +gRNA during replication or the synthesis of +sgRNAs during transcription (in the case of other PQSs not selected in this work and overlapped with structural proteins coding region). Although G4 blockage of RdRp has not been experimentally established, it was reported that a stable G4 located at the 3′ end of the hepatitis C virus negative-sense strand could inhibit the RNA synthesis by reducing the RdRp activity [75]. These observations position the PQSs found in the −gRNA as additional putative targets for exploring antiviral strategies targeted to increase (or decrease) the stabilities of viral G4s. However, G4s may not only act as negative elements in nucleic acids metabolism, since there are evidences that these structures may have positive effects in transcription and translation [76], by acting as specific anchoring sites for protein factors or by competing with alternative nucleic acid structures with inhibitory effects. In addition, G4s could be thought as specific anchoring elements for viral RNA encapsidation by structural proteins.

Virus–host cell interplay may involve viral proteins interacting with viral and host G4s, as well as cellular proteins interacting with viral G4s. Recent reports about the specific interaction of viral proteins with SARS-CoV-2 G4s support the relevance of these structures for viral replication. These viral proteins may target not only viral but also cellular G4s. For instance, SUD of nsp3 in SARS-CoV has been shown to bind G4 through a specific macrodomain [69], which is essential for the activity of the RTC of this virus [77]. SARS-CoV SUD binding to viral and/or cellular RNAs with G4s could affect their stability and translation, thus controlling the host cell's response to the viral infection [69]. Nsp3 of SARS-CoV-2 contains a similar SUD predicted to conserve critical amino acids for G4 binding [32,33,36,78], thus probably sharing with SARS-CoV the viral pathogenic mechanism dependent on nsp3-SUD interaction with G4 structures. Another viral protein, the nsp13 with RNA helicase activity, was also informed as able to bind and probably unfold viral RNA G4s, thus favoring viral translation, transcription, and replication processes [30]. Of notice, besides serving as potential targets for antiviral treatment against SARS-CoV-2, RNA G4s could also be used for the design of biosensors in the detection of viral particles through G4 interaction proteins of SARS-CoV-2 and other coronaviruses, with the potential to replace the antibody-based detection methods and to improve the diagnosis of SARS-CoV-2 and other coronaviruses [79].

With a focus on host proteins, a cellular RNA helicase (Asp-Glu-Ala-Asp (DEAD)-box polypeptide 5 or DDX5) was detected to interact with nsp13 of SARS-CoV, and viral replication was significantly inhibited by knocking down the expression of DDX5 [80]. This suggests that host DDX5 or other DEAD-box helicases could be hijacked by CoVs to enhance the transcription and proliferation of viral genome through G4 unfolding. Other cellular host RNA-binding proteins (SRSF7, HNRNPA1 and TRA2A) were proposed from a predictive analysis as promising candidates for binding and resolving G4s formed in SARS-CoV-2 RNA genome [33]. The recent report about CNBP as the main host cellular protein interacting with SARS-CoV-2 genome, the observation that CNBP expression is induced in response to host cells infection [19], and the fact that CNBP knock-out improves virus-induced cell death [62], highlights the pivotal role of CNBP as an important host protein for controlling viral infection. CNBP is involved in the transcriptional activation of pro-immflamatory cytokines required for activating and maintaining the immune response [67,68], probably acting as a SARS-CoV-2 antiviral regulator [19]. In addition, CNBP is a single-stranded nucleic acid binding protein with chaperone activity that may control gene expression (at transcriptional and translational levels) through G4 unfolding [65,66]. Considering these evidences, together with the variety of putative functions of G4s in SARS-CoV-2 replication cycle and the G4 unfolding capacity of CNBP showed here, we hypothesize that CNBP may act as a direct regulator of viral gene expression, transcription, and replication, adding relevant evidences for understanding the role of this protein in the control of SARS-CoV-2 infection. Although previous data has shown that CNBP binds to +gRNA within infected cells and there is still no evidence of interaction with −gRNA or −sgRNAs, our results show for the first time that CNBP is capable of binding and unfolding in vitro G4s present in the −gRNA. This opens the hypothesis that this protein may interact with both viral RNA strands, favoring the unwinding of G4s and controlling viral replication, transcription, and gene expression in those steps where G4 structures may be acting. To date, no pharmacological products with therapeutic potential have been designed for modifying CNBP activity. Future experimental work is needed for assessing the function of G4s present in +gRNA and −gRNA in viral host cell infection, as well as the action of cellular proteins such as CNBP in these processes, with perspectives of understanding useful molecular mechanisms for the design of new antiviral strategies.

3. Materials and Methods

3.1. Bioinformatics

The linear genomes of the six viruses used here were downloaded from the genome database of the National Center for Biotechnology Information (NCBI, https://www.ncbi.nlm.nih.gov/, accessed on 2 January 2021) [81]. Full names and accession numbers are indicated in Table 1.

PQSs prediction was performed by the web-based server QGRS mapper (https://bioinformatics.ramapo.edu/QGRS/index.php, accessed on 2 January 2021) [45]. PQSs found by QGRS mapper were analyzed also by PQS Finder [46] and G4RNA screener [47]. G4RNA screener is a web algorithm that identifies regions in RNA sequences prone to fold into G4 based on three scoring systems: cGcC (Consecutive G over consecutive C ratio) that addresses the issue of competition in between G4 and Watson-Crick structures [48], G4H (G4Hunter) which is similar to the cGcC but was built to analyze DNA sequences [49], and G4NN (G4 Neural Network) that is based on sequences of the G4RNA database converted into vectors of their trinucleotide content to train an artificial neural network [50]. The parameters used for the algorithms are detailed in Supplementary Tables S1 and S2.

The nucleotide and amino acid variations of SARS-CoV-2 genome and their associated data were searched by using the Nextstrain tool for analysis and visualization of 325,005 SARS-CoV-2 full genomic sequences sampled by different laboratories worldwide between 10 January 2020 and 11 January 2021, available in GISAID database (https://www.gisaid.org/, accessed on 11 January 2021) [52].

The predictions of RNA secondary structures were performed using the web based software RNAfold (http://rna.tbi.univie.ac.at/cgi-bin/RNAWebSuite/RNAfold.cgi, accessed on 12 February 2021) [82], and NUPACK (http://www.nupack.org/, accessed on 12 February 2021) [83], using the settings indicated in Supplementary Table S4.

3.2. Oligonucleotides

Synthetic single-stranded desalted oligoribonucleotides (Table 2) were purchased from Sigma-Aldrich, dissolved in bidistilled water and stored at −20 °C until use. Concentrations were determined by spectrophotometry using extinction coefficients provided by the manufacturer.

3.3. Circular Dichroism (CD) Spectroscopy

Intramolecular G4s were folded by dissolving 2 μM RNA oligonucleotides in 10 mM Tris-HCl pH 7.5 and different KCl or LiCl concentrations, as indicated in each figure, heating for 5 min at 95 °C and slowly cooling to 20 °C. For analysis of CNBP and BSA effect, prior to CD spectroscopy, proteins dissolved in CNBP buffer were added to the pre-folded G4 (8 μM) at equimolar concentrations and incubated at 37 °C for 30 min. CD spectra were recorded at 20 °C over a wavelength range of 220–320 nm with a Jasco-1500 spectropolarimeter (10 mm quartz cell, 100 nm/min scanning speed, 1 s response time, 1 nm data pitch, 1 nm band width, average of four scans). The spectral contribution of buffers, salts, and proteins were appropriately subtracted by using the software supplied with the spectropolarimeter. Three qualitative rules-of-thumb exist for CD spectral features associated with particular G4 topologies, namely parallel (with an important positive band around 264 nm and a relative shallow negative band at 245 nm), antiparallel (with a positive band around 295 nm, and a negative band around 265 nm), or hybrid (with two positive bands around 295 nm and 264 nm, and a negative band around 245 nm) [84,85]. The CD melting curves were recorded by ellipticity measurements between 20 °C and 95 °C at the wavelength corresponding to the maximum observed at the initial temperature (20 °C) for positive band around 264 nm, using the same parameters set for the spectra, except for 5 nm band width, a temperature increase speed of 1 °C/min, and a sampling interval of 0.1 °C. Data was analyzed in SigmaPlot 11.0 with a nonlinear least squares fitting procedure assuming a two-state transition of a monomer from a folded (G4) to an unfolded state with no change in heat capacity upon unfolding [86,87]. Melting curves were plotted as the fractional population of the G4-folded oligonucleotides (FG4 = [θ(T) − θU]/[θG4 − θu]) vs. temperature, where θf and θu are the ellipticities of the fully folded and unfolded states, respectively. The reported Tm represent the temperature at which both states are equally populated (FG4 = Fu = 0.5). Tm, ΔH, θf, and θu are those which provide the best fit of experimental melting data and the shown spectra and melting curves are representative of at least three independent experiments. $\Delta G_{37\,°C}$ were estimated following the procedures indicated elsewhere [86].

3.4. 1D ^1H Nuclear Magnetic Resonance (NMR)

NMR spectroscopy provides information about the type of base associations in the nucleic acid oligonucleotides by imino protons signals in the spectral region between 9 and 16 ppm. For instance, Watson-Crick typically presents signals clustered around 13–14 ppm, G4 around 11–12 ppm and i-motifs around 15–16 ppm [60]. In this work, NMR spectra were acquired at 20 °C on a 700MHz Bruker Avance III spectrometer (Bruker Biospin, MA, USA) equipped with a triple resonance inverse NMR probe (5 mm $^1H/D$-$^{13}C/^{15}N$ TXI). Samples contained 50 μM RNA oligonucleotides folded in 10 mM Tris at pH 7.5 supplemented with 100 or 200 mM KCl as described for CD spectroscopy. We used 5 mm Shigemi tubes that were previously treated with HCl 1M and washed extensively with water and ethanol to remove RNAses. 1D ^1H NMR spectra were registered using a pulse sequence with excitation sculpting (zgesgp) for water suppression [88]. We used 8K points, 4096 scans, a recycling delay of 1.4 s and a sweep width of 22 ppm. Experimental time for each NMR spectrum was 1 h 56 min. We repeated the spectra at different time points to discard degradation of the RNA oligonucleotides. Processing was done using an exponential window function multiplication with a line broadening of 10 Hz and baseline correction. We used Topspin 3.5 software (Bruker, Biospin, MA, USA) for acquisitions, processing, and analysis of the NMR spectra.

3.5. Thermal Difference Spectroscopy (TDS)

Two μM RNA oligonucleotides folded in 10 mM Tris at pH 7.5 supplemented with 100 or 200 mM KCl as described for CD spectroscopy were scanned to measure absorbance over the wavelength range of 220−320 nm using a scan speed of 100 nm/min and a data interval of 1 nm and a 10 mm quartz cell. Spectra were recorded at 20 °C and then at

70 °C using a Jasco V-630BIO spectrophotometer with peltier temperature control. The absorbance spectra obtained at these two temperatures were subtracted (A 70 °C–A 20 °C) and plotted on a graph to obtain TDS according to [61]. Typical G4 signature presents two positive peaks around 243 and 273 nm and a negative peak at 295 nm.

3.6. ThT Fluorescence Assays

ThT (3,6-Dimethyl-2-(4-dimethylaminophenyl) benzothiazolium cation, Sigma-Aldrich T3516) fluorescence assays were performed as previously described [89]. Briefly, 100 µL of 2 µM RNA oligonucleotides folded in 10 mM Tris at pH 7.5 supplemented with 100 or 200 mM KCl as described for CD spectroscopy were mixed with 100 µL of 1 µM ThT and loaded into 96-well black microplates (Greiner, NC, USA). Fluorescence emission measurements were performed using a microplate reader (Synergy 2 Multi-Mode Microplate Reader, BioTek, VT, USA) with excitation filter of 485 ± 20 nm and an emission filter of 528 ± 20 nm. Each sample was tested by triplicate and fluorescence values were relativized to ThT fluorescence in the absence of oligonucleotides (F_0). A threshold of 10-fold increase was used for considering G4 formation. NRAS RNA oligonucleotide representing a PQS from human NRAS 5′ UTR [90] was used as positive control for the assay (Supplementary Figure S4).

3.7. CNBP Expression and Purification

The pET-32a-TEV-CNBP plasmid [66] was used for recombinant expression and purification of tag-free human CNBP following guidelines detailed elsewhere [91]. CNBP was obtained in CNBP buffer (50 mM Tris–HCl pH 7.5; 300 mM NaCl; 1 mM DTT, 5 mM Imidazole and 0.1 mM $ZnCl_2$), which was used in several in vitro assays as a control.

3.8. Electrophoretic Mobility Shift Assay (EMSA)

EMSAs were performed as described previously [92] with some modifications for non-radioactive detection. Binding reactions were performed in 20 mM HEPES pH 8.0, 10 mM $MgCl_2$, 1 mM EDTA, 0.5 µg/µL Heparin, 1 mM DTT, 1 µg/µL BSA, and 10% glycerol. Probes were added to a final concentration of 0.75 µM. 100 mM KCl or LiCl were added, depending on the folding condition of the probe, as indicated in Figure 3a. Final reaction volumes were 20 µL. Binding reactions were incubated for 30 min at 37 °C and then loaded in 14% polyacrylamide gels containing 5% glycerol in TBE 0.5X. After electrophoresis, gels were exposed for 10 min to SYBR Gold stain [93] to detect bands, the fluorescence of which was subsequently registered in a Typhoon FLA 7000 Scanner (GE Healthcare, NJ, USA) using ImageQuant 5.2 software. RNA oligonucleotides used as probes were previously folded by thermal denaturation and slow renaturation in a buffer (10 mM Tris-HCl pH 8.0, 1 mM EDTA) in the presence of LiCl or KCl at concentrations of 100 or 200 mM, as indicated for each case in the figure caption (Figure 3). The CNBP buffer was used for the CNBP dilutions and for the reactions with no added CNBP.

Supplementary Materials: The following are available online at https://www.mdpi.com/1422-0067/22/5/2614/s1, Figure S1: Multiple sequence alignments of the position-conserved selected SARS-CoV-2 PQSs identified in this study with their homologous PQSs in other members of the *Coronaviridae* family, Figure S2: CD melting curves obtained for selected PQSs in the positive and negative-sense RNA genomes of SARS-CoV-2, Figure S3: TDS obtained for selected PQSs in the positive and negative-sense RNA genomes of SARS-CoV-2, Figure S4: ThT fluorescence assay for selected PQSs in the positive and negative-sense RNA genomes of SARS-CoV-2., Table S1: PQS predicted in +gRNA of the SARS-CoV-2 and other members of the *Coronaviridae* family, Table S2: PQS predicted in −gRNA of the SARS-CoV-2 and other members of the *Coronaviridae* family, Table S3: Analysis of variations within selected PQSs in +gRNA and −gRNA of the SARS-CoV-2, Table S4: Prediction of secondary structures and minimum free energy (MFE) of the selected PQSs in +gRNA and −gRNA of the SARS-CoV-2.

Author Contributions: P.A. and G.B. performed the conceptualization and design of the work. G.B. carried on most bioinformatics analyses and in vitro experiments. E.J.P. participated in CD spectroscopy, ThT fluorescence assays, and analysis of mutations in SARS-CoV-2 selected PQSs. A.B. performed and analyzed NMR spectroscopies. P.A. was responsible for supervision, funding acquisition, project administration, and obtaining of resources. P.A. and G.B. conducted the original draft writing and visualization, while P.A., G.B., E.J.P. and A.B. performed the writing review and editing. All authors have read and agreed to the published version of the manuscript.

Funding: This research was funded by Agencia Nacional de Promoción Científica y Tecnológica and Fondo para la Investigación Científica y Tecnológica, grant number PICT 2017-0976 to P.A.; and Universidad Nacional de Rosario, grant number BIO573 to P.A. The APC was not funded.

Institutional Review Board Statement: Not applicable.

Informed Consent Statement: Not applicable.

Data Availability Statement: The data presented in this study are available in the article and supplementary material.

Acknowledgments: G.B. is Fellow of ANPCyT-FONCyT and E.J.P. is Fellow of CONICET. P.A. and A.B. are Career Researchers of CONICET, and P.A. is Professor of UNR. We are thankful to Franco Biglione for help with CD melting temperatures calculations and to Nora B. Calcaterra for mentorship during work execution and writing.

Conflicts of Interest: The authors declare no conflict of interest. The funders had no role in the design of the study; in the collection, analyses, or interpretation of data; in the writing of the manuscript, or in the decision to publish the results.

References

1. Zhou, P.; Yang, X.L.; Wang, X.G.; Hu, B.; Zhang, L.; Zhang, W.; Si, H.R.; Zhu, Y.; Li, B.; Huang, C.L.; et al. A pneumonia outbreak associated with a new coronavirus of probable bat origin. *Nature* **2020**, *579*, 270–273. [CrossRef] [PubMed]
2. Gorbalenya, A.E.; Baker, S.C.; Baric, R.S.; de Groot, R.J.; Drosten, C.; Gulyaeva, A.A.; Haagmans, B.L.; Lauber, C.; Leontovich, A.M.; Neuman, B.W.; et al. The species Severe acute respiratory syndrome-related coronavirus: Classifying 2019-nCoV and naming it SARS-CoV-2. *Nat. Microbiol.* **2020**, *5*, 536–544. [CrossRef]
3. Malik, Y.A. Properties of coronavirus and SARS-CoV-2. *Malays. J. Pathol.* **2020**, *42*, 3–11. [PubMed]
4. Rota, P.A.; Oberste, M.S.; Monroe, S.S.; Nix, W.A.; Campagnoli, R.; Icenogle, J.P.; Peñaranda, S.; Bankamp, B.; Maher, K.; Chen, M.; et al. Characterization of a novel coronavirus associated with severe acute respiratory syndrome. *Science* **2003**, *300*, 1394–1399. [CrossRef]
5. Ramadan, N.; Shaib, H. Middle east respiratory syndrome coronavirus (MERS-COV): A review. *GERMS* **2019**, *9*, 35–42. [CrossRef]
6. Lu, R.; Zhao, X.; Li, J.; Niu, P.; Yang, B.; Wu, H.; Wang, W.; Song, H.; Huang, B.; Zhu, N.; et al. Genomic characterisation and epidemiology of 2019 novel coronavirus: Implications for virus origins and receptor binding. *Lancet* **2020**, *395*, 565–574. [CrossRef]
7. Chu, H.; Chan, J.F.-W.; Yuen, T.T.-T.; Shuai, H.; Yuan, S.; Wang, Y.; Hu, B.; Yip, C.C.-Y.; Tsang, J.O.-L.; Huang, X.; et al. Comparative tropism, replication kinetics, and cell damage profiling of SARS-CoV-2 and SARS-CoV with implications for clinical manifestations, transmissibility, and laboratory studies of COVID-19: An observational study. *Lancet Microbe* **2020**, *1*, e14–e23. [CrossRef]
8. Mousavizadeh, L.; Ghasemi, S. Genotype and phenotype of COVID-19: Their roles in pathogenesis. *J. Microbiol. Immunol. Infect.* **2020**, in press. [CrossRef] [PubMed]
9. Fehr, A.R.; Perlman, S. Coronaviruses: An overview of their replication and pathogenesis. In *Coronaviruses: Methods and Protocols*; Springer: New York, NY, USA, 2015; Volume 1282, pp. 1–23. ISBN 9781493924387. [CrossRef]
10. Sola, I.; Almazán, F.; Zúñiga, S.; Enjuanes, L. Continuous and discontinuous RNA synthesis in coronaviruses. *Annu. Rev. Virol.* **2015**, *2*, 265–288. [CrossRef] [PubMed]
11. Chan, Y.K.; Gack, M.U. Viral evasion of intracellular DNA and RNA sensing. *Nat. Rev. Microbiol.* **2016**, *14*, 360–373. [CrossRef]
12. Blanco-Melo, D.; Nilsson-Payant, B.E.; Liu, W.C.; Uhl, S.; Hoagland, D.; Møller, R.; Jordan, T.X.; Oishi, K.; Panis, M.; Sachs, D.; et al. Imbalanced host response to SARS-CoV-2 drives development of COVID-19. *Cell* **2020**, *181*, 1036–1045.e9. [CrossRef]
13. Emanuel, W.; Kirstin, M.; Vedran, F.; Asija, D.; Theresa, G.L.; Roberto, A.; Filippos, K.; David, K.; Salah, A.; Christopher, B.; et al. Bulk and single-cell gene expression profiling of SARS-CoV-2 infected human cell lines identifies molecular targets for therapeutic intervention. *bioRxiv* **2020**. [CrossRef]
14. Bouhaddou, M.; Memon, D.; Meyer, B.; White, K.M.; Rezelj, V.V.; Correa Marrero, M.; Polacco, B.J.; Melnyk, J.E.; Ulferts, S.; Kaake, R.M.; et al. The global phosphorylation landscape of SARS-CoV-2 infection. *Cell* **2020**, *182*, 685–712.e19. [CrossRef]
15. Bojkova, D.; Klann, K.; Koch, B.; Widera, M.; Krause, D.; Ciesek, S.; Cinatl, J.; Münch, C. Proteomics of SARS-CoV-2-infected host cells reveals therapy targets. *Nature* **2020**, *583*, 469–472. [CrossRef] [PubMed]
16. Klann, K.; Bojkova, D.; Tascher, G.; Ciesek, S.; Münch, C.; Cinatl, J. Growth factor receptor signaling inhibition prevents SARS-CoV-2 replication. *Mol. Cell* **2020**, *80*, 164–174.e4. [CrossRef] [PubMed]

17. Gordon, D.E.; Jang, G.M.; Bouhaddou, M.; Xu, J.; Obernier, K.; White, K.M.; O'Meara, M.J.; Rezelj, V.V.; Guo, J.Z.; Swaney, D.L.; et al. A SARS-CoV-2 protein interaction map reveals targets for drug repurposing. *Nature* **2020**, *583*, 459–468. [CrossRef] [PubMed]
18. Gordon, D.E.; Hiatt, J.; Bouhaddou, M.; Rezelj, V.V.; Ulferts, S.; Braberg, H.; Jureka, A.S.; Obernier, K.; Guo, J.Z.; Batra, J.; et al. Comparative host-coronavirus protein interaction networks reveal pan-viral disease mechanisms. *Science* **2020**, *370*. [CrossRef]
19. Schmidt, N.; Lareau, C.A.; Keshishian, H.; Ganskih, S.; Schneider, C.; Hennig, T.; Melanson, R.; Werner, S.; Wei, Y.; Zimmer, M.; et al. The SARS-CoV-2 RNA–protein interactome in infected human cells. *Nat. Microbiol.* **2020**. [CrossRef]
20. Smyth, R.P.; Negroni, M.; Lever, A.M.; Mak, J.; Kenyon, J.C. RNA structure-a neglected puppet master for the evolution of virus and host immunity. *Front. Immunol.* **2018**, *9*, 2097. [CrossRef] [PubMed]
21. Varshney, D.; Spiegel, J.; Zyner, K.; Tannahill, D.; Balasubramanian, S. The regulation and functions of DNA and RNA G-quadruplexes. *Nat. Rev. Mol. Cell Biol.* **2020**, *21*, 459–474. [CrossRef]
22. Marsico, G.; Chambers, V.S.; Sahakyan, A.B.; McCauley, P.; Boutell, J.M.; Di Antonio, M.; Balasubramanian, S. Whole genome experimental maps of DNA G-quadruplexes in multiple species. *Nucleic Acids Res.* **2019**, *47*, 3862–3874. [CrossRef]
23. Bartas, M.; Cutová, M.; Brázda, V.; Kaura, P.; Št'Astný, J.; Kolomazník, J.; Coufal, J.; Goswami, P.; Červeň, J.; Pečinka, P. The presence and localization of G-quadruplex forming sequences in the domain of bacteria. *Molecules* **2019**, *24*, 1711. [CrossRef] [PubMed]
24. Brázda, V.; Luo, Y.; Bartas, M.; Kaura, P.; Porubiaková, O.; Šťastný, J.; Pečinka, P.; Verga, D.; Da Cunha, V.; Takahashi, T.S.; et al. G-Quadruplexes in the archaea domain. *Biomolecules* **2020**, *10*, 1349. [CrossRef] [PubMed]
25. Seifert, S. Above and beyond watson and crick: Guanine quadruplex structures and microbes. *Annu. Rev. Microbiol.* **2018**, *72*, 49–69. [CrossRef] [PubMed]
26. Saranathan, N.; Vivekanandan, P. G-Quadruplexes: More than just a kink in microbial genomes. *Trends Microbiol.* **2018**, *27*, 148–163. [CrossRef] [PubMed]
27. Lavezzo, E.; Berselli, M.; Frasson, I.; Perrone, R.; Palù, G.; Brazzale, A.R.; Richter, S.N.; Toppo, S. G-Quadruplex forming sequences in the genome of all known human viruses: A comprehensive guide. *PLoS Comput. Biol.* **2018**, *14*, e1006675. [CrossRef]
28. Ruggiero, E.; Richter, S.N. Survey and summary G-quadruplexes and G-quadruplex ligands: Targets and tools in antiviral therapy. *Nucleic Acids Res.* **2018**, *46*, 3270–3283. [CrossRef]
29. Ruggiero, E.; Richter, S.N. Viral G-quadruplexes: New frontiers in virus pathogenesis and antiviral therapy. In *Annual Reports in Medicinal Chemistry*; Academic Press Inc.: San Diego, CA, USA, 2020; Volume 54, pp. 101–131. [CrossRef]
30. Ji, D.; Juhas, M.; Tsang, C.M.; Kwok, C.K.; Li, Y.; Zhang, Y. Discovery of G-quadruplex-forming sequences in SARS-CoV-2. *Brief. Bioinform.* **2020**, bbaa114. [CrossRef]
31. Panera, N.; Tozzi, A.E.; Alisi, A. The G-quadruplex/helicase world as a potential antiviral approach against COVID-19. *Drugs* **2020**, *80*, 941–946. [CrossRef]
32. Zhang, R.; Xiao, K.; Gu, Y.; Liu, H.; Sun, X. Whole genome identification of potential G-quadruplexes and analysis of the G-quadruplex binding domain for SARS-CoV-2. *Front. Genet.* **2020**, *11*, 587829. [CrossRef]
33. Bartas, M.; Brázda, V.; Bohálová, N.; Cantara, A.; Volná, A.; Stachurová, T.; Malachová, K.; Jagelská, E.B.; Porubiaková, O.; Červeň, J.; et al. In-Depth bioinformatic analyses of nidovirales including human SARS-CoV-2, SARS-CoV, MERS-CoV viruses suggest important roles of non-canonical nucleic acid structures in their lifecycles. *Front. Microbiol.* **2020**, *11*, 1583. [CrossRef] [PubMed]
34. Belmonte-Reche, E.; Serrano-Chacón, I.; Gonzalez, C.; Gallo, J.; Bañobre-López, M. Exploring G and C-quadruplex structures as potential targets against the severe acute respiratory syndrome coronavirus 2. *bioRxiv* **2020**. [CrossRef]
35. Zhao, C.; Qin, G.; Niu, J.; Wang, Z.; Wang, C.; Ren, J.; Qu, X. Targeting RNA G-quadruplex in SARS-CoV-2: A promising therapeutic target for COVID-19? *Angew. Chem. Int. Ed.* **2021**, *60*, 432–438. [CrossRef] [PubMed]
36. Cui, H.; Zhang, L. G-Quadruplexes are present in human coronaviruses including SARS-CoV-2. *Front. Microbiol.* **2020**, *11*, 567317. [CrossRef] [PubMed]
37. Zaccaria, F.; Fonseca Guerra, C. RNA versus DNA G-quadruplex: The origin of increased stability. *Chem. Eur. J.* **2018**, *24*, 16315–16322. [CrossRef]
38. Perrone, R.; Nadai, M.; Poe, J.A.; Frasson, I.; Palumbo, M.; Palù, G.; Smithgall, T.E.; Richter, S.N. Formation of a unique cluster of G-quadruplex structures in the HIV-1 nef coding region: Implications for antiviral activity. *PLoS ONE* **2013**, *8*, e73121. [CrossRef] [PubMed]
39. Murat, P.; Zhong, J.; Lekieffre, L.; Cowieson, N.P.; Clancy, J.L.; Preiss, T.; Balasubramanian, S.; Khanna, R.; Tellam, J. G-Quadruplexes regulate Epstein-Barr virus-encoded nuclear antigen 1 mRNA translation. *Nat. Chem. Biol.* **2014**, *10*, 358–364. [CrossRef] [PubMed]
40. Fleming, A.M.; Ding, Y.; Alenko, A.; Burrows, C.J. Zika virus genomic RNA possesses conserved G-quadruplexes characteristic of the flaviviridae family. *ACS Infect. Dis.* **2016**, *2*, 674–681. [CrossRef]
41. Wang, S.R.; Zhang, Q.Y.; Wang, J.Q.; Ge, X.Y.; Song, Y.Y.; Wang, Y.F.; Li, X.D.; Fu, B.S.; Xu, G.H.; Shu, B.; et al. Chemical targeting of a G-quadruplex RNA in the Ebola Virus L. Gene. *Cell Chem. Biol.* **2016**, *23*, 1113–1122. [CrossRef]
42. Zahin, M.; Dean, W.L.; Ghim, S.; Joh, J.; Gray, R.D.; Khanal, S.; Bossart, G.D.; Mignucci-Giannoni, A.A.; Rouchka, E.C.; Jenson, A.B.; et al. Identification of G-quadruplex forming sequences in three manatee papillomaviruses. *PLoS ONE* **2018**, *13*, e0195625. [CrossRef] [PubMed]

43. Majee, P.; Kumar Mishra, S.; Pandya, N.; Shankar, U.; Pasadi, S.; Muniyappa, K.; Nayak, D.; Kumar, A. Identification and characterization of two conserved G-quadruplex forming motifs in the Nipah virus genome and their interaction with G-quadruplex specific ligands. *Sci. Rep.* **2020**, *10*, 1477. [CrossRef]
44. Zhang, Y.; Liu, S.; Jiang, H.; Deng, H.; Dong, C.; Shen, W.; Chen, H.; Gao, C.; Xiao, S.; Liu, Z.F.; et al. G2-Quadruplex in the 3'UTR of IE180 regulates pseudorabies virus replication by enhancing gene expression. *RNA Biol.* **2020**, *17*, 816–827. [CrossRef]
45. Kikin, O.; D'Antonio, L.; Bagga, P.S. QGRS mapper: A web-based server for predicting G-quadruplexes in nucleotide sequences. *Nucleic Acids Res.* **2006**, *34*, W676–W682. [CrossRef] [PubMed]
46. Labudova, D.; Hon, J.; Lexa, M.; Lexa, M. Pqsfinder web: G-quadruplex prediction using optimized pqsfinder algorithm. *Bioinformatics* **2020**, *36*, 2584–2586. [CrossRef] [PubMed]
47. Garant, J.M.; Perreault, J.P.; Scott, M.S. G4RNA screener web server: User focused interface for RNA G-quadruplex prediction. *Biochimie* **2018**, *151*, 115–118. [CrossRef] [PubMed]
48. Beaudoin, J.D.; Jodoin, R.; Perreault, J.P. New scoring system to identify RNA G-quadruplex folding. *Nucleic Acids Res.* **2014**, *42*, 1209–1223. [CrossRef] [PubMed]
49. Bedrat, A.; Lacroix, L.; Mergny, J.L. Re-Evaluation of G-quadruplex propensity with G4Hunter. *Nucleic Acids Res.* **2016**, *44*, 1746–1759. [CrossRef]
50. Garant, J.M.; Perreault, J.P.; Scott, M.S. Motif independent identification of potential RNA G-quadruplexes by G4RNA screener. *Bioinformatics* **2017**, *33*, 3532–3537. [CrossRef]
51. Rangan, R.; Zheludev, I.N.; Hagey, R.J.; Pham, E.A.; Wayment-Steele, H.K.; Glenn, J.S.; Das, R. RNA genome conservation and secondary structure in SARS-CoV-2 and SARS-related viruses: A first look. *RNA* **2020**, *26*, 937–959. [CrossRef]
52. Elbe, S.; Buckland-Merrett, G. Data, disease and diplomacy: GISAID's innovative contribution to global health. *Glob. Chall.* **2017**, *1*, 33–46. [CrossRef]
53. Nakken, S.; Rognes, T.; Hovig, E. The disruptive positions in human G-quadruplex motifs are less polymorphic and more conserved than their neutral counterparts. *Nucleic Acids Res.* **2009**, *37*, 5749–5756. [CrossRef] [PubMed]
54. Baral, A.; Kumar, P.; Halder, R.; Mani, P.; Yadav, V.K.; Singh, A.; Das, S.K.; Chowdhury, S. Quadruplex-single nucleotide polymorphisms (Quad-SNP) influence gene expression difference among individuals. *Nucleic Acids Res.* **2012**, *40*, 3800–3811. [CrossRef]
55. Inagaki, H.; Ota, S.; Nishizawa, H.; Miyamura, H.; Nakahira, K.; Suzuki, M.; Nishiyama, S.; Kato, T.; Yanagihara, I.; Kurahashi, H. Obstetric complication-associated ANXA5 promoter polymorphisms may affect gene expression via DNA secondary structures. *J. Hum. Genet.* **2019**, *64*, 459–466. [CrossRef] [PubMed]
56. Beaudoin, J.D.; Perreault, J.P. 5′-UTR G-quadruplex structures acting as translational repressors. *Nucleic Acids Res.* **2010**, *38*, 7022–7036. [CrossRef] [PubMed]
57. Zeraati, M.; Moye, A.L.; Wong, J.W.H.; Perera, D.; Cowley, M.J.; Christ, D.U.; Bryan, T.M.; Dinger, M.E. Cancer-Associated noncoding mutations affect RNA G-quadruplex-mediated regulation of gene expression. *Sci. Rep.* **2017**, *7*, 1–11. [CrossRef]
58. Lee, D.S.M.; Ghanem, L.R.; Barash, Y. Integrative analysis reveals RNA G-quadruplexes in UTRs are selectively constrained and enriched for functional associations. *Nat. Commun.* **2020**, *11*. [CrossRef]
59. Bhattacharyya, D.; Mirihana Arachchilage, G.; Basu, S. Metal cations in G-quadruplex folding and stability. *Front. Chem.* **2016**, *4*, 1–14. [CrossRef]
60. Phan, A.T.; Mergny, J.L. Human telomeric DNA: G-quadruplex, i-motif and Watson-Crick double helix. *Nucleic Acids Res.* **2002**, *30*, 4618–4625. [CrossRef]
61. Mergny, J.L.; Li, J.; Lacroix, L.; Amrane, S.; Chaires, J.B. Thermal difference spectra: A specific signature for nucleic acid structures. *Nucleic Acids Res.* **2005**, *33*, 1–6. [CrossRef] [PubMed]
62. Wei, J.; Alfajaro, M.M.; DeWeirdt, P.C.; Hanna, R.E.; Lu-Culligan, W.J.; Cai, W.L.; Strine, M.S.; Zhang, S.M.; Graziano, V.R.; Schmitz, C.O.; et al. Genome-Wide CRISPR screens reveal host factors critical for SARS-CoV-2 infection. *Cell* **2020**, *184*, 76–91.e13. [CrossRef]
63. Calcaterra, N.B.; Armas, P.; Weiner, A.M.J.; Borgognone, M. CNBP: A multifunctional nucleic acid chaperone involved in cell death and proliferation control. *IUBMB Life* **2010**, *62*, 707–714. [CrossRef]
64. Armas, P.; Margarit, E.; Mouguelar, V.S.; Allende, M.L.; Calcaterra, N.B. Beyond the binding site: In vivo identification of tbx2, smarca5 and wnt5b as molecular targets of CNBP during embryonic development. *PLoS ONE* **2013**, *8*, e63234. [CrossRef]
65. Benhalevy, D.; Gupta, S.K.; Danan, C.H.; Ghosal, S.; Sun, H.-W.; Kazemier, H.G.; Paeschke, K.; Hafner, M.; Juranek, S.A. The human CCHC-type zinc finger nucleic acid-binding protein binds g-rich elements in target mRNA coding sequences and promotes translation. *Cell Rep.* **2017**, *18*, 2979–2990. [CrossRef] [PubMed]
66. David, A.P.; Pipier, A.; Pascutti, F.; Binolfi, A.; Weiner, A.M.J.; Challier, E.; Heckel, S.; Calsou, P.; Gomez, D.; Calcaterra, N.B.; et al. CNBP controls transcription by unfolding DNA G-quadruplex structures. *Nucleic Acids Res.* **2019**, *47*. [CrossRef] [PubMed]
67. Lee, E.; Lee, T.A.; Kim, J.H.; Park, A.; Ra, E.A.; Kang, S.; Choi, H.; Choi, J.L.; Huh, H.D.; Lee, J.E.; et al. CNBP acts as a key transcriptional regulator of sustained expression of interleukin-6. *Nucleic Acids Res.* **2017**, *45*, 3280–3296. [CrossRef] [PubMed]
68. Chen, Y.; Sharma, S.; Assis, P.A.; Jiang, Z.; Elling, R.; Olive, A.J.; Hang, S.; Bernier, J.; Huh, J.R.; Sassetti, C.M.; et al. CNBP controls IL-12 gene transcription and Th1 immunity. *J. Exp. Med.* **2018**, *215*, 3136–3150. [CrossRef]

69. Tan, J.; Vonrhein, C.; Smart, O.S.; Bricogne, G.; Bollati, M.; Kusov, Y.; Hansen, G.; Mesters, J.R.; Schmidt, C.L.; Hilgenfeld, R. The SARS-Unique Domain (SUD) of SARS coronavirus contains two macrodomains that bind G-quadruplexes. *PLoS Pathog.* **2009**, *5*, e1000428. [CrossRef]
70. Endoh, T.; Kawasaki, Y.; Sugimoto, N. Suppression of gene expression by G-quadruplexes in open reading frames depends on G-quadruplex stability. *Angew. Chem. Int. Ed.* **2013**, *52*, 5522–5526. [CrossRef]
71. Wang, S.R.; Min, Y.Q.; Wang, J.Q.; Liu, C.X.; Fu, B.S.; Wu, F.; Wu, L.Y.; Qiao, Z.X.; Song, Y.Y.; Xu, G.H.; et al. A highly conserved G-rich consensus sequence in hepatitis C virus core gene represents a new anti–hepatitis C target. *Sci. Adv.* **2016**, *2*, e1501535. [CrossRef] [PubMed]
72. Pontier, D.B.; Kruisselbrink, E.; Guryev, V.; Tijsterman, M. Isolation of deletion alleles by G4 DNA-induced mutagenesis. *Nat. Methods* **2009**, *6*, 655–657. [CrossRef]
73. Tateishi-Karimata, H.; Isono, N.; Sugimoto, N. New insights into transcription fidelity: Thermal stability of non-canonical structures in template DNA regulates transcriptional arrest, pause, and slippage. *PLoS ONE* **2014**, *9*, e90580. [CrossRef] [PubMed]
74. Hagihara, M.; Yoneda, K.; Yabuuchi, H.; Okuno, Y.; Nakatani, K. A reverse transcriptase stop assay revealed diverse quadruplex formations in UTRs in mRNA. *Bioorg. Med. Chem. Lett.* **2010**, *20*, 2350–2353. [CrossRef]
75. Jaubert, C.; Bedrat, A.; Bartolucci, L.; Di Primo, C.; Ventura, M.; Mergny, J.L.; Amrane, S.; Andreola, M.L. RNA synthesis is modulated by G-quadruplex formation in Hepatitis C virus negative RNA strand. *Sci. Rep.* **2018**, *8*, 8120. [CrossRef]
76. Agarwala, P.; Pandey, S.; Mapa, K.; Maiti, S. The G-quadruplex augments translation in the 5′ untranslated region of transforming growth factor β2. *Biochemistry* **2013**, *52*, 1528–1538. [CrossRef]
77. Kusov, Y.; Tan, J.; Alvarez, E.; Enjuanes, L.; Hilgenfeld, R. A G-quadruplex-binding macrodomain within the "SARS-unique domain" is essential for the activity of the SARS-coronavirus replication-transcription complex. *Virology* **2015**, *484*, 313–322. [CrossRef]
78. Hognon, C.; Miclot, T.; García-Iriepa, C.; Francés-Monerris, A.; Grandemange, S.; Terenzi, A.; Marazzi, M.; Barone, G.; Monari, A. Role of RNA guanine quadruplexes in favoring the dimerization of SARS unique domain in coronaviruses. *J. Phys. Chem. Lett.* **2020**, *11*, 5661–5667. [CrossRef] [PubMed]
79. Xi, H.; Juhas, M.; Zhang, Y. G-quadruplex based biosensor: A potential tool for SARS-CoV-2 detection. *Biosens. Bioelectron.* **2020**, *167*, 112494. [CrossRef]
80. Chen, J.Y.; Chen, W.N.; Poon, K.M.V.; Zheng, B.J.; Lin, X.; Wang, Y.X.; Wen, Y.M. Interaction between SARS-CoV helicase and a multifunctional cellular protein (Ddx5) revealed by yeast and mammalian cell two-hybrid systems. *Arch. Virol.* **2009**, *154*, 507–512. [CrossRef] [PubMed]
81. Sayers, E.W.; Beck, J.; Bolton, E.E.; Bourexis, D.; Brister, J.R.; Canese, K.; Comeau, D.C.; Funk, K.; Kim, S.; Klimke, W.; et al. Database resources of the national center for biotechnology information. *Nucleic Acids Res.* **2021**, *49*, D10–D17. [CrossRef]
82. Lorenz, R.; Bernhart, S.H.; Höner zu Siederdissen, C.; Tafer, H.; Flamm, C.; Stadler, P.F.; Hofacker, I.L. ViennaRNA Package 2.0. *Algorithms Mol. Biol.* **2011**, *6*, 26. [CrossRef] [PubMed]
83. Zadeh, J.N.; Steenberg, C.D.; Bois, J.S.; Wolfe, B.R.; Pierce, M.B.; Khan, A.R.; Dirks, R.M.; Pierce, N.A. NUPACK: Analysis and design of nucleic acid systems. *J. Comput. Chem.* **2011**, *32*, 170–173. [CrossRef] [PubMed]
84. Del Villar-Guerra, R.; Gray, R.D.; Chaires, J.B. Characterization of quadruplex DNA structure by circular dichroism. *Curr. Protoc. Nucleic Acid Chem.* **2017**, *68*, 17.8.1–17.8.16. [CrossRef] [PubMed]
85. Del Villar-Guerra, R.; Trent, J.O.; Chaires, J.B. G-Quadruplex secondary structure obtained from circular dichroism spectroscopy. *Angew. Chem. Int. Ed.* **2018**, *57*, 7171–7175. [CrossRef]
86. Petraccone, L.; Erra, E.; Esposito, V.; Randazzo, A.; Galeone, A.; Barone, G.; Giancola, C.; Scienze, D.; Cintia, V.; Chimica, D. Biophysical properties of quadruple helices of modified human telomeric DNA. *Biopolymers* **2005**, *77*, 75–85. [CrossRef] [PubMed]
87. Greenfield, N.J. Using circular dichroism collected as a function of temperature to determine the thermodynamics of protein unfolding and binding interactions. *Nat. Protoc.* **2006**, *1*, 2527–2535. [CrossRef] [PubMed]
88. Hwang, T.L.; Shaka, A.J. Water suppression that works. Excitation sculpting using arbitrary wave-forms and pulsed-field gradients. *J. Magn. Reson. Ser. A* **1995**, *112*, 275–279. [CrossRef]
89. David, A.P.; Margarit, E.; Domizi, P.; Banchio, C.; Armas, P.; Calcaterra, N.B. G-Quadruplexes as novel cis-elements controlling transcription during embryonic development. *Nucleic Acids Res.* **2016**, *44*, 4163–4173. [CrossRef]
90. Kumari, S.; Bugaut, A.; Huppert, J.L.; Balasubramanian, S. An RNA G-quadruplex in the 5′ UTR of the NRAS proto-oncogene modulates translation. *Nat. Chem. Biol.* **2007**, *3*, 218–221. [CrossRef]
91. Challier, E.; Lisa, M.-N.; Nerli, B.B.; Calcaterra, N.B.; Armas, P. Novel high-performance purification protocol of recombinant CNBP suitable for biochemical and biophysical characterization. *Protein Expr. Purif.* **2014**, *93*, 23–31. [CrossRef] [PubMed]
92. Armas, P.; Nasif, S.; Calcaterra, N.B. Cellular nucleic acid binding protein binds G-rich single-stranded nucleic acids and may function as a nucleic acid chaperone. *J. Cell. Biochem.* **2008**, *103*, 1013–1036. [CrossRef]
93. Tuma, R.S.; Beaudet, M.P.; Jin, X.; Jones, L.J.; Cheung, C.; Yue, S.; Singer, V.L. Characterization of SYBR gold nucleic acid gel stain: A dye optimized for use with 300-nm ultraviolet transilluminators. *Anal. Biochem.* **1999**, *288*, 278–288. [CrossRef] [PubMed]

International Journal of *Molecular Sciences*

MDPI

Article

Tracing dsDNA Virus–Host Coevolution through Correlation of Their G-Quadruplex-Forming Sequences

Natália Bohálová [1,2], Alessio Cantara [1,2], Martin Bartas [3], Patrik Kaura [4], Jiří Šťastný [4,5], Petr Pečinka [3], Miroslav Fojta [1] and Václav Brázda [1,*]

[1] Institute of Biophysics of the Czech Academy of Sciences, Královopolská 135, 612 65 Brno, Czech Republic; natalia.bohalova@ibp.cz (N.B.); alexcantara41@gmail.com (A.C.); fojta@ibp.cz (M.F.)
[2] Department of Experimental Biology, Faculty of Science, Masaryk University, Kamenice 5, 625 00 Brno, Czech Republic
[3] Department of Biology and Ecology, Institute of Environmental Technologies, Faculty of Science, University of Ostrava, 710 00 Ostrava, Czech Republic; martin.bartas@osu.cz (M.B.); petr.pecinka@osu.cz (P.P.)
[4] Faculty of Mechanical Engineering, Brno University of Technology, Technická 2896/2, 616 69 Brno, Czech Republic; 160702@vutbr.cz (P.K.); stastny@fme.vutbr.cz (J.Š.)
[5] Department of Informatics, Mendel University in Brno, Zemědělská 1, 613 00 Brno, Czech Republic
* Correspondence: vaclav@ibp.cz; Tel.: +420-541517231; Fax: +420-541211293

Abstract: The importance of gene expression regulation in viruses based upon G-quadruplex may point to its potential utilization in therapeutic targeting. Here, we present analyses as to the occurrence of putative G-quadruplex-forming sequences (PQS) in all reference viral dsDNA genomes and evaluate their dependence on PQS occurrence in host organisms using the G4Hunter tool. PQS frequencies differ across host taxa without regard to GC content. The overlay of PQS with annotated regions reveals the localization of PQS in specific regions. While abundance in some, such as repeat regions, is shared by all groups, others are unique. There is abundance within introns of Eukaryota-infecting viruses, but depletion of PQS in introns of bacteria-infecting viruses. We reveal a significant positive correlation between PQS frequencies in dsDNA viruses and corresponding hosts from archaea, bacteria, and eukaryotes. A strong relationship between PQS in a virus and its host indicates their close coevolution and evolutionarily reciprocal mimicking of genome organization.

Keywords: G-quadruplex; virus; bioinformatics; coevolution; host; dsDNA; G4Hunter

Citation: Bohálová, N.; Cantara, A.; Bartas, M.; Kaura, P.; Šťastný, J.; Pečinka, P.; Fojta, M.; Brázda, V. Tracing dsDNA Virus–Host Coevolution through Correlation of Their G-Quadruplex-Forming Sequences. *Int. J. Mol. Sci.* **2021**, *22*, 3433. https://doi.org/10.3390/ijms22073433

Academic Editor: James K. Bashkin

Received: 2 March 2021
Accepted: 23 March 2021
Published: 26 March 2021

Publisher's Note: MDPI stays neutral with regard to jurisdictional claims in published maps and institutional affiliations.

1. Introduction

Viruses are intracellular parasites closely coevolving with their host organisms and thus shaping genotypic diversity [1,2]. The interplay between a virus and its host constitutes a powerful mechanism of reciprocal selection pressure. Coevolution of the two can be traced by nucleic acid sequence, protein tertiary structure, and also at the whole-function level. For example, hosts' antiviral defense mechanisms often originate from viruses [2–4]. The study of reciprocal coevolutionary adaptations between a virus and the host immune system could provide new insights and potential strategies in developing antiviral treatments [5].

G-quadruplexes (G4s) are noncanonical, local secondary structures of nucleic acids that have been identified as having regulatory roles within cells in gene expression, replication, and telomere maintenance [6–8]. A G4 consists of stacked planar G-quartets, which are built by Hoogsteen hydrogen bond-based pairing of four guanines. It has been demonstrated that G4s are very often targets for various cellular proteins [9–11], and a specific domain for G4 recognition has been shown [12,13]. Moreover, several proteins are also capable of stabilizing the G4 structure [14,15]. Recently, it has been demonstrated that the G4 binding domain is also conserved between SARS-CoV and SARS-CoV-2 genomes [16], and it was proven to be crucial for the SARS-CoV life cycle [17]. G4s can be found in all domains of life [18–21], and they have been described as constituting an important structural genomic feature with various functions in several viral classes [21,22]. The G4 formation was shown to limit the replication and transcription of the Ebola virus, hepatitis B virus (HBV), hepatitis C virus (HCV), human immunodeficiency virus (HIV), and several viruses from the *Herpesviridae* family [23–26]. In the life cycles of the Epstein–Barr virus (EBV) and Kaposi sarcoma herpesvirus (KSHV), moreover, RNA G4 has been described as a cis-acting regulatory element that downregulates the translation of highly antigenic proteins and thus influences the immune evasion of the virus and eases transit and persistence in the latent period of infection [27,28]. Importantly, the functions of G4 may be modified via their stabilization by proteins or small-molecule ligands [29,30]. Therefore, stabilizing G4 ligands are considered promising antiviral and antibacterial drugs [29]. Coevolution of viral and host loop sequences of the G-quadruplex-forming sequences in human *Herpesviridae* viruses was recently proposed [31].

G4Hunter is one of several software programs available for predicting putative G-quadruplex forming sequences (PQS) [32–34]. The G4Hunter algorithm searches for Gs/Cs and sums up the scores for the groups of bases. The final score is thus a combination of G-richness and G-skewness and the presence of G-blocks. The default threshold is set to a G4Hunter score of 1.2, which has proven to be a reasonable compromise between false-negative and false-positive results. The higher the score, the higher the probability for a G4 structure to form [33]. G4Hunter provides the benefit of targeting even atypical G4s that could not be found by pattern-based algorithms [35,36].

Here, we present an extensive analysis of 2903 viruses across a diverse range of host organisms. Our goals were to identify PQS occurrence and localization in the genomes of viruses infecting a given host group, study the evolutionary differences related to PQS, and describe the potential dependence of PQS frequency between a virus and the corresponding host.

2. Results and Discussion

2.1. Variation in Frequency for G4-Forming Sequences in dsDNA Viruses Grouped by Host

Using the National Center for Biotechnology Information (NCBI) taxonomy classification, the analyzed viruses were divided into three domains according to their host organisms: Archaea, Bacteria, and Eukaryota. The domains were further divided into 23 groups (12 with five or more sequenced genomes) as shown in the phylogenetic tree in Figure 1. Phylogenetic classification of the viruses and corresponding host is presented in Supplementary Materials S1. All hosts were assigned by the Virus-Host database without further modification [37], which could have limited the potential host range, especially in arboviruses. Whereas 95% of all known bacteriophages and archaea-infecting viruses have dsDNA genomes [38], eukaryotes could be infected by all classes of the Baltimore virus classification, which means, in addition to dsDNA viruses, also ssDNA viruses, dsRNA viruses, ssRNA viruses, ssRNA reverse-transcribing viruses, and dsDNA reverse-transcribing viruses. We therefore restricted the analyses of PQS occurrence to only dsDNA viruses, although they have not been found to infect higher plants belonging to the Streptophyta but only lower species of plants belonging to Chlorophyta [39,40].

For further statistical analyses, only those groups with five or more sequenced genomes were included. We analyzed the PQS occurrence in all 2903 reference dsDNA viral genomes divided according to their host organisms (Supplementary Materials S2). A summary of all PQS found in ranges of the G4Hunter score intervals (1.2–1.4, 1.4–1.6, 1.6–1.8, 1.8–2.0, and 2.0 and higher) and precomputed PQS frequencies per 1000 nt is shown in Table 1.

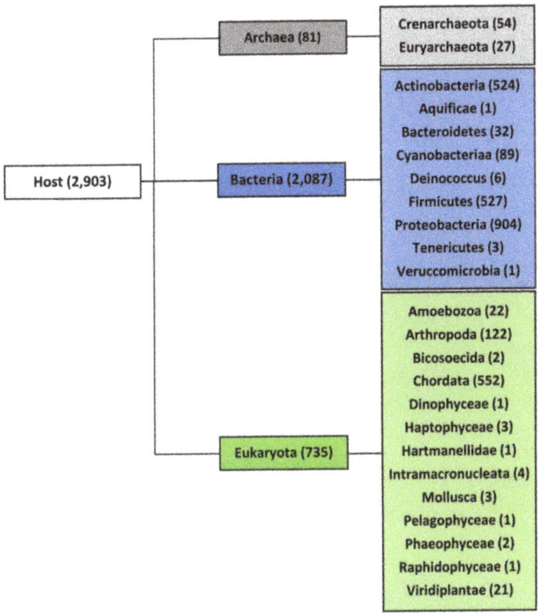

Figure 1. Full set of viral genomes divided according to host. The number of accessible unique genomes for each domain and group is shown in brackets.

Table 1. Total number of putative G-quadruplex-forming sequences (PQS) and their resulting frequencies per 1000 nt in all 2903 viral genomes and host categories, grouped by G4Hunter score. Frequencies were calculated as the total number of PQS in each category divided by the total length of all analyzed sequences, multiplied by 1000 and normalized by the number of viruses infecting one genus.

G4Hunter Score	PQS Frequency per 1000 nt			
	All	Archaea	Bacteria	Eukaryota
1.2–1.4	1.27	1.74	0.88	1.46
1.4–1.6	0.039	0.025	0.026	0.047
1.6–1.8	0.0042	0	0.00088	0.0062
1.8–2.0	0.00025	0	0.000041	0.00038
2.0 and more	0.00021	0	0.000050	0.00031

The mean frequency for all viral genomes in G4Hunter score interval 1.2–1.4 was 1.27 PQS per 1000 nt (see above). The lowest frequency in the same interval was observed for bacteriophages (0.88 PQS per 1000 nt), whereas the highest frequency was detected for archaea-infecting viruses (1.74). Surprisingly, there was not one PQS with a G4Hunter score higher than 1.4 found in the archaea host domain. In the Bacteria and Eukaryota host domains, by contrast, there were some PQS found even with G4Hunter scores higher than 2.0.

The numbers of analyzed viral sequences, grouped by their host phylogenetic categories; median genome length; mean, minimum, and maximum observed frequency of PQS per 1000 nt; and total PQS counts are summarized in Table 2. Just four viral groups (viruses infecting Euryarchaeota, Actinobacteria, *Deinococcus*, and Proteobacteria) showed >50% GC content. On the other hand, three groups (viruses infecting Bacteroidetes, Firmicutes, and Arthropoda) showed <40% GC content. Detailed statistical characteristics for PQS frequencies per 1000 nt (including mean, variance, and outliers) are depicted in boxplots for all inspected host groups in Figure 2. The mean frequency for all viral genomes was 1.32 PQS per 1000 nt. Detailed statistical analyses of host inter-domain and intergroup comparisons are presented in Supplementary Materials S3. We observed the highest mean frequency in the archaea host domain (1.76), followed by viruses infecting Eukaryota (1.52), whereas the lowest was noted in bacteriophages (0.89). At the group level, the most extreme values were found within the viruses infecting the bacteria domain. The lowest mean frequency was found in viruses infecting the Firmicutes (0.32) followed by Bacteroidetes (0.41). The highest PQS frequency was observed in the *Deinococcus* host group (4.21), followed by Actinobacteria (2.27). In viruses infecting the archaea domain, notable enrichment relative to the average was found for both the Euryarchaeota (1.69) and Crenarchaeota (1.85) groups. Within the Eukaryota host domain, the highest PQS frequency was observed for Chordata (2.18), the lowest for Arthropoda (0.30). The mean PQS frequency found in viruses infecting humans was lower (1.75) than the average PQS in the Chordata host group as normalized by the number of viruses infecting one host genus (2.18). We created a cluster dendrogram, as shown in Figure 3, to further reveal and graphically depict similarities among host groups. The input data and R code are listed in Supplementary Materials S4. Viruses infecting humans are notably clustered together with other viruses infecting Chordata on the left side of the dendrogram. Other viruses infecting eukaryotes are clustered in the second branch on the right.

Table 2. Distribution of PQS frequencies in viruses according to host organisms. Genomic length, PQS frequencies, and total counts. Seq (total number of sequences), Median (median length of sequences), GC% (average GC content), PQS (total number of predicted PQS), Mean f (mean frequency of predicted PQS per 1000 nt normalized by the number of viruses infecting one genus), Min f (the lowest frequency of predicted PQS per 1000 nt), Max f (the highest frequency of predicted PQS per 1000 nt), and Cov (% of genome covered by PQS).

All	Seq	Median	GC%	PQS	Mean f	Min f	Max f	Cov
All	3134	44,746.5	44.94	220,569	1.32	0	11.51	3.34
Domain	**Seq**	**Median**	**GC%**	**PQS**	**Mean f**	**Min f**	**Max f**	**Cov**
Archaea	81	33,356	48.92	3137	1.76	0	4.80	4.32
Bacteria	2087	49,639	48.10	112,664	0.89	0	11.51	2.11
Eukaryota	966	7951.5	43.09	104,768	1.52	0	11.44	3.93
Group	**Seq**	**Median**	**GC%**	**PQS**	**Mean f**	**Min f**	**Max f**	**Cov**
Crenarchaeota	54	32,047.5	40.91	1012	1.85	0	4.80	4.76
Euryarchaeota	27	49,107	54.92	2125	1.69	0.28	3.75	3.99
Actinobacteria	524	53,403.5	60.90	61,313	2.27	0.33	7.02	5.12
Bacteroidetes	32	47,060	38.12	477	0.41	0.03	1.14	1.01
Cyanobacteria	89	174,079	43.33	3875	0.82	0.06	3.88	2.10
Deinococcus	6	61,150	50.26	726	4.21	0.33	11.51	10.45
Firmicutes	527	41,843	38.14	7886	0.32	0	1.39	0.78
Proteobacteria	904	49,035	50.07	38,334	0.80	0	4.55	1.90
Amoebozoa	22	495,022	42.47	21,931	0.66	0	1.89	1.60
Arthropoda	345	7276	38.77	4957	0.30	0	1.92	0.73
Chordata	561	7852	45.48	72,420	2.18	0	11.44	5.65
Viridiplantae	21	193,301	46.91	3542	1.06	0	2.01	2.54
Subgroup	**Seq**	**Median**	**GC%**	**PQS**	**Mean f**	**Min f**	**Max f**	**Cov**
Humans	120	7344	42.55	15,996	1.75	0	11.44	4.48

The colors correspond to phylogenetic tree depiction in Figure 1 (Grey—Archaea, Blue—Bacteria, Green—Eukaryota as host organisms).

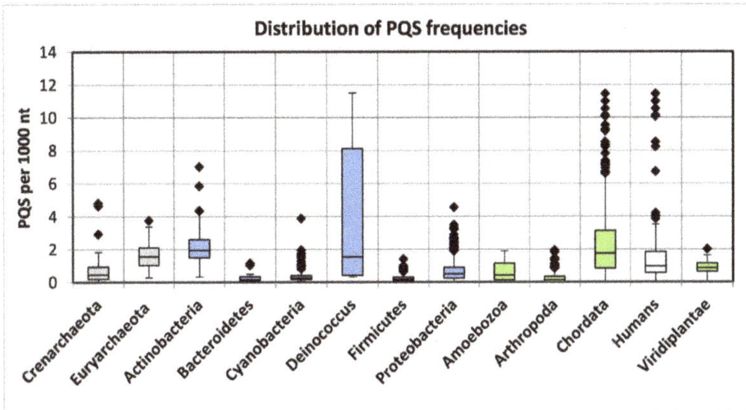

Figure 2. Frequencies of PQS in host groups of the analyzed viral genomes. Data within boxes span the interquartile range and whiskers show the lowest and highest values within the 1.5 interquartile range. Black diamonds denote outliers. The colors correspond to phylogenetic tree depiction.

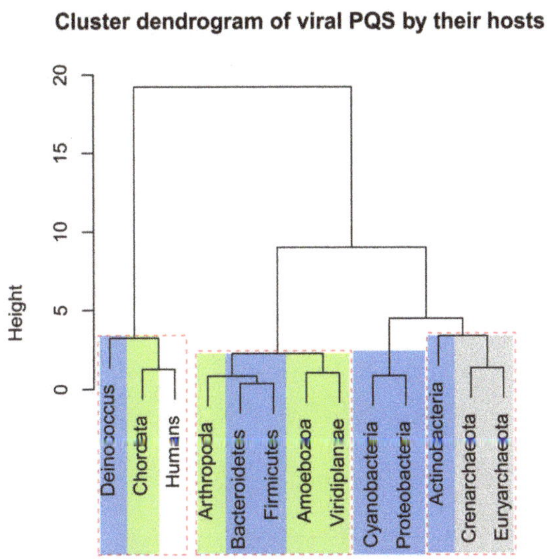

Figure 3. Cluster dendrogram based on PQS characteristics in all viral species by their host. Input data are listed in Supplementary Materials S4. Statistically significant clusters (based upon approximately unbiased *p*-values above 95, equivalent to *p*-values lower than 0.05) are highlighted by rectangles drawn with broken red lines. The colors correspond to phylogenetic tree depiction.

2.2. Features Characteristic for Hosts Are Enriched for PQS in Corresponding dsDNA Viral Genomes

To evaluate the localization of PQS within viral genomes, we overlapped PQS with annotation regions extracted from the NCBI database (Supplementary Materials S5). We took the PQS frequency per 1000 nt in genes as a reference and plotted the ratio of the PQS frequency in features to that in genes (Figure 4). PQS frequencies differ depending on the annotated motif and across different hosts. In the archaea domain, the most notable enrichment was found inside and 100 bp after *stem_loops* (4.2× and 10.2× enrichment) and *mobile_elements* (3.5× and 3.4× enhancement). Predictably, abundance was also found in the archaea-infecting viruses' *repeat_regions* (2.9×). The *repeat_regions* were also enriched for

PQS inside bacteria-infecting (3.3×) and Eukaryota-infecting viruses (4.4×). The highest relative frequency inside bacteria was found in *tmRNA* (3.13) and *ncRNA* (1.9×), followed by a region 100 bp long before *misc_RNA* and *misc_recomb* (1.8 and 1.5× abundance). In addition to *repeat_regions*, we noted PQS enrichment in *misc_RNA* (6.7×), *variation* (5.2×), *protein_bind* (3.8×), and *introns* (1.9×) of Eukaryota-infecting viruses. Notably, the PQS frequency was increased in comparison to *genes* inside *introns* only in Eukaryota-infecting viruses (1.9× enrichment), whereas *introns* in bacteria-infecting viruses were depleted for PQS presence (0.14× lower PQS frequency in comparison to genes). This indicates that the prevalence of PQS in specific viral features is important for the host's cellular machinery. A G4 located in an intron could affect the expression profile; it was shown to regulate the splicing of alternative isoforms of a p53 protein in the human genome [41].

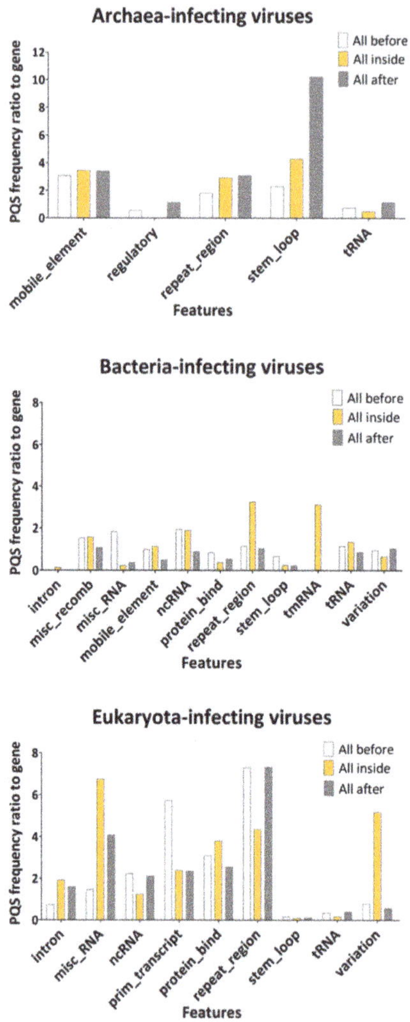

Figure 4. The ratio of PQS frequencies per 1000 nt between gene annotation and other annotated locations from the NCBI database. PQS frequencies within (inside), before (100 nt), and after (100 nt) annotated locations were analyzed. Detailed results are summarized in Supplementary Materials S5.

2.3. PQS Frequencies of dsDNA Viruses Correlate with Their Hosts' Genomes

Next, to evaluate the relationship between PQS frequencies of the virus and the host, we analyzed selected genomes of host organisms for PQS presence. For hosts from archaea and bacteria, we utilized the previously published results of our group on PQS occurrence in all accessible archaeal and bacterial genomes [19,42]. In all analyses, we used the same workflow and same parameters for G4Hunter and data processing. Reference genomes of the Eukaryota hosts were retrieved from the NCBI database, and their list together with correlation analyses is available in Supplementary Materials S1. We selected all available reference genomes of hosts belonging to Viridiplantae, and for the remaining Eukaryota groups (Arthropoda and Chordata), we selected the 10 most frequent hosts in each corresponding category. There is no reference genome, however, for the *Acanthamoeba* genus, the only host of Amoebozoa-infecting viruses. We always compared a single eukaryotic host genome to all corresponding dsDNA viruses. The overall results of the correlation analyses are presented in Figure 5 and in the Supplementary Material S6. Spearman's correlation coefficient for the average of PQS frequencies in all investigated virus-host pairs was determined to be 0.7677, with a statistically significant p-value of 7×10^{-7} (Figure 5A). To exclude the GC content as a bias factor, we plotted also the average PQS/GC per 1000 nt. The correlation coefficient for the average of PQS/GC per 1000 nt then increased to 0.822, with p-value of 3×10^{-8} (Figure 5B).

The strongest correlation was found between virus–bacteria pairs. Spearman's correlation coefficient for PQS frequency of bacteria-infecting viruses and their hosts showed a strong, statistically significant (p-value < 0.01) positive correlation (Figure 5E,F). The correlation coefficient was 0.9429 for PQS frequency and 0.9985 for GC/PQS, with p-value < 0.01. Our previously published PQS frequencies of all known bacterial genomes [19] correspond to the frequencies determined here for PQS in bacteriophage genomes. In all virus–bacteria pairs, the mean PQS frequency was higher in the bacterial group than in the viral host group. A statistically significant positive correlation (p-value < 0.05) was observed, also with PQS frequencies grouped by G4Hunter score intervals and PQS frequencies identified by the Tetraplex Finder module of QuadBase2 software with default low stringency parameters (Supplementary Materials S7). Dispersion of PQS frequencies among bacteriophages was more diverse than inside other viruses, and the same observation (higher diversity in PQS frequencies) has been reported for bacteria compared to other hosts [18]. The corresponding frequencies of virus and bacteria hosts confirmed by correlation analyses pointed to their having close coevolutionary processes. The second highest correlation coefficient was found for the Archaea subgroup, with a value of 0.9 for PQS average frequency and PQS/GC (p-value < 0.05) (Figure 5C,D). To distinguish several different phyla, we further divided Crenarchaeota into two subgroups (*Sulfolobus*, Thermoproteales) and Euryarchaeota into three subgroups (*Acidianus*, Halobacteriales, Haloferacales). Because of the high diversity and the low number of sequenced genomes in several categories, the minimum number of viral or host genomes for statistical analyses was set to four.

Inside the eukaryote domain, we noted lower but still statistically significant (p-value < 0.01) correlation coefficients of 0.6509 for PQS frequencies and 0.7737 for PQS/GC. This finding could be attributed to two main causes (Figure 5G,H). First, the statistical sample for Eukaryota host genomes was significantly smaller, in comparison to that for bacteria and archaea host domains; an average of six host genomes were analyzed for each group of Eukaryota, shown in Table 2. Second, with the increasing complexity of the organisms, genomic duplications, and extensive repetitions, the correlation could be less obvious and significant.

Recently, coevolution of G4 sequence composition between dsDNA viruses from the *Herpesviridae* family and host has been proposed, as herpesviruses are often enriched for C-rich looped G4s, which are binding sites for host transcription factors, and TTA-looped G4s identical to the telomeric motif of vertebrates [31]. Mimicking the genome organization of the host could influence the PQS prevalence in dsDNA viral genomes and vice versa.

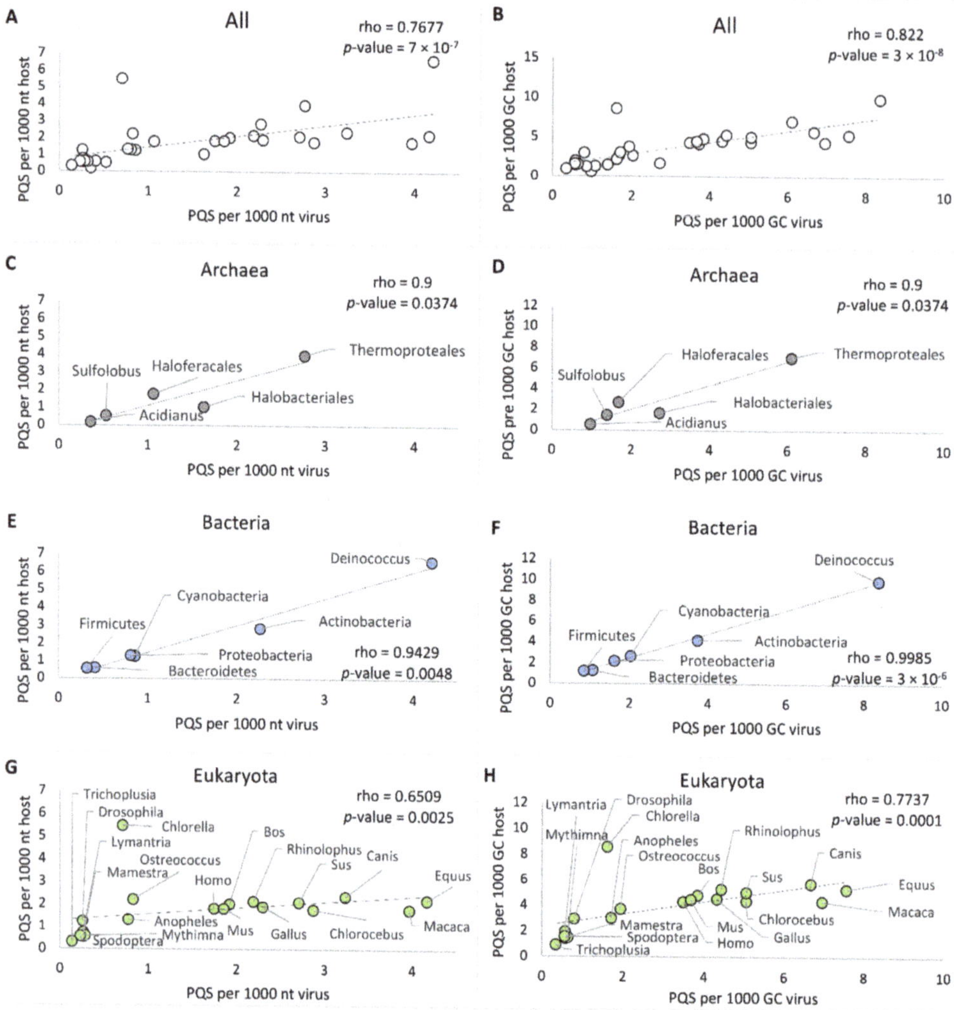

Figure 5. Relationships between virus and various hosts as measured by observed PQS frequency per 1000 nt and PQS frequency per 1000 GC. (**A**) All host-virus pairs, PQS frequencies; (**B**) All host–virus pairs, PQS per 1000 GC; (**C**) Archaea–virus pairs, PQS frequencies; (**D**) Archaea–virus pairs, PQS per 1000 GC; (**E**) Bacteria–virus pairs, PQS frequencies; (**F**) Bacteria–virus pairs, PQS per 1000 GC; (**G**) Eukaryota–virus pairs, PQS frequencies; (**H**) Eukaryota–virus pairs, PQS per 1000 GC of the archaea–virus pairs.

3. Materials and Methods

3.1. Viral and Host Sequences

A total of 3134 sequences of 2903 unique viral genomes were downloaded from the genome database of the National Center for Biotechnology Information (NCBI). Where more than one sequence was available, only the reference genome was used in the analyses. Hosts were assigned according to the NCBI and the Virus-Host database [37]. Subviral agents were assigned to the host of the coinfected virus as stated in the database [37]. A full list of NCBI IDs and host assignments are presented in Supplementary Materials S1. For hosts from archaea and bacteria, we utilized the previously published results of our group on PQS occurrence in all accessible archaeal and bacterial genomes [19,42]. For eukaryote groups, we selected the 10 most frequented hosts for each viral group, and these are also listed in Supplementary Materials S1.

3.2. *PQS Analyses*

Analyses were run using the computational core of DNA analyzer software written in Java [43] with G4Hunter algorithm implementation [32] and default parameters (25 nt for window size, G4 score threshold 1.2). The overall results as to the number of PQS found together with the size of genomic DNA, GC content, PQS frequency normalized for 1000 nt, and lengths of sequences covered with PQS are summarized in Supplementary Materials S2. The average PQS frequency of host groups, shown in Table 1, was normalized by the number of viruses infecting each genus to avoid sampling bias due to the overabundance of viruses infecting specific species (such as a human). PQS were also classified into the five intervals of the G4Hunter score: 1.2–1.4, 1.4–1.6, 1.6–1.8, 1.8–2.0, and 2.0 and more. To confirm the results acquired by G4Hunter, bacteriophages and corresponding hosts were selected and analyzed by the Tetraplex Finder module of QuadBase2 software with default low stringency parameters (nonoverlapping PQS with minimum two-tracked PQS and loop length of 1–12 nt), and the results are listed in Supplementary Materials S7 [44].

3.3. *Statistical Analysis*

The normality of data was tested by the Shapiro–Wilk test. The nonparametric Kruskal–Wallis test was utilized for statistical evaluations of the differences in PQS among host phylogenetic groups. Post hoc multiple pairwise comparisons, using Dunn's test with Bonferroni correction of the significance level, were applied with the p-value cutoff set at 0.05; data are available in Supplementary Materials S3 for sequences grouped by their host organism and their comparison to the PQS frequency of host groups. For correlation analyses, the two-tailed Spearman's correlation coefficient was considered. To further reveal and graphically depict similarities between viral hosts, we constructed a cluster dendrogram of PQS characteristics in program R, version 3.6.3, using the pvclust package. The following values were used as input data: Mean f (mean of predicted PQS per 1000 nt), Min f (the lowest frequency of predicted PQS per 1000 nt), Max f (the highest frequency of predicted PQS per 1000 nt), and Cov % (% of genome covered by PQS) (Supplementary Materials S4). The following parameters were used for both analyses, the cluster method "ward.D2", distance "euclidean", and the number of bootstrap resampling 10,000. Statistically significant clusters (based on approximately unbiased p-values values above 95, equivalent to p values less than 0.05) are highlighted in Figure 5 by rectangles marked with broken red lines. R code is provided in Supplementary Materials S4.

3.4. *Overlay of PQS with Annotated Features from NCBI*

Annotated feature tables of all viral genomes were downloaded from the NCBI database. Features were grouped by their names as stated in the feature table file. PQS occurrence was analyzed inside and around (before and after) a predefined featured neighborhood (±100 nt). From this analysis, we obtained a file with feature names and numbers of PQS found inside and around features. Further processing was performed in Microsoft Excel and the data are available as Supplementary Material S5.

4. Conclusions

PQS frequencies in viral genomes differ across host taxa and correspond to the PQS frequencies of the host organism. The overlay of PQS with annotated regions revealed nonrandom localization of G4 sequences and their abundance in various regions, such as repeat regions, stem-loops, mobile elements, protein-binding regions, RNA, etc. While abundance and depletion in some locations are shared by viruses of different hosts, others are unique. For example, there is an abundance of PQS in introns of Eukaryota-infecting viruses, but depletion of PQS in introns of bacteria-infecting viruses. Our study revealed the correlation between PQS frequencies of dsDNA viruses and corresponding hosts from archaea, bacteria, and even eukaryotes, which indicate their close coevolution and evolutionarily reciprocal mimicking of genome organization.

Supplementary Materials: Supplementary Materials can be found at https://www.mdpi.com/article/10.3390/ijms22073433/s1.

Author Contributions: Conceptualization, V.B.; methodology, V.B.; software, P.K. and J.Š.; validation, V.B.; formal analysis, N.B., A.C. and M.B.; investigation, N.B. and A.C.; resources, P.K. and J.Š.; data curation, N.B. and A.C.; writing—original draft preparation, N.B., A.C., M.B. and V.B.; writing—review and editing, P.P. and M.F.; visualization, N.B., A.C. and M.B.; supervision, V.B., M.F. and P.P.; project administration, V.B.; and funding acquisition, V.B. and M.F. All authors have read and agreed to the published version of the manuscript.

Funding: This work was supported by The Czech Science Foundation (18-15548S) and by the SYMBIT project Reg. no. CZ.02.1.01/0.0/0.0/15_003/0000477 financed from the ERDF.

Institutional Review Board Statement: Not applicable.

Informed Consent Statement: Not applicable.

Data Availability Statement: Not applicable.

Acknowledgments: We thank Gale A. Kirking for proofreading and providing feedback on the manuscript.

Conflicts of Interest: The authors declare no competing interests.

Abbreviations

EBV	Epstein–Barr virus
G4s	G-quadruplexes
HBV	hepatitis B virus
HCV	hepatitis C virus
HIV	human immunodeficiency virus
KSHV	Kaposi sarcoma herpes virus
PQS	putative G-quadruplex-forming sequences
NCBI	National Center for Biotechnology Information

References

1. McLaughlin, R.N.; Malik, H.S. Genetic conflicts: The usual suspects and beyond. *J. Exp. Biol.* **2017**, *220*, 6–17. [CrossRef]
2. Kaján, G.L.; Doszpoly, A.; Tarján, Z.L.; Vidovszky, M.Z.; Papp, T. Virus–Host Coevolution with a Focus on Animal and Human DNA Viruses. *J. Mol. Evol.* **2020**, *88*, 41–56. [CrossRef]
3. Charpentier, E.; Doudna, J.A. Rewriting a genome. *Nat. Cell Biol.* **2013**, *495*, 50–51. [CrossRef]
4. Moelling, K.; Broecker, F.; Russo, G.; Sunagawa, S. RNase H as Gene Modifier, Driver of Evolution and Antiviral Defense. *Front. Microbiol.* **2017**, *8*, 1745. [CrossRef] [PubMed]
5. Woolhouse, M.E.J.; Webster, J.P.; Domingo, E.; Charlesworth, B.; Levin, B.R. Biological and biomedical implications of the co-evolution of pathogens and their hosts. *Nat. Genet.* **2002**, *32*, 569–577. [CrossRef] [PubMed]
6. Lemarteleur, T.; Gomez, D.; Paterski, R.; Mandine, E.; Mailliet, P.; Riou, J.-F. Stabilization of the c-myc gene promoter quadruplex by specific ligands' inhibitors of telomerase. *Biochem. Biophys. Res. Commun.* **2004**, *323*, 802–808. [CrossRef] [PubMed]
7. Patel, D.J.; Phan, A.T.; Kuryavyi, V. Human telomere, oncogenic promoter and 5′-UTR G-quadruplexes: Diverse higher order DNA and RNA targets for cancer therapeutics. *Nucleic Acids Res.* **2007**, *35*, 7429–7455. [CrossRef]
8. Rhodes, D.; Lipps, H.J. G-quadruplexes and their regulatory roles in biology. *Nucleic Acids Res.* **2015**, *43*, 8627–8637. [CrossRef] [PubMed]
9. Mishra, S.K.; Tawani, A.; Mishra, A.; Kumar, A. G4IPDB: A database for G-quadruplex structure forming nucleic acid interacting proteins. *Sci. Rep.* **2016**, *6*, 38144. [CrossRef] [PubMed]
10. Brázda, V.; Hároníková, L.; Liao, J.C.C.; Fojta, M. DNA and RNA Quadruplex-Binding Proteins. *Int. J. Mol. Sci.* **2014**, *15*, 17493–17517. [CrossRef] [PubMed]
11. Alavi, S.; Ghadiri, H.; Dabirmanesh, B.; Moriyama, K.; Khajeh, K.; Masai, H. G-quadruplex binding protein Rif1, a key regulator of replication timing. *J. Biochem.* **2021**, *169*, 1–14. [CrossRef] [PubMed]
12. Brázda, V.; Červeň, J.; Bartas, M.; Mikysková, N.; Coufal, J.; Pečinka, P. The Amino Acid Composition of Quadruplex Binding Proteins Reveals a Shared Motif and Predicts New Potential Quadruplex Interactors. *Molecules* **2018**, *23*, 2341. [CrossRef]

13. Masuzawa, T.; Oyoshi, T. Roles of the RGG Domain and RNA Recognition Motif of Nucleolin in G-Quadruplex Stabilization. *ACS Omega* **2020**, *5*, 5202–5208. [CrossRef]
14. Brázda, V.; Coufal, J.; Liao, J.C.; Arrowsmith, C.H. Preferential binding of IFI16 protein to cruciform structure and superhelical DNA. *Biochem. Biophys. Res. Commun.* **2012**, *422*, 716–720. [CrossRef] [PubMed]
15. Tosoni, E.; Frasson, I.; Scalabrin, M.; Perrone, R.; Butovskaya, E.; Nadai, M.; Palù, G.; Fabris, D.; Richter, S.N. Nucleolin stabilizes G-quadruplex structures folded by the LTR promoter and silences HIV-1 viral transcription. *Nucleic Acids Res.* **2015**, *43*, 8884–8897. [CrossRef] [PubMed]
16. Bartas, M.; Brázda, V.; Bohálová, N.; Cantara, A.; Volná, A.; Stachurová, T.; Malachová, K.; Jagelská, E.B.; Porubiaková, O.; Červeň, J.; et al. In-Depth Bioinformatic Analyses of Nidovirales Including Human SARS-CoV-2, SARS-CoV, MERS-CoV Viruses Suggest Important Roles of Non-canonical Nucleic Acid Structures in Their Lifecycles. *Front. Microbiol.* **2020**, *11*, 1583. [CrossRef]
17. Kusov, Y.; Tan, J.; Alvarez, E.; Enjuanes, L.; Hilgenfeld, R. A G-quadruplex-binding macrodomain within the "SARS-unique domain" is essential for the activity of the SARS-coronavirus replication–transcription complex. *Virology* **2015**, *484*, 313–322. [CrossRef]
18. Ding, Y.; Fleming, A.M.; Burrows, C.J. Case studies on potential G-quadruplex-forming sequences from the bacterial orders Deinococcales and Thermales derived from a survey of published genomes. *Sci. Rep.* **2018**, *8*, 15679. [CrossRef]
19. Bartas, M.; Čutová, M.; Brázda, V.; Kaura, P.; Šťastný, J.; Kolomazník, J.; Coufal, J.; Goswami, P.; Červeň, J.; Pečinka, P. The Presence and Localization of G-Quadruplex Forming Sequences in the Domain of Bacteria. *Molecules* **2019**, *24*, 1711. [CrossRef]
20. Brazda, V.; Fojta, M.; Bowater, R.P. Structures and stability of simple DNA repeats from bacteria. *Biochem. J.* **2020**, *477*, 325–339. [CrossRef]
21. Ruggiero, E.; Richter, S.N. Viral G-quadruplexes: New frontiers in virus pathogenesis and antiviral therapy. *Annu. Rep. Med. Chem.* **2020**, *54*, 101–131. [CrossRef] [PubMed]
22. Saranathan, N.; Vivekanandan, P. G-Quadruplexes: More Than Just a Kink in Microbial Genomes. *Trends Microbiol.* **2019**, *27*, 148–163. [CrossRef]
23. Wang, S.-R.; Zhang, Q.-Y.; Wang, J.-Q.; Ge, X.-Y.; Song, Y.-Y.; Wang, Y.-F.; Li, X.-D.; Fu, B.-S.; Xu, G.-H.; Shu, B.; et al. Chemical Targeting of a G-Quadruplex RNA in the Ebola Virus L Gene. *Cell Chem. Biol.* **2016**, *23*, 1113–1122. [CrossRef] [PubMed]
24. Jaubert, C.; Bedrat, A.; Bartolucci, L.; Di Primo, C.; Ventura, M.; Mergny, J.-L.; Amrane, S.; Andreola, M.-L. RNA synthesis is modulated by G-quadruplex formation in Hepatitis C virus negative RNA strand. *Sci. Rep.* **2018**, *8*, 1–9. [CrossRef]
25. Frasson, I.; Nadai, M.; Richter, S.N. Conserved G-Quadruplexes Regulate the Immediate Early Promoters of Human Alphaherpesviruses. *Molecules* **2019**, *24*, 2375. [CrossRef] [PubMed]
26. Liu, Y.; Le, C.; Tyrrell, D.L.; Le, X.C.; Li, X.-F. Aptamer Binding Assay for the E Antigen of Hepatitis B Using Modified Aptamers with G-Quadruplex Structures. *Anal. Chem.* **2020**, *92*, 6495–6501. [CrossRef] [PubMed]
27. Murat, P.; Zhong, J.; Lekieffre, L.; Cowieson, N.P.; Clancy, J.L.; Preiss, T.; Balasubramanian, S.; Khanna, R.; Tellam, J. G-quadruplexes regulate Epstein-Barr virus–encoded nuclear antigen 1 mRNA translation. *Nat. Chem. Biol.* **2014**, *10*, 358–364. [CrossRef]
28. Dabral, P.; Babu, J.; Zareie, A.; Verma, S.C. LANA and hnRNP A1 Regulate the Translation of LANA mRNA through G-Quadruplexes. *J. Virol.* **2020**, *94*, 94. [CrossRef] [PubMed]
29. Ruggiero, E.; Richter, S.N. G-quadruplexes and G-quadruplex ligands: Targets and tools in antiviral therapy. *Nucleic Acids Res.* **2018**, *46*, 3270–3283. [CrossRef]
30. De Cian, A.; Gros, J.; Guédin, A.; Haddi, M.; Lyonnais, S.; Guittat, L.; Riou, J.-F.; Trentesaux, C.; Saccà, B.; Lacroix, L.; et al. DNA and RNA Quadruplex ligands. *Nucleic Acids Symp. Ser.* **2008**, *52*, 7–8. [CrossRef]
31. Lombardi, E.P.P.; Londoño-Vallejo, A.; Nicolas, A. Relationship Between G-Quadruplex Sequence Composition in Viruses and Their Hosts. *Molecules* **2019**, *24*, 1942. [CrossRef]
32. Brázda, V.; Kolomazník, J.; Lýsek, J.; Bartas, M.; Fojta, M.; Šťastný, J.; Mergny, J.-L. G4Hunter web application: A web server for G-quadruplex prediction. *Bioinformatics* **2019**, *35*, 3493–3495. [CrossRef]
33. Bedrat, A.; Lacroix, L.; Mergny, J.-L. Re-evaluation of G-quadruplex propensity with G4Hunter. *Nucleic Acids Res.* **2016**, *44*, 1746–1759. [CrossRef] [PubMed]
34. Lombardi, E.P.; Londoño-Vallejo, A. A guide to computational methods for G-quadruplex prediction. *Nucleic Acids Res.* **2020**, *48*, 1–15. [CrossRef]
35. Lightfoot, H.L.; Hagen, T.; Tatum, N.J.; Hall, J. The diverse structural landscape of quadruplexes. *FEBS Lett.* **2019**, *593*, 2083–2102. [CrossRef]
36. Guédin, A.; Gros, J.; Alberti, P.; Mergny, J.-L. How long is too long? Effects of loop size on G-quadruplex stability. *Nucleic Acids Res.* **2010**, *38*, 7858–7868. [CrossRef] [PubMed]
37. Mihara, T.; Nishimura, Y.; Shimizu, Y.; Nishiyama, H.; Yoshikawa, G.; Uehara, H.; Hingamp, P.; Goto, S.; Ogata, H. Linking Virus Genomes with Host Taxonomy. *Viruses* **2016**, *8*, 66. [CrossRef] [PubMed]
38. Ofir, G.; Sorek, R. Contemporary Phage Biology: From Classic Models to New Insights. *Cell* **2018**, *172*, 1260–1270. [CrossRef] [PubMed]
39. Van Etten, J.L.; Dunigan, D.D. Chloroviruses: Not your everyday plant virus. *Trends Plant. Sci.* **2012**, *17*, 1–8. [CrossRef] [PubMed]
40. Hull, R. *Comparative Plant. Virology*, 2nd ed.; Elsevier: Amsterdam, The Netherlands; Academic Press: Boston, MA, USA, 2009; ISBN 9780123741547.

41. Marcel, V.; Tran, P.L.; Sagne, C.; Martel-Planche, G.; Vaslin, L.; Teulade-Fichou, M.-P.; Hall, J.; Mergny, J.-L.; Hainaut, P.; Van Dyck, E. G-quadruplex structures in TP53 intron 3: Role in alternative splicing and in production of p53 mRNA isoforms. *Carcinogenesis* **2010**, *32*, 271–278. [CrossRef]
42. Brázda, V.; Luo, Y.; Bartas, M.; Kaura, P.; Porubiaková, O.; Šťastný, J.; Pečinka, P.; Verga, D.; Da Cunha, V.; Takahashi, T.S.; et al. G-Quadruplexes in the Archaea Domain. *Biomolecules* **2020**, *10*, 1349. [CrossRef] [PubMed]
43. Brázda, V.; Kolomazník, J.; Lýsek, J.; Hároníková, L.; Coufal, J.; Šť'Astný, J. Palindrome analyser—A new web-based server for predicting and evaluating inverted repeats in nucleotide sequences. *Biochem. Biophys. Res. Commun.* **2016**, *478*, 1739–1745. [CrossRef] [PubMed]
44. Dhapola, P.; Chowdhury, S. QuadBase2: Web server for multiplexed guanine quadruplex mining and visualization. *Nucleic Acids Res.* **2016**, *44*, 277–283. [CrossRef] [PubMed]

International Journal of
Molecular Sciences

MDPI

Article

Rhodamine 6G-Ligand Influencing G-Quadruplex Stability and Topology

Lukáš Trizna [1], Ladislav Janovec [2], Andrea Halaganová [1] and Viktor Víglaský [1,*]

[1] Department of Biochemistry, Institute of Chemistry, Faculty of Sciences, Pavol Jozef Šafárik University, 04001 Košice, Slovakia; lukas.trizna@yahoo.com (L.T.); andreahalaganova@gmail.com (A.H.)

[2] Department of Organic Chemistry, Institute of Chemistry, Faculty of Sciences, Pavol Jozef Šafárik University, 04001 Košice, Slovakia; ladislav.janovec@upjs.sk

[*] Correspondence: viktor.viglasky@upjs.sk; Tel.: +421-55-2341262

Abstract: The involvement of G-quadruplex (G4) structures in nucleic acids in various molecular processes in cells such as replication, gene-pausing, the expression of crucial cancer-related genes and DNA damage repair is well known. The compounds targeting G4 usually bind directly to the G4 structure, but some ligands can also facilitate the G4 folding of unfolded G-rich sequences and stabilize them even without the presence of monovalent ions such as sodium or potassium. Interestingly, some G4-ligand complexes can show a clear induced CD signal, a feature which is indirect proof of the ligand interaction. Based on the dichroic spectral profile it is not only possible to confirm the presence of a G4 structure but also to determine its topology. In this study we examine the potential of the commercially available Rhodamine 6G (RhG) as a G4 ligand. RhG tends to convert antiparallel G4 structures to parallel forms in a manner similar to that of Thiazole Orange. Our results confirm the very high selectivity of this ligand to the G4 structure. Moreover, the parallel topology of G4 can be verified unambiguously based on the specific induced CD profile of the G4-RhG complex. This feature has been verified on more than 50 different DNA sequences forming various non-canonical structural motifs.

Keywords: G-quadruplex; ligand; rhodamine; thiazole orange; thioflavin T

Citation: Trizna, L.; Janovec, L.; Halaganová, A.; Víglaský, V. Rhodamine 6G-Ligand Influencing G-Quadruplex Stability and Topology. *Int. J. Mol. Sci.* **2021**, *22*, 7639. https://doi.org/10.3390/ijms22147639

Academic Editor: Vaclav Brazda

Received: 17 June 2021
Accepted: 15 July 2021
Published: 16 July 2021

1. Introduction

G-quadruplexes (G4s) are relatively common in the genomes of all living cells, including viruses, but their frequency differs from species to species [1,2]. The G4 motif may be an integral part of some artificially developed DNA and RNA aptamers [3–6]. The dispersion of putative G4 sequences in genomes is not random, however, and their localization is closely correlated with specific gene functions [7]. An investigation of different genomes using various searching algorithms indicated that at least 3.10^5 and up to 3.10^6 G4-putative sequences can be formed in the human genome [8,9]. Therefore, in the past decade, considerable efforts have been made with the aim of developing small molecular probes capable of selectively recognizing G4s in therapeutic drug screening and biosensor construction since DNA molecules are not readily visible in such assays [10–14].

An extremely wide range of fluorophores which target nucleic acids have been identified to date, and several of these optical probes are routinely used in fluorescent microscopy studies to stain genetic material in the nucleus (e.g., DAPI and Hoechst) [15,16]. Frequently, π-π interactions between polyaromatic systems and nucleobases play a crucial role in determining the binding mode, typically through intercalation and insertion in between base pairs of duplex DNA or in end-stacking on the G-quartets of G4s. In addition to these binding modes, ligands can also bind to the grooves of DNA or through direct coordination. Additionally, electrostatic interactions play an important role in increasing the affinity between positively charged optical probes and the negatively charged phosphates found in nucleic acids.

In this study we investigate the interaction of series of known G4-DNA forming oligonucleotides with rhodamine dyes containing a fluorescent xanthene core; specifically, Rhodamine B (RhB) and Rhodamine 6G (RhG). The results were compared with those obtained for other well-characterized G4 ligands, primarily Thiazole Orange (TO) and Thioflavin T (ThT), Figure 1.

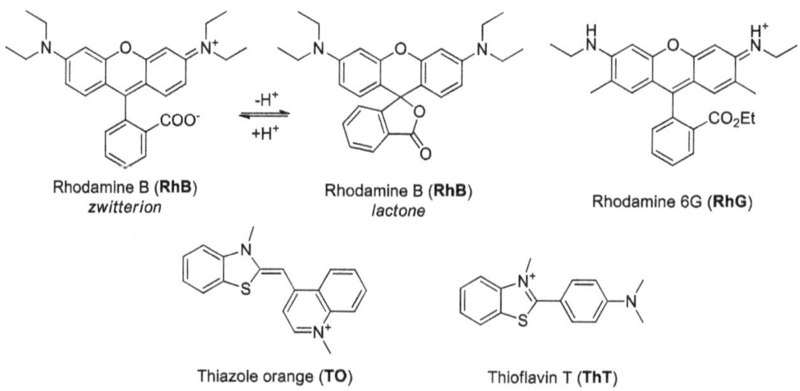

Figure 1. Structure of ligands directly used in this study.

RhB is a widely available dye which is commonly used as a water tracer or as a colorant in textiles and foodstuffs, but it can also serve as a fluorescent biomarker [17,18]. Nanoparticles consisting of RhB derivatives have shown considerable potential for applications in the field of biomedical sciences [19]. RhG is an organic laser dye which is suitable for use in studying the probes as it has a high quantum yield for fluorescence. As with RhB, this agent has a wide range of potential applications, ranging from use as a fluorescence tracker which can help in defining the spectroscopic characteristics with a high conversion efficiency and precision to its use as a leukocyte marker [20]. No cytotoxicity has been detected for RhG at μM concentrations in vivo [21]. ThT is an effective fluorescence probe in the detection of DNA and RNA G4s. Nucleolar G-quadruplexes in living cells have been visualized by using ThT and the high selectivity of the ligand allows researchers to distinguish between G4 and non-G4 structures [22]. The cyanine dye TO is widely used as a fluorescent probe which becomes illuminated upon binding to almost all forms of DNA, but the dye exhibits poor selectivity in differentiating G-DNA from other structural forms of DNA [23].

The adopted structure of each oligonucleotide was verified using UV absorption, CD spectroscopy and electrophoretic separation in the presence of either sodium or potassium ions. Circular dichroism (CD) spectroscopy has been used to monitor spectral profiles of non-canonical structural motifs structure formation under different conditions, mainly the presence of ligands and cations. This method was also combined with other techniques to identify other properties of the folded structures such as multimerization and stability. For this purpose, various types of electrophoreses, UV-Vis absorption and fluorescent spectroscopies were performed. Parallel and antiparallel G4 topologies can typically be identified by determining the position of the positive and negative peaks in CD spectra in the UV range of 230–320 nm [24]. In order to eliminate the false confirmation of conformation on the basis of CD spectra profiles alone, CD melting curves and temperature gradient-gel electrophoresis (TGGE) were used because, as is generally known, the stability and melting temperature of G4s are significantly higher in the presence of potassium than in the presence of sodium ions [25]. Other non-canonical forms are significantly less sensitive to the presence of potassium. In addition, the ligand gradient-gel electrophoresis (LGGE) was also used in this study to monitor the influence of ligand to G4 topology.

The main goal of this study is to demonstrate that the fluorophore RhG selectively binds to parallel forms of G4s. In order to verify the relatively high selectivity, other sequences capable of forming non-canonical structures were also analyzed.

2. Results and Discussion

2.1. The Spectral Properties of DNA-Ligand Complexes

Parallel G-quadruplex structures exhibit a clear positive band at ~265 nm and a negative peak at ~240 nm, while antiparallel G4 structures exhibit a positive CD signal at ~295 nm and a negative signal at ~265 nm. In contrast, the so-called (3 + 1) conformer, in which three strands are in the same alignment with another strand oriented in the opposite direction, exhibits a positive shoulder at 265—270 nm, but it should be noted that a mix of parallel and antiparallel structures can show signatures which are close to the topology of (3 + 1) conformers [26,27]. However, other structural forms may display a positive peak close to 265 nm, but this spectrum does not necessarily indicate the presence of a G-quadruplex [28,29]. CD spectra can also be used for the detection of i-motifs; the maximum and minimum Cotton effects at 288 and 258 nm are indicative of the formation of this structure and the peaks at 275 and 249 nm are indicative of unstructured DNA [30]. Interestingly, some achiral ligands binding to DNA form a chiral complex which shows an induced CD (ICD) signal close to the wavelength region of absorption, but it should be noted here that ICD can also be observed in the UV region in which G4s show a characteristic CD signal with the UV-ICD signal interfering with an original signal corresponding to G4 formation. The results show the unique ICD profile in visible region caused by G4-ligand interaction [31,32]. A ligand causing ICD in a G4-ligand complex usually stabilizes the G4 motif, but it can also induce topological changes and facilitate G4 multimerization [27]. On this basis, ICD signatures can be used to determine whether a sequence forms G4 motif.

More than 50 different oligomeric sequences which form different non-canonical structures have been analyzed; see the Material and Methods section for more details.

The UV-Vis spectra of the studied ligands are shown in Figure 2. In the absence of DNA, each ligand shows an absorption signal in the visible wavelength in the range of 350–560 nm, and an ICD signal is therefore expected in this region.

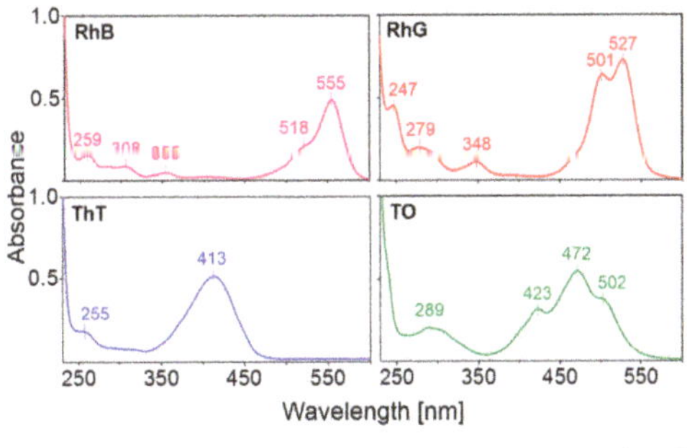

Figure 2. Absorption spectra of 130 μM ligands used in this study in a 25 mM mRB buffer, pH 7.0.

The representative spectral results for various non-canonical motifs in the presence of various ligands are shown in Figure 3. All DNA sequences summarized in the Table 1 were analyzed using CD and absorption spectroscopy.

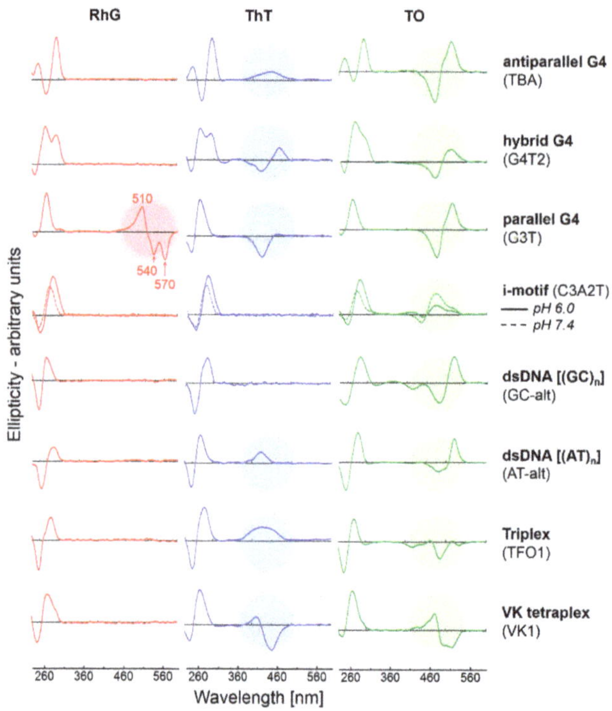

Figure 3. Representative CD spectra of DNA sequences able to adopt different non-canonical structures. The concentration of DNA and ligand was 27 and 130 μM (~1:5 eqv.), respectively, in each measurement. ICD signals are highlighted with colored circles. The mBR contains 50 mM NaCl. The positive and two negative peaks are observed at ~510, ~540 and ~570 nm, respectively.

Table 1. Selected melting temperatures obtained using CD spectroscopy.

Oligo	Wavelength [nm]	Melting Temperature [°C]			
		50 mM NaCl		50 mM KCl	
		No Ligand	RhG	No Ligand	RhG
HTR	294	52.0	52.0	63.5	64.0
Scle	264	62.0	77.0	81.5	>100
TBA	294	20.0	26.8	46.5	48.5
Hema	264	ND	54.3	72.0	86.0
STAT	264	54.5	76.5	92.8	>100
HCV	264	44.5	60.0	72.6	86.0
ionK	294	46.2	48.5	59.1	58.0
VEGF	264	47.5	82.0	85.6	87.9
		no ligand [a]		RhG [a]	
C3A2T [b]	286	28.3		24.3	
TFO1	282	20.4		17.7	

[a] obtained in absence of salt, [b] obtained in pH 6.0 and ND—not determined.

The results clearly demonstrate that, with the exception of RhB, ICD signals only occur when G4s are formed regardless of the ligand used. However, the results also suggest the poor selectivity of ThT and TO, with ICDs being observed for many different non-canonical forms, including that of the dsDNA-ThT complex; the presence of ICD is a signature of DNA-ligand interaction. Nevertheless, the profile of the G4-TO complexes shows some common features as has been demonstrated in our previous study [31,32]. The profile of G4-TO complexes is very similar to those of other G4 putative sequences. However, the ICD signal obtained with TO does not allow us to distinguish between parallel and antiparallel G4 topologies. In contrast, the ICD signal with RhG is observed only in the case of parallel G4 topologies, and we can therefore suggest that the selectivity of RhG is restricted for the determination of parallel G4s. In addition, the presence of salts was found to have interfered only slightly with the shape of ICD profiles, Figure 3. Any parallel G4 structure exhibits almost the same CD spectral features as those obtained with the G3T oligonucleotide in the presence of RhG.

RhG versus RhB

In contrast to RhB, RhG is positively charged in a neutral condition, and a significantly higher affinity with DNA would therefore be expected. In addition to its above-mentioned affinity G4s, its significant advantage of RhG over RhB is the presence of an ethyl ester group in its structure, Figure 1. The presence of an ester protects the carboxyl group and blocks the formation of a lactone cyclic structure, resulting in the greater stability of the RhG structure. RhB is also able to form cyclic forms, a factor which increases its structural variability, and which may also explain why the affinity for G4 structures is not as pronounced as that recorded for RhG. In addition, the carboxyl group forms a COO-anionic structural form under certain environmental conditions (e.g., high pH), which also has an adverse effect on its affinity for negatively charged DNA [33].

2.2. *RhG: Influence on Polymorphism and Stability*

The Scle core sequence (d[TGGGGGGGTGGTGGGT]) derived from the sclerostin binding aptamer [34] adopts a clear parallel G4 structure in the presence of 50 mM potassium; a positive CD signal is observed at 265 nm, Figure 4. The results also show an influence of increasing concentration of DNA in the presence of 130 μM of RhB and RhG. The RhG ICD signal strength is dependent on Scle concentration. The electrophoretic separation shows unambiguously that at least three different folds of Scle can be formed under the given conditions. A clear isosbestic point at ~539 nm in the absorption spectra was also observed in the RhG spectra, panel C. However, the RhB spectra indicate a different pattern of behavior; a negligible effect on spectral shift and an unclear isosbestic point were detected, panel F. Nevertheless, the influence of both RhB and RhG on the distribution of topological forms are also clear from the results; the intensity and position of bands are different from those observed in the PAGE experiment without the presence of the ligands. An intensive ICD signal was also observed for the Scle-RhG complex. G3T showed a very similar sequence to that of Scle is G3T, with only one G base being substituted by T.

The results shown in Figure 5 demonstrate the effect of RhG on a series of sequences $d(G_3NG_3)G_3$, where N represents either C, T or A, respectively. Any of these sequences adopts parallel G4 in either the presence or absence of RhG ligand regardless of whether sodium or potassium cations are present; CD positive peaks are observed at 265 nm. CD spectral features for this set of oligonucleotides are almost identical to those observed for the core of Scle sequence. The electrophoretic results show the effect of RhG on electrophoretic mobility and on the elimination of the number of folds, primarily for the less stable G3A oligomer which shows the highest level of polymorphism. These set of oligonucleotides preferentially form dimeric structure. These results suggest that RhG has a significant effect on the folding process, with multimeric topological forms being facilitated. The electrophoreses of other sequences adopting different non-canonical motifs are summarized in Supplementary Figure S1. These results also confirm the effect of ligand-induced multimerization.

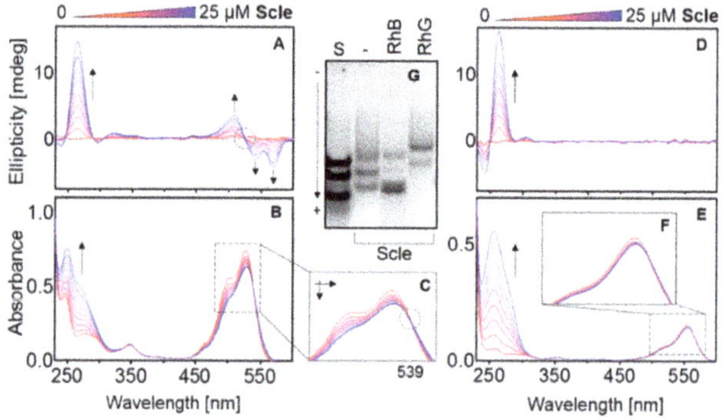

Figure 4. CD and UV-Vis spectra of 130 μM RhG (**A**,**B**) and RhB (**D**,**E**) in mRB, supplemented with 50 mM KCl, pH 7.0. The increment of the Scle oligomer is 3.35 μM. The final concentration of DNA is ~27 μM (0.2 ekv). The enlarged UV-Vis region of the RhG absorption spectra shows a clear isosbestic point at 539 nm (**C**), but not in (**F**). 12% PAGE (**G**) in corresponding conditions; the concentration of the ligand in the two columns is 130 μM. The standard is a mixture of oligonucleotides AC9, AC18 and AC28.

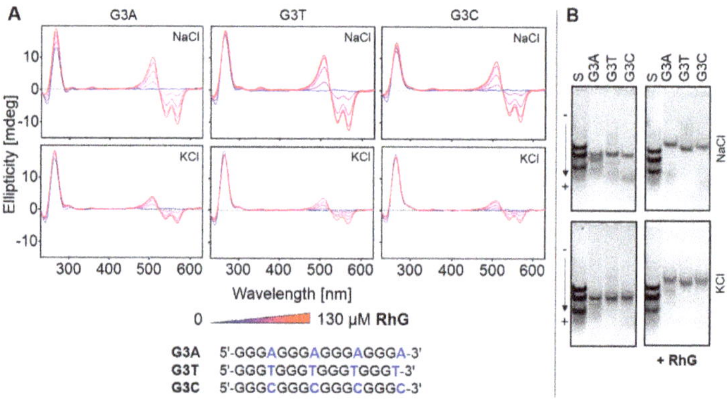

Figure 5. CD titration spectral measurements and PAGE of $d(G_3NG_3)G_3$ sequences (~27 μM) at different ionic conditions in the presence of increasing concentrations of RhG up to 260 μM (**A**). The increment of RhG is ~33 μM. The left and right PAGE panels (**B**) represent electrophoretic records in the absence and presence of 130 μM of ligand, respectively. Electrophoresis was performed in the presence of both 50 mM NaCl and KCl. The mixture of AC9, AC18 and AC28 is used as standard.

2.3. *Temperature and Concentration Measurements*

CD and UV melting analyses were performed using the method described in our previous studies [27,28]. However, as has already been mentioned, DNA sequences rarely adopt only a single well defined and stable conformation, and instead typically form a wide range of different topological isoforms. This feature may explain why the spectral measurements display a melting curve which represents the average melting of a mix of topological forms which have occurred in the solution. It is therefore not appropriate to apply van't Hoff analysis in the case of this type of melting curve as this approach is intended for use with two-state systems [35,36]. As a result, it should be noted that the selected values of melting temperature of G4s and one triplex and i-motif obtained using CD spectroscopy which are summarized in Table 1 represent only indicative values. In addition, spectral melting curves of this type cannot offer an unambiguous explanation of declination from the two-state mechanism. The corresponding electrophoreses are shown in Supplementary Figure S1.

Although the values are only indicative, it is clear that antiparallel and hybrid G4 conformers are only slightly stabilized with RhG, but the melting temperature of parallel G4s shows a more significant increase. The triplexes and i-motifs have been destabilized with RhG. In order to provide clearer evidence, G4 forming sequences were also examined using TGGE. The results shown in Figure 6 include TGGE results for HTR in both the presence and absence of RhG in the gel. In order to eliminate the occurrence of a high electric current only 2.5 mM of KCl was used. The corresponding CD measurements under identical conditions may help to clarify the melting mechanism and the influence of RhG on this process. Parallel G4 structures were not found to be the dominant form at lower temperatures, even with the presence of RhG, but a temperature increase resulted in an increasing population of parallel G4 structures at the expense of antiparallel hybrid conformations. This change can be seen clearly in the dotted CD melting curve obtained at 264 nm. However, the parallel topology was also found to be more stable than the hybrid structure.

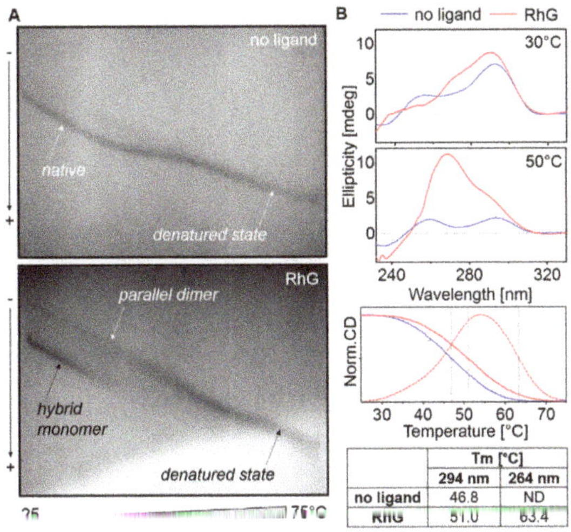

Figure 6. (**A**) TGGE record of HTR sequence in 25 mM mBR, pH 7.0 supplemented with 2.5 mM KCl (up). The corresponding electrophoretic result contained 260 μM of Rh6G (down). (**B**) CD spectra under the same conditions as the TGGE assay in the presence and absence of RhG. The temperature dependences were obtained at 264 (red dashed line) and 294 (solid lines) nm. CD melting temperatures are shown in the enclosed table. These temperatures agree with those obtained with TGGE: 46.5 °C and 50.6 °C for antiparallel G4 in the absence and presence of RhG, respectively, and >62 ± 2 °C for parallel G4 with RhG.

The TGGE profiles show clearly that the mobility of the parallel G4 dimer differs only slightly from that of the denatured form, and therefore this method is not generally applicable in melting analyses of any G4 structures. Another interesting example which demonstrates the influence of the ligand to G4 structure is the unorthodox arrangement found in the LGGE electrophoresis, in which the nonlinear gradient of the ligand is applied in a perpendicular direction to the sample movement, Figure 7. This methodology allows some details concerning the interaction of DNA with the ligand and the folding process to be clarified. The results demonstrate the influence of the studied rhodamines on the multimerization of G4 and on the occurrence of other topologies. Increasing concentrations of RhG resulted in an abundance of parallel G4 structures but also of multimeric forms.

The isothermal analysis performed at laboratory temperature enables the visualization of the occurrence of native conformers at various concentrations of the ligand. A sequence which occurs in the HIV genome was used for this purpose [37]. The effect observed in the case of RhG was not recorded with RhB, a result which is probably due to the weak interaction of the ligand with G4. No conversion to parallel topology or multimerization occurred and, of course, no ICD signal was detected, not even in the presence of other oligonucleotides. Nonetheless, RhB did exert a slight stabilization effect on G4 structures, an effect which was primarily observed in conditions in which sodium was present, but potassium was absent; the T_m value was seen to have increased by approximately 0.5–3 °C (not shown). The continuous mobility profile of the electrophoretic band allowed the entire topological conversion connected with the multimerization of the appropriate sequence to be monitored. Using a combination of LGGE with, for example, CD titrating analysis, it is possible to determine many of the "hidden" details of ligand-DNA interaction. This methodology is introduced here for the first time, and we believe that it will prove to be a useful tool in future DNA-ligand interaction studies.

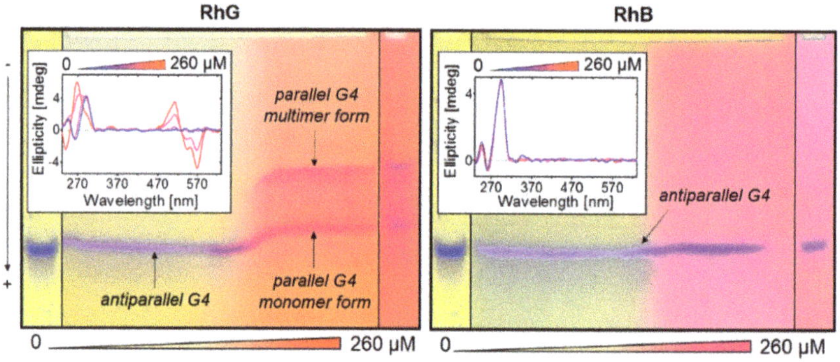

Figure 7. LGGE of HIV-M27 (d-[GTGGCCTGGGCGGGACTTGGGA]) performed in a 25 mM mRB buffer, pH 7.0 supplemented with 50 mM KCl and 0 to 260 µM rhodamines RhG (**left**) and RhB **right**). The concentration of polyacrylamide was 12%. The inset represents a corresponding CD spectrum under the same conditions. The concentration of ligand in CD was 0–260 µM, the increment is 65 µM. The G4 conversion from antiparallel to parallel monomer and dimer is highlighted with arrows. The left and right columns represent standard PAGE of HIV-M27 performed in gels containing 0 and 260 µM of ligands, respectively.

2.4. Fluorescence Spectroscopic Properties of RhB and RhG

The two reference fluorescent ligands ThT and TO which target G4 structures have been studied in depth as fluorescent G4 ligands, with both dyes displaying considerable fluorescence yields when bound to G4 in comparison to their independent fluorescence in solution without the target structures [38,39]. The selectivity of TO was lower than that of ThT, and TO also displays a strong illumination effect upon association with various topological forms of nucleic acids [38]. The binding constants of the two agents are in the micromolar range [40,41]. As was mentioned above in the introduction, rhodamines are also fluorophores, and it is therefore appropriate to analyze their interactions with DNA using fluorescence spectroscopy, thereby allowing the DNA-rhodamine complexes to be determined in more detail. The most relevant results are shown in Figure 8. The fluorescence of RhG was quenched when the ligand interacted with DNA, but the strongest effect was observed for parallel G4 structures, panel A, although other structural forms of DNA were found to quench RhG fluorescence to a less significant degree (not shown). However, RhB quenching was almost negligible for all the DNA sequences used, including those featuring G4 motifs, panel B. As evidence of the affinity of RhG, the promising G4 ligand of ThT was displaced from the ThT-G4 complex by RhG and RhB, panel C and D, respectively. It is evident that the signal corresponding to the ThT-G4 complex at ~485 nm was eliminated at increasing amounts of RhG and the signal at 555 nm corresponding to RhG was found to increase; ThT is displaced by RhG. However, this effect was not observed in the case of RhB. A crossover point analogical to the isosbestic point was also observed which indicates that these spectra are coupled. Even though the concentration of RhB was 10-fold higher, no light-down corresponding to ThT-G4 complex was observed. Although the determination of the binding constant of RhG to DNA was not a primary aim of this study, it was possible to estimate this value based on the results of the experiments monitoring G4 ligand displacement by rhodamines. The binding constant of RhG falls in the same region as that of ThT because an equivalent amount of RhG can displace ThT. However, this constant varies, and it is dependent on the G4 topology and the presence of cationic molecules. Due to the complexity of this relationship, a deeper analysis of the binding constant lies beyond the scope of this study.

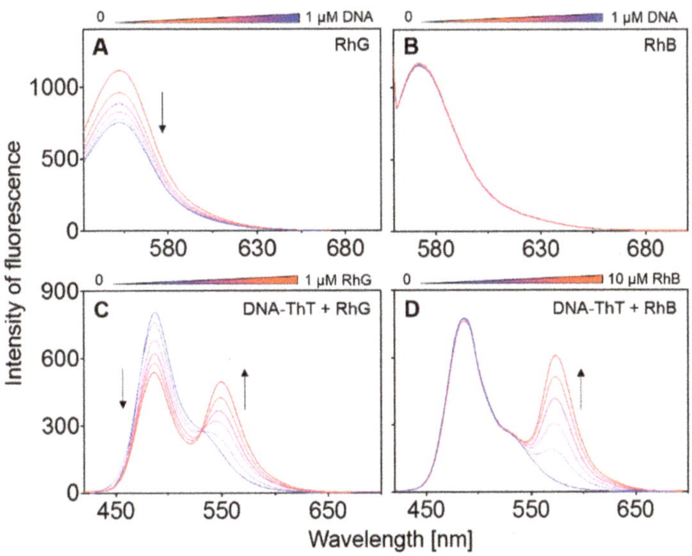

Figure 8. Fluorescence titration of 25 nM RhG (**A**) and RhB (**B**) with the stepwise addition of the G3T oligonucleotide (0, 0.2, 0.4, 0.6, 0.8 and 1.0 μM) corresponding to 0–40 molar equivalents. Increasing concentrations of DNA are highlighted with arrows. Measurements were performed in a mRB buffer supplemented with 50 mM KCl at a pH of 7.0; excitation of RhG and RhB at 527 nm and 555 nM, respectively, the excitation and emission slits were 2.5 nm (5 nm in B) and the scan speed was 240 nm/min. 1 μM G4-ThT mixture (1: 1 molar eqv.) titrated with 0–1 μM RhG (**C**) and 0–10 μM RhB (**D**), the excitation was at 413 nm.

2.5. Molecular Modeling of Ligand-G4 Interactions

Docking simulations were also carried out to demonstrate the putative binding of RhG within G4 structures. The simulations of ligand binding were performed with structures representing parallel dimer (2le6) and hybrid (2jpz) G4 structures because RhG can be shown to affect these topologies, Figure 9. 2le6 [5′-d(GIGTGGGTGGGTGGGT)-3′] and 2jpz [5′-d(TTAGGGTTAGGGTTAGGGTTAGGGTT)-3′] represent structural analogs to G3T and HTR structures, respectively. The docking simulation may not represent the true structure of the DNA-ligand complex because the ligand interaction may slightly alter the initial coordinate values of the atoms in the G4 structure, and this declination may continue until the complex reaches its most stable form. The docking simulation identifies the best configuration for a fixed structure in terms of the given data. Nevertheless, the 2le6 structure could represent a structure which is close to that induced with RhG. The most populated binding clusters of the five anchored RhGs could be a source of the strong ICD signals, e.g. Figure 4a. The best matches were obtained for structures in which RhG was bound into the G4 grooves close to cavity formed by G4 loop. Although this type of molecular modeling did not confirm the presence of stacking interactions with terminal G-quartets, an arrangement which is typically observed with NMR or crystallographic data, we cannot rule out the possibility that such an interaction occurs in G4-RhG complexes [41,42]. Nevertheless, this set of results demonstrates possible places where the initial attachments of RhG with the folded G4 structure occur and not a consequent G4 structure modification driven by the ligand. We realize that PDB sequences are not the same as studied oligomers, but results obtained with docking simulation show that G4 structures contain the binding sites tailored for RhG.

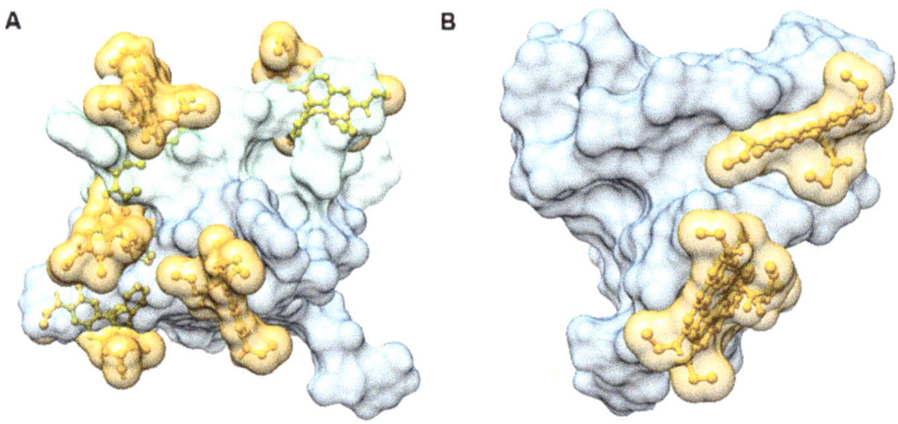

Figure 9. Putative binding of the RhG ligand within the quadruplex structure PDB 2le6 (**A**) and 2jpz (**B**) obtained from docking simulations. Only the leading structures of the most populated binding clusters are depicted. The quadruplex is drawn in a solvent-accessible surface representation. The ligand is shown in a ball and stick representation. The solvent-accessible surface of the ligand is also shown. The ligand is shown fitting into the quadruplex grooves. The subunits of 2le6 are colored with different hues, pale green and blue grey. Images were prepared using Chimera software [43].

3. Materials and Methods

All chemicals and reagents were obtained from commercial sources. DNA oligonucleotides were obtained from Metabion, Germany (Table 2). PAGE purified DNA was dissolved in double distilled water prior to use. Thiazole Orange, Thioflavin T, Rhodamine B and Rhodamine 6G were purchased form Merk, Slovakia (390062, T3516, R6626, 252441). Single strand concentrations were determined precisely by measuring absorbance (~260 nm) at 95 °C using molar extinction coefficients [44]. DNA concentrations were determined using UV measurements carried out on a Jasco J-810 spectropolarimeter (Easton, MD, USA).

Table 2. Sequences of oligonucleotides used in the present study.

No.	Name	Sequence in 5′→3′ Direction	Category and Preferred Motif	
1	G3A	GGGAGGGAGGGAGGGA		
2	G3C	GGGCGGGCGGGCGGGC		
3	G3T	GGGTGGGTGGGTGGGT		
4	G3T2	GGGTTGGGTTGGGTTGGG		
5	G3T3	GGGTTTGGGTTTGGGTTTGGG		
6	G3T4	GGGTTTTGGGTTTTGGGTTTTGGG		
7	HTR	GGGTTAGGGTTAGGGTTAGGG		
8	HTR2	AGGGTTAGGGTTAGGGTTAGGGT	G_3N_n [31]	
9	HTR-T	GGGTTAGGGTTAGGGTTAGGGT		
10	G3T2C	GGGTTCGGGTTCGGGTTCGGG		
11	8G3	GGGTTAGGGTTAGGGTTAGGGTTAGGGTTAGGGTTAGGGTTAGGG		
12	8G3T2	GGGTTGGGTTGGGTTGGGTTGGGTTGGGTTGGGTTGGG		
13	8G3T3	GGGTTTGGGTTTGGGTTTGGGTTTGGGTTTGGGTTTGGGTTTGGG		
14	G3-3-A20	GGGTTAGGGTTAGGGTTAGGGAAAAAAAAAAAAAAAAAAAA		
15	G3-5-T20	TTTTTTTTTTTTTTTTTTTTGGGTTAGGGTTAGGGTTAGGG		
16	G4T	GGGGTGGGGTGGGGTGGGG		
17	G4T2	GGGGTTGGGGTTGGGGTTGGGG		
18	G4T3	GGGGTTTGGGGTTTGGGGTTTGGGG	G_4N_n [31]	
19	G4T4	GGGGTTTTGGGGTTTTGGGGTTTTGGGG		
20	G4T2A	GGGGTTAGGGGTTAGGGGTTAGGGG		
21	HCV	GGGCGTGGTGGGTGGGGT		
22	Hema	GGGGTCGGGCGGGCCGGGTG		
23	HIV	GGGGTGGGAGGAGGGT		G-quadruplex
24	Insu	GGTGGTGGGGGGGGTTGGTAGGGT		
25	ionK	GGGTTAGGGTTAGGGTAGGG		
26	OCH-A	CGGGTGTGGGTGGCGTAAAGGGA	Aptamers [45]	
27	Scle	TGGGGGGGTGGGTGGGT		
28	STAT	GGGCGGGCGGGCGGG		
29	TBA	GGTTGGTGTGGTTGG		
30	TBA-5T	GGTTGGTGTGGTTGGTTTTTGGTTGGTGTGGTTGG		
31	VEGF	GGGGCGGGCCGGGGGCGGG		
32	HIV1-K02	GTGGCCTGGGCGGGACTGGGGA		
33	HIV1-K03	CGGGGTTGGGAGGTGGGT	HIV [37]	
34	HIV1-L20	TGGGAGGGATAAGGGGCGGTTCGGGGA		
35	HIV1-M27	GTGGCCTGGGCGGGACTTGGGA		
36	E-Cote2	TGGGGAGGGTGGGGAGGGTGGGGAAGG		
37	E-Cote4	TGGGATGGGTGGGGTGCTTGTCTGGGGC	Ebola virus [32]	
38	MarRavn	GTGGTCGGCGTGGGGGGGAGGGT		
39	c-myc	TGGGGAGGGTGGGGAGGGTGGGGAAGG		
40	N-myc	TAGGGCGGGAGGGAGGGAA		
41	pUC-G1	GGGGTGTTGGCGGGTGTCGGGGC	Others	
42	RAN	TGGGGGTGGGGTTGGGTGGTGT		
43	RAN-del	TGGGGGTGGGGTTGGGTGGT		
44	Z-G4	TGGTGGTGGTGTGGTGGTGGTGGTGTT		
45	i-HTR	CCCAATCCCAATCCCAATCCC		
46	i-HTR2	TCCCAATCCCAATCCCAATCCCA	i-motif	
47	C3-Msl1	CCCTAACCCTAAACCCTAACCC		
48	AC9	ACACACACA		
49	AC12	ACACACACACAC	ssDNA	
50	AC18	ACACACACACACACACAC		
51	AC28	ACACACACACACACACACACACACACAC		
52	AT-alt	ATATATATATATCCCATATATATATAT		
53	GC-alt	GCGCGCGCGCGCTTTGCGCGCGCGCGC	dsDNA	
54	ctDNA	Unspecified calf thymus DNA		
55	TFO1	AAAAAAAACCCCTTTTTTTTCCCCTTTTTTTT	triplex	
56	TFO2	AGAGAGAACCCCTTCTCTCTTATATCTCTCTT		
57	VK1	GGGAGCGAGGGAGCG	AG-tetraplex [29]	

3.1. Circular Dichroism Spectroscopy

CD spectra were recorded on a Jasco J-810 spectropolarimeter equipped with a PTC-423L temperature controller using a quartz cell of 1 mm optical path length in a reaction volume of ~150 μL; instrument scanning speed of 100 nm/min, 1 nm pitch and 1 nm bandwidth, with a response time of 2 s. CD data represents three averaged scans taken at a temperature range of 0–100 °C. Scans were performed over a range of 220–700 nm. All other parameters and conditions were the same as those which were described previously [27]. A modified Britton–Robinson buffer (mBR) in which TRIS was used instead of potassium/sodium hydroxide (25 mM phosphoric acid, 25 mM boric acid and 25mM acetic acid) was used in all spectral analyses and was supplemented by either 2.5–50 mM potassium chloride or 50 mM sodium chloride; pH was adjusted by TRIS to a final value of 7.0. I-motif was also measured in acidic conditions (pH 4–6). DNA titration was performed with increasing concentrations of the ligand. Each ligand was solubilized in DMSO or ethanol to reach a final concentration of 10 mM in the stock solution. The concentrations of DNA and ligand in the 1 mm quartz cell were 30 μM and 0–200 μM, respectively, and the increment of the ligand was ~67 μM. Each sample was mixed vigorously for 3 min following the addition of ligand; CD/UV spectra were performed immediately.

3.2. CD Melting Curves

CD melting profiles were collected at ~295 and ~265 nm as a function of temperature using a procedure which has been published previously [27,36]. The temperature ranged from 0 to 100 °C, and the heating rate was 0.25 °C per minute. The melting temperature (Tm) was defined as the temperature of the mid-transition point.

3.3. Electrophoresis

Samples consisting of 0.3 μL of 1mM stock solutions were separated using nondenaturing PAGE in a temperature-controlled electrophoretic apparatus (Z375039-1EA; Sigma-Aldrich, San Francisco, CA, USA) on 12% acrylamide (19: 1 acrylamide/bisacrylamide) gels. DNA was loaded onto 13 by 16 by 0.1 cm gels. Electrophoresis was run at 10 °C for 2 h at 125V (~8 $V \cdot cm^{-1}$). Each gel was stained with StainsAll (Sigma-Aldrich). All electrophoretic measurements were performed in a mBR buffer at pH 7.0. Temperature gradient gel electrophoresis (TGGE) equipment was used according to a method which has been described previously [44,45]. The gel concentration was 12%. Electrophoreses were run perpendicularly to the temperature-gradient (20–80 °C) for 3 h at 160 V (~8 $V \cdot cm^{-1}$). Approximately 12 μg of DNA was loaded into the electrophoretic well. DNA oligomers were visualized with Stains-all after the electrophoresis. Ligand gradient gel electrophoresis (LGGE) is similar to denaturing gradient gel electrophoresis (DGGE), but in place of a denaturing agent, a concentration gradient of ligand (0–260 μM) is applied perpendicularly to the movement of the DNA sample. The same apparatus used for standard PAGE analyses was used in this assay. The technique was developed and applied in our laboratory to monitor the folding and multimerization effect of the ligands on G4 structures.

3.4. Fluorescence Spectroscopy

Fluorescence spectra were acquired at 20 ± 1 °C with a Jasco FP-8300 Spectrofluorometer which was equipped with a Peltier temperature controller ETC-815. A quartz cuvette with a 10 mm path length was used in all experiments. In the fluorescence measurements, the excitation and emission slits were 2.5–5 nm, and the scan speed was 240 nm/min. Then, 25 nM of ligand was titrated with DNA (0–1 μM) in a mRB buffer in both the presence and absence of monovalent metal cations. The molar ratios between DNA and ligand were 1:40, 1:32, 1:24, 1:16 and 1:8. The excitation wavelength was adjusted to 527 and 413 nm for RhG and ThT, respectively.

3.5. Docking Studies

Molecular models of RhG was created using the building options in an ACD/ChemSketch (ACD/ChemSketch package 2020.2.0 www.acdlabs.com (accessed on 17 June 2021). The models were built as 3D structures and saved as *Mopac* input files using the ACD/3D Viewer (ACD/3D Viewer package 2020.2.0 www.acdlabs.com (accessed on 17 June 2021). MOPAC2016 was used to optimize the ligand geometry [46]. Chimera software [43] was used to extract coordinates of a G4 structure from a pdb file id: *2jpz* [47]; *2le6* [48]. As NMR spectroscopy was used to determine the coordinates of a G4 structure, only the coordinates of the first model were selected for a docking simulation. In the case of the quadruplex id: *2le6*, the position of the last residue dT(16) was changed in order to allow stacking interactions with a ligand during the docking run. The position of the last nucleotide was changed using the Structure Measurements module in the Chimera software program; the torsion angle defined as dG(15).A C3′–dG(15).A O3′–dT(16) P–dT(16) O5′ was changed from −74 to +74 degrees. MGL TOOLS 1.5.6 software was used to assign Gasteiger partial atomic charges to the G4 structure [49]. The Antachamber module of the Ambertools 18 software package was used to derive charges for the ligands via the AM1-bcc method. Docking simulations were carried out using Autodock ver. 4.2, while MGL TOOLS 1.5.6 was used to prepare the input files [50]. United atom representations were used for the ligands and G4 structures. The grid for energy for G4s pdb id: *2jpz* and *2le6* was set at the center of the macromolecule with the dimensions of 120 × 120 × 120 points (x,y,z) and a spacing of 0.375 Å. Docking runs were performed using a Larmarckian genetic algorithm. Docking began with a population of random ligand conformations in a random orientation and at a random translation. Each docking experiment was derived from 100 different runs which were set to terminate after a maximum of 25×10^5 energy evaluations or 27×10^3 generations, yielding 100 docked conformations. The population size was set to 300. For other parameters, the default values were used. Five docking runs were performed for the ligand.

4. Conclusions

In conclusion, the results of the study indicate that RhG acts as a promising stabilizer of G4 structures. The ligand was found to bind preferentially to parallel G4 topologies and to promote G4 multimerization, while the fluorescence quenching induced with G4, and the resultant ICD values are highly significant in comparison with other DNA structures. The findings of the computer modeling predict that RhG binds to the grooves of G4 structures. In addition, LGGE is the first application to our knowledge that can demonstrate the concentration effect of the ligand on the G4 topology. Our results regarding the selectivity of RhG to G4s could serve as a starting point for the development and synthesis of novel fluorescent organic and metalloorganic G4-probes derived from the basic skeleton of RhG. Highly selective optical probes are frequently required for the construction of functionalized-nanoparticles and drug delivery systems, and therefore these types of small molecules show great potential for future applications in molecular biology and in a wide range of biomedical fields.

Supplementary Materials: The following are available online at https://www.mdpi.com/article/10.3390/ijms22147639/s1, Figure S1: PAGE of DNA oligonucleotides used in this study.

Author Contributions: L.T. and V.V. performed the conceptualization and design of the work. L.J. carried out the majority of the bioinformatic simulation analyses. L.T. performed most of the experimental measurements. A.H. initially participated in the spectral analyses of some of the DNA-ThT complexes. V.V. was responsible for the supervision, funding application, project administration, and sourcing of resources. All authors have read and agreed to the published version of the manuscript.

Funding: This work was supported by the Slovak Grant Agency (1/0138/20) and internal university grant (VVGS-PF-2020-1431).

Institutional Review Board Statement: Not applicable.

Informed Consent Statement: Not applicable.

Data Availability Statement: The data presented in this study are available in the article and Supplementary Materials.

Conflicts of Interest: The authors declare no conflict of interest. The funders had no role in the design of the study; in the collection, analyses, or interpretation of data; in the writing of the manuscript, or in the decision to publish the results.

Abbreviations

CD, circular dichroism; G4, G-quadruplex; G-rich sequence, guanine-rich sequence; ICD, induced circular dichroism; LGGE, ligand gradient gel electrophoresis; mBR, modified Britton-Robinson buffer; RhG, rhodamine 6G; RhB, rhodamine B; TGGE, temperature gradient gel electrophoresis; ThT, thioflavin T; TO, thiazole orange.

References

1. Bohálová, N.; Cantara, A.; Bartas, M.; Kaura, P.; Šťastný, J.; Pečinka, P.; Fojta, M.; Mergny, J.L.; Brázda, V. Analyses of viral genomes for G-quadruplex forming sequences reveal their correlation with the type of infection. *Biochimie* **2021**, *186*, 13–27. [CrossRef]
2. Ruggiero, E.; Richter, S.N. G-quadruplexes and G-quadruplex ligands: Targets and tools in antiviral therapy. *Nucleic Acids Res.* **2018**, *46*, 3270–3283. [CrossRef] [PubMed]
3. Viglasky, V.; Hianik, T. Potential uses of G-quadruplex-forming aptamers. *Gen. Physiol. Biophys.* **2013**, *32*, 149–172. [CrossRef]
4. Roxo, C.; Kotkowiak, W.; Pasternak, A. G-Quadruplex-Forming Aptamers-Characteristics, Applications, and Perspectives. *Molecules* **2019**, *24*, 3781. [CrossRef] [PubMed]
5. Banco, M.T.; Ferré-D'Amaré, A.R. The emerging structural complexity of G-quadruplex RNAs. *RNA* **2021**, *27*, 390–402. [CrossRef] [PubMed]
6. Mitra, J.; Ha, T. Nanomechanics and co-transcriptional folding of Spinach and Mango. *Nat. Commun.* **2019**, *10*, 4318. [CrossRef] [PubMed]
7. Eddy, J.; Maizels, N. Gene function correlates with potential for G4 DNA formation in the human genome. *Nucleic Acids Res.* **2006**, *34*, 3887–3896. [CrossRef]
8. Bedrat, A.; Lacroix, L.; Mergny, J.L. Re-evaluation of G-quadruplex propensity with G4Hunter. *Nucleic Acids Res.* **2016**, *44*, 1746–1759. [CrossRef] [PubMed]
9. Huppert, J.L.; Balasubramanian, S. Prevalence of quadruplexes in the human genome. *Nucleic Acids Res.* **2005**, *33*, 2908–2916. [CrossRef]
10. Huppert, J.L. Four-stranded nucleic acids: Structure, function and targeting of G-quadruplexes. *Chem. Soc. Rev.* **2008**, *37*, 1375–1384. [CrossRef]
11. Bidzinska, J.; Cimino-Reale, G.; Zaffaroni, N.; Folini, M. G-quadruplex structures in the human genome as novel therapeutic targets. *Molecules* **2013**, *18*, 12368–12395. [CrossRef] [PubMed]
12. Neidle, S. Quadruplex nucleic acids as targets for anticancer therapeutics. *Nat. Rev. Chem.* **2017**, *1*, 95. [CrossRef]
13. Ou, T.M.; Lu, Y.J.; Tan, J.H.; Huang, Z.S.; Wong, K.Y.; Gu, L.Q. G-quadruplexes: Targets in anticancer drug design. *ChemMedChem* **2008**, *3*, 690–713. [CrossRef] [PubMed]
14. Tao, Y.; Zheng, Y.; Zhai, Q.; Wei, D. Recent advances in the development of small molecules targeting RNA G-quadruplexes for drug discovery. *Bioorg. Chem.* **2021**, *110*, 804. [CrossRef]
15. Berrones Reyes, J.; Kuimova, M.K.; Vilar, R. Metal complexes as optical probes for DNA sensing and imaging. *Curr. Opin. Chem. Biol.* **2021**, *61*, 179–190. [CrossRef]
16. Guan, L.; Zhao, J.; Sun, W.; Deng, W.; Wang, L. Meso-Substituted Thiazole Orange for Selective Fluorescence Detection to G-Quadruplex DNA and Molecular Docking Simulation. *ACS Omega* **2020**, *5*, 26056–26062. [CrossRef] [PubMed]
17. Zhang, H.; Wang, L.; Jiang, W. Label free DNA detection based on gold nanoparticles quenching fluorescence of Rhodamine B. *Talanta* **2011**, *85*, 725–729. [CrossRef] [PubMed]
18. Xu, J.; Li, Y.; Wang, L.; Huang, Y.; Liu, D.; Sun, R.; Luo, J.; Sun, C. A facile aptamer-based sensing strategy for dopamine through the fluorescence resonance energy transfer between rhodamine B and gold nanoparticles. *Dyes Pigments* **2015**, *123*, 55–63. [CrossRef]
19. Kennedy, J.; Larrañeta, E.; McCrudden, M.T.C.; McCrudden, C.M.; Brady, A.J.; Fallows, S.J.; McCarthy, H.O.; Kissenpfennig, A.; Donnelly, R.F. In vivo studies investigating biodistribution of nanoparticle-encapsulated rhodamine B delivered via dissolving microneedles. *J. Control. Release* **2017**, *265*, 57–65. [CrossRef]
20. Jbeily, N.; Claus, R.A.; Dahlke, K.; Neugebauer, U.; Bauer, M.; Gonnert, F.A. Comparative suitability of CFDA-SE and rhodamine 6G for in vivo assessment of leukocyte-endothelium interactions. *J. Biophotonics* **2014**, *7*, 369–375. [CrossRef]
21. Thaler, S.; Haritoglou, C.; Choragiewicz, T.J.; Messias, A.; Baryluk, A.; May, C.A.; Rejdak, R.; Fiedorowicz, M.; Zrenner, E.; Schuettauf, F. In vivo toxicity study of rhodamine 6G in the rat retina. *Investig. Ophthalmol. Vis. Sci.* **2008**, *49*, 2120–2126. [CrossRef] [PubMed]
22. Xu, S.; Li, Q.; Xiang, J.; Yang, Q.; Sun, H.; Guan, A.; Wang, L.; Liu, Y.; Yu, L.; Shi, Y.; et al. Thioflavin T as an efficient fluorescence sensor for selective recognition of RNA G-quadruplexes. *Sci. Rep.* **2016**, *6*, 24793. [CrossRef]
23. Largy, E.; Granzhan, A.; Hamon, F.; Verga, D.; Teulade-Fichou, M.P. Visualizing the quadruplex: From fluorescent ligands to light-upprobes. *Top. Curr. Chem.* **2013**, *330*, 111–177.
24. Vorlíčková, M.; Kejnovská, I.; Bednářová, K.; Renčiuk, D.; Kypr, J. Circular dichroism spectroscopy of DNA: From duplexes to quadruplexes. *Chirality* **2012**, *24*, 691–698. [CrossRef]

25. Hud, N.V.; Smith, F.W.; Anet, F.A.; Feigon, J. The selectivity for K+ versus Na+ in DNA quadruplexes is dominated by relative free energies of hydration: A thermodynamic analysis by 1H NMR. *Biochemistry* **1996**, *35*, 15383–15390. [CrossRef]
26. Luu, K.N.; Phan, A.T.; Kuryavyi, V.; Lacroix, L.; Patel, D.J. Structure of the human telomere in K+ solution: An intramolecular (3 + 1) G-quadruplex scaffold. *J. Am. Chem. Soc.* **2006**, *128*, 9963–9970. [CrossRef]
27. Tóthová, P.; Krafčíková, P.; Víglaský, V. Formation of highly ordered multimers in G-quadruplexes. *Biochemistry* **2014**, *53*, 7013–7027. [CrossRef]
28. Tlučková, K.; Marusič, M.; Tóthová, P.; Bauer, L.; Sket, P.; Plavec, J.; Víglasky, V. Human papillomavirus G-quadruplexes. *Biochemistry* **2013**, *52*, 7207–7216. [CrossRef] [PubMed]
29. Kocman, V.; Plavec, J. A tetrahelical DNA fold adopted by tandem repeats of alternating GGG and GCG tracts. *J. Nat. Commun.* **2014**, *5*, 5831. [CrossRef] [PubMed]
30. Iaccarino, N.; Di Porzio, A.; Amato, J.; Pagano, B.; Brancaccio, D.; Novellino, E.; Leardi, R.; Randazzo, A. Assessing the influence of pH and cationic strength on i-motif DNA structure. *Anal. Bioanal. Chem.* **2019**, *411*, 7473–7479. [CrossRef] [PubMed]
31. Demkovičová, E.; Bauer, L'.; Krafčíková, P.; Tlučková, K.; Tóthova, P.; Halaganová, A.; Valušová, E.; Víglaský, V. Telomeric G-Quadruplexes: From Human to Tetrahymena Repeats. *J. Nucleic Acids* **2017**, *2017*, 9170371. [CrossRef] [PubMed]
32. Krafčíková, P.; Demkovičová, E.; Víglaský, V. Ebola virus derived G-quadruplexes: Thiazole orange interaction. *Biochim. Biophys. Acta Gen. Subj.* **2017**, *1861*, 1321–1328. [CrossRef] [PubMed]
33. Helttunen, K.; Prus, P.; Luostarinen, M.; Nissinen, M. Interaction of aminomethylated resorcinarenes with rhodamine B. *New J. Chem.* **2009**, *33*, 1148–1154. [CrossRef]
34. Shum, K.T.; Chan, C.; Leung, C.M.; Tanner, J.A. Identification of a DNA aptamer that inhibits sclerostin's antagonistic effect on Wnt signalling. *Biochem. J.* **2011**, *434*, 493–501. [CrossRef] [PubMed]
35. Víglaský, V.; Antalík, M.; Bagel'ová, J.; Tomori, Z.; Podhradský, D. Heat-induced conformational transition of cytochrome c observed by temperature gradient gel electrophoresis at acidic pH. *Electrophoresis* **2000**, *21*, 850–858. [CrossRef]
36. Víglaský, V.; Bauer, L.; Tlucková, K. Structural features of intra- and intermolecular G-quadruplexes derived from telomeric repeats. *Biochemistry* **2010**, *49*, 2110–2120. [CrossRef]
37. Krafčíková, P.; Demkovičová, E.; Halaganová, A.; Víglaský, V. Putative HIV and SIV G-Quadruplex Sequences in Coding and Noncoding Regions Can Form G-Quadruplexes. *J. Nucleic Acids* **2017**, *2017*, 6513720. [CrossRef]
38. Gabelica, V.; Maeda, R.; Fujimoto, T.; Yaku, H.; Murashima, T.; Sugimoto, N.; Miyoshi, D. Multiple and cooperative binding of fluorescence light-up probe thioflavin T with human telomere DNA G-quadruplex. *Biochemistry* **2013**, *52*, 5620–5628. [CrossRef]
39. Lubitz, I.; Zikich, D.; Kotlyar, A. Specific high-affinity binding of thiazole orange to triplex and G-quadruplex DNA. *Biochemistry* **2010**, *49*, 3567–3574. [CrossRef] [PubMed]
40. Mohanty, J.; Barooah, N.; Dhamodharan, V.; Harikrishna, S.; Pradeepkumar, P.I.; Bhasikuttan, A.C. Thioflavin T as an efficient inducer and selective fluorescent sensor for the human telomeric G-quadruplex DNA. *J. Am. Chem. Soc.* **2013**, *135*, 367–376. [CrossRef]
41. Sjekloca, L.; Ferre-D'Amare, A.R. Binding between G Quadruplexes at the Homodimer Interface of the Corn RNA Aptamer Strongly Activates Thioflavin T Fluorescence. *Cell Chem. Biol.* **2019**, *26*, 1159. [CrossRef]
42. Campbell, N.H.; Parkinson, G.N.; Reszka, A.P.; Neidle, S. Structural basis of DNA quadruplex recognition by an acridine drug. *J. Am. Chem. Soc.* **2008**, *130*, 6722–6724. [CrossRef]
43. Pettersen, E.F.; Goddard, T.D.; Huang, C.C.; Couch, G.S.; Greenblatt, D.M.; Meng, E.C.; Ferrin, T.E. UCSF Chimera—A visualization system for exploratory research and analysis. *J. Comput. Chem.* **2004**, *25*, 1605–1612. [CrossRef] [PubMed]
44. Bauer, L.; Tlučková, K.; Tóhová, P.; Viglaský, V. G-quadruplex motifs arranged in tandem occurring in telomeric repeats and the insulin-linked polymorphic region. *Biochemistry* **2011**, *50*, 7484–7492. [CrossRef] [PubMed]
45. Poniková, S.; Tlučková, K.; Antalík, M.; Víglaský, V.; Hianik, T. The circular dichroism and differential scanning calorimetry study of the properties of DNA aptamer dimers. *Biophys. Chem.* **2011**, *155*, 29–35. [CrossRef]
46. Stewart, J.J.P. Optimization of parameters for semiempirical methods I. Method. *J. Comput. Chem.* **1989**, *10*, 209–220. [CrossRef]
47. Dai, J.; Carver, M.; Punchihewa, C.; Jones, R.A.; Yang, D. Structure of the hybrid-2 type intramolecular human telomeric G-quadruplex in K+ solution: Insights into structure polymorphism of the human telomeric sequence. *Nucleic Acids Res.* **2007**, *35*, 4927–4940. [CrossRef]
48. Do, N.Q.; Lim, K.W.; Teo, M.H.; Heddi, B.; Phan, A.T. Stacking of G-quadruplexes: NMR structure of a G-rich oligonucleotide with potential anti-HIV and anticancer activity. *Nucleic Acids Res.* **2011**, *39*, 9448–9457. [CrossRef]
49. Sanner, M.F. Python: A programming language for software integration and development. *J. Mol. Graphics Mod.* **1999**, *17*, 57–61.
50. Morris, G.M.; Ruth, H.; Lindstrom, W.; Sanner, M.F.; Belew, R.K.; Goodsell, D.S.; Olson, A.J. Software news and updates AutoDock4 and AutoDockTools4: Automated docking with selective receptor flexibility. *J. Comput. Chem.* **2009**, *30*, 2785–2791. [CrossRef] [PubMed]

International Journal of *Molecular Sciences*

MDPI

Communication

G-Quadruplex in Gene Encoding Large Subunit of Plant RNA Polymerase II: A Billion-Year-Old Story

Adriana Volná [1], Martin Bartas [2], Václav Karlický [1,3], Jakub Nezval [1], Kristýna Kundrátová [2], Petr Pečinka [2], Vladimír Špunda [1,3,*] and Jiří Červeň [2,*]

1 Department of Physics, Faculty of Science, University of Ostrava, 710 00 Ostrava, Czech Republic; adriana.volna@osu.cz (A.V.); vaclav.karlicky@osu.cz (V.K.); jakub.nezval@osu.cz (J.N.)
2 Department of Biology and Ecology, Institute of Environmental Technologies, Faculty of Science, University of Ostrava, 710 00 Ostrava, Czech Republic; martin.bartas@osu.cz (M.B.); P18116@student.osu.cz (K.K.); petr.pecinka@osu.cz (P.P.)
3 Global Change Research Institute, Czech Academy of Sciences, Bělidla 4a, 603 00 Brno, Czech Republic
* Correspondence: vladimir.spunda@osu.cz (V.Š.); jiri.cerven@osu.cz (J.Č.)

Abstract: G-quadruplexes have long been perceived as rare and physiologically unimportant nucleic acid structures. However, several studies have revealed their importance in molecular processes, suggesting their possible role in replication and gene expression regulation. Pathways involving G-quadruplexes are intensively studied, especially in the context of human diseases, while their involvement in gene expression regulation in plants remains largely unexplored. Here, we conducted a bioinformatic study and performed a complex circular dichroism measurement to identify a stable G-quadruplex in the gene *RPB1*, coding for the RNA polymerase II large subunit. We found that this G-quadruplex-forming locus is highly evolutionarily conserved amongst plants sensu lato (Archaeplastida) that share a common ancestor more than one billion years old. Finally, we discussed a new hypothesis regarding G-quadruplexes interacting with UV light in plants to potentially form an additional layer of the regulatory network.

Keywords: evolution; plant science; nucleic acids; circular dichroism; UV light

Citation: Volná, A.; Bartas, M.; Karlický, V.; Nezval, J.; Kundrátová, K.; Pečinka, P.; Špunda, V.; Červeň, J. G-Quadruplex in Gene Encoding Large Subunit of Plant RNA Polymerase II: A Billion-Year-Old Story. *Int. J. Mol. Sci.* **2021**, *22*, 7381. https://doi.org/10.3390/ijms22147381

Academic Editor: Laurence Lins

Received: 28 May 2021
Accepted: 5 July 2021
Published: 9 July 2021

1. Introduction

G-quadruplexes (G4s) in nucleic acids are noncanonical four-stranded structures, which are different from classical double-stranded DNA (B-DNA form) described in 1953 by James Watson, Francis Crick, and Rosalind Franklin [1,2]. The basic building block for a G4 is a so-called guanine quartet formed by a G:G Hoogsteen base pairing, a structure which was first proposed by Gellert and his colleagues in 1962 [3,4]. G4 formation usually requires monovalent cations with a positive charge, such as potassium (K^+) and sodium (Na^+) ions [5]. It has been demonstrated that G4s have the potential to frequently occur in specific genomic loci, generally called G4-forming sequences or putative G4 sites. These regions are widely found in various eukaryotes [6–8], prokaryotes [9–11], and even viruses [12–15]. There is direct evidence of the functional relevance of such a structure; that is, G4s generally slow the replication process and induce instability during leading-strand replication [16,17], affect transcription by arresting RNA polymerase [18–20], and stop translation of the protein if a stable G4 is formed in the transcribed mRNA [19,21]. There are also studies suggesting that G4s are sensitive to UV light in vitro [22,23]. However, G4s are still a neglected area in plant research when compared to humans and model animals. Limited knowledge on the topic was reviewed in [24] stating the unknown function of most plant G4s. In this study, we inspected the *RPB1* gene, which encodes the large subunit of RNA polymerase II. RNA polymerase II (DNA directed RNA polymerase II) is usually associated with transcription of most structural genes. Eukaryotic RNA polymerase II consists of 12 subunits encoded by different genes [25,26], and in 2005, an exact 3D structure of RNA-polymerase II from *Saccharomyces cerevisiae* was resolved [27]. The activity of RNA polymerase II is precisely regulated on several levels, but there are large differences among species, especially between mammals and plants. For example, mammalian RNA polymerase II is localized to the so-called transcription factories in the nucleus, while in plants, RNA polymerase II is rather more evenly distributed in the nucleoplasm [28]. Moreover, in plants, specialized polymerases have evolved (RNA polymerases IV and V, and organelle-specific polymerases), so the specificity of substrates slightly differs [29]. Recently, a novel role of plant RNA polymerase II has been described—they silence retrotransposons and, thus, maintain genome stability [30]. RNA polymerase II is a multi-subdomain complex; the number of domains, as well as their positions, differs between plant species, which points to a complicated evolution for this enzyme, as reviewed in [26]. RNA polymerase protein complexes are considered to be one of the main regulators of gene expression processes in all living organisms [31,32]. Thus, we decided to carry out our G4 analysis in the coding regions of the *RPB1* gene in different evolutionarily distant organisms belonging to the plant kingdom. Although regulation of gene expression is more complex, levels of active RNA polymerase II are important for the overall level of transcription. Therefore, stable G4(s) in the coding sequence of the large subunit of RNA polymerase II could significantly reduce the level of transcription.

2. Results and Discussion

At first, we decided to study the conservation of potential G4-forming RBP1 sequences in evolutionarily distant plants. We aligned the *RPB1* coding sequences of various representative green plants (Viridiplantae) and their closest relative groups, Rhodophyta and Glaucophyta, which all belong to the Archaeplastida supergroup [33]; see Table 1. The alignment revealed a single highly conserved G4 site (Figure 1a), which we inspected further using in silico G4 predictions. It was found that at least 18 of the 20 sequences analyzed had a G4-forming potential.

Table 1. The most important species inspected in this study (for more species, see Supplementary Material S1). Columns contain their Latin and common names (if applicable, higher and lower taxonomy units, NCBI accession numbers of *RPB1* coding regions, and the presence of predicted conserved G4 sites).

Latin Name/Common Name	Higher Taxonomy Unit	Lower Taxonomy Unit	*RPB1* CDS NCBI ID	Predicted Conserved G4 Site [1]
Amborella trichopoda/ amborella	Viridiplantae	Amborella	XM_006828781.3	Yes
Arabidopsis thaliana/ thale cress	Viridiplantae	Brassicales	NM_119746.4	Yes
Bathycoccus prasinos	Viridiplantae	Chlorophyta	XM_007510177.1	Yes
Chondrus crispus	Rhodophyta	Gigartinales	XM_005718456.1	Yes
Cyanidioschyzon merolae	Rhodophyta	Cyanidiales	XM_005538436.1	No
Cyanophora paradoxa	Glaucophyta	Cyanophoraceae	DQ223186.1	Yes
Coccomyxa subellipsoidea	Viridiplantae	Chlorophyta	XM_005651715.1	Yes
Galdieria sulphuraria	Rhodophyta	Cyanidiales	XM_005707370.1	Yes
Helianthus annuus/ common sunflower	Viridiplantae	Asterales	XM_035988906.1	Yes
Juglans regia/ English walnut	Viridiplantae	Fagales	XM_035687192.1	Yes
Micromonas pusilla	Viridiplantae	Chlorophyta	XM_003055558.1	Yes
Ostreococcus tauri	Viridiplantae	Chlorophyta	XM_022983138.1	Yes
Papaver somniferum/ opium poppy	Viridiplantae	Ranunculales	XM_026541050.1	Yes
Populus alba/ white poplar	Viridiplantae	Malpighiales	XM_035033875.1	Yes
Prunus dulcis/ almond	Viridiplantae	Rosales	XM_034368569.1	Yes
Setaria viridis/ bristlegrass	Viridiplantae	Poales	XM_034744839.1	No

Table 1. *Cont.*

Latin Name/Common Name	Higher Taxonomy Unit	Lower Taxonomy Unit	*RPB1* CDS NCBI ID	Predicted Conserved G4 Site [1]
Solanum tuberosum/ potato	Viridiplantae	Solanales	XM_006340725.2	Yes
Vitis riparia/ riverbank grape	Viridiplantae	Vitales	XM_034819617.1	Yes
Volvox carteri	Viridiplantae	Chlorophyta	XM_002949413.1	Yes
Zea mays/ maize	Viridiplantae	Poales	NM_001305817.1	Yes

[1] G4 sites were predicted using four independent approaches. For more details, see the Supplementary Material S1.

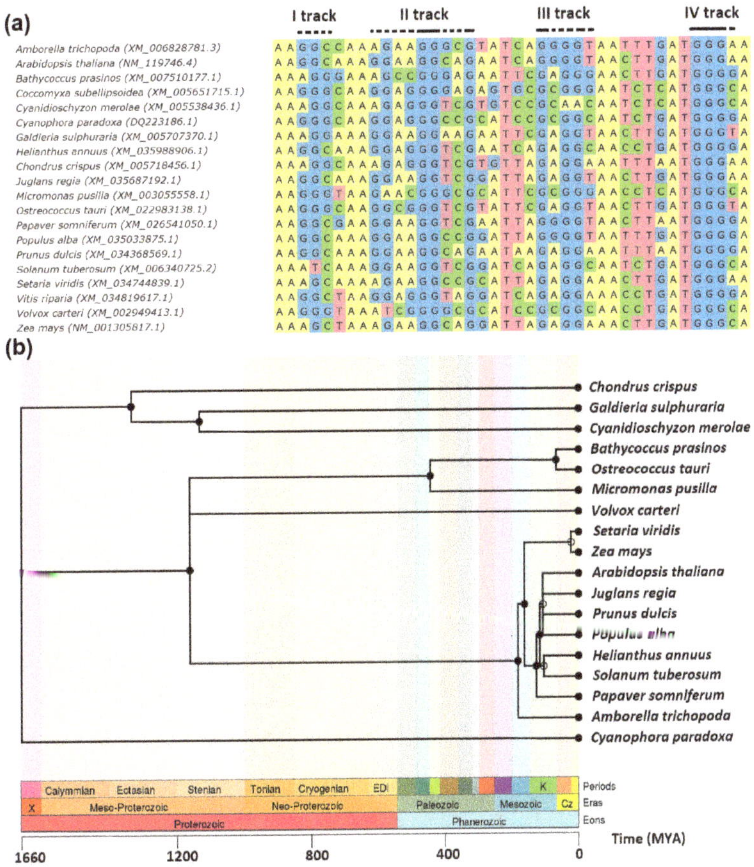

Figure 1. Conserved G4 locus in the *RPB1* gene. (**a**) Multiple sequence alignment of conserved G4-forming loci inside the gene coding for the large subunit of RNA polymerase II (RPB1). The alignment was constructed using the MUSCLE algorithm [34] via UGENE workflow [35]. Mainly the second (II) and fourth (IV) guanine tracks show strong conservation of guanine residues. (**b**) Taxonomic tree with the time of branching estimations (MYA) constructed using the TimeTree tool [36]. *Coccomyxa subellipsoidea* is omitted here because of its unclear phylogeny.

The identified region was approximately 40 nucleotides long and contained four well-conserved guanine tracks. Their G4 forming potential was verified by four different methods, including QGRS mapper [37]; the G4Hunter [38] algorithm; and the G4RNAscreener web server [39], which comprises cGcC [40] and neural network approaches [41]. Considering the *RPB1* CDS sequence from *Arabidopsis thaliana* (NM_119746.4), the G4 locus started at nucleotide position 1257 and ended at nucleotide position 1296. For the whole genome, the identified region occupies the coordinates of chromosome 4, 16,966,308–16,966,347. The fourth guanine track is 100% conserved, while the other tracks contain a certain plasticity, which can play an important role in the resulting conformation of the formed G4 (Figure 1a). From an evolutionary perspective, it is remarkable that this G4 locus has remained preserved for more than one billion years (Figure 1b). This finding was highly unexpected, because the vast majority of G4 loci are highly divergent and non-conserved, even between closely related species [42,43].

It can be hypothesized that the coding region of the *RPB1* CDS is a priori conserved to maintain the unaltered amino acid sequence of the RNA polymerase II large subunit. Therefore, we inspected the whole *RPB1* CDS (app. 6000 bp) and found that the 40-bp-long potential G4 locus is the most conserved (based on a multiple sequence alignment of *RPB1* gene in plants; details are enclosed in the Supplementary Material S2). Although the G4 locus of the *RPB1* gene is perfectly conserved among evolutionarily distant plant species, its paralogs in *Arabidopsis thaliana* (*rpa1*, *rpc1*, *rpd1a*, and *rpd1b*) have this locus modified by deletions and/or substitutions that disrupt G4-forming potential (see Supplementary Material S3). The largest and catalytic component of RNA polymerase II (RPB1) synthesizes mRNA precursors and many functional non-coding RNAs. RPB1 forms the polymerase active center [44]. Therefore, it is possible that the G4 characterized in our study plays an important regulatory role in vivo by affecting transcription of the *RPB1* gene, thus forming a negative feedback loop, because it is generally accepted that G4s inhibit transcription rates [45–47]. Currently, there is also a whole-genome experimental map of G4s in multiple species, including *Arabidopsis thaliana* [48], but no signal for the whole *RPB1* gene was identified via this analysis. We suggest that this could possibly stem from the low-sequencing coverage of this particular site. In addition, only 2407 G4s were mapped across the five *Arabidopsis thaliana* chromosomes using this approach. More specifically, the total number of putative G4 sites in *Arabidopsis thaliana* is supposed to be at minimum five times higher if the strict threshold (1.4) of the G4Hunter prediction algorithm [49] is used. When using data from the *Arabidopsis thaliana* isoform sequencing [50], we found that there is probably an *RPB1* gene isoform comprising exons 1–8 (*RPB1* has 13 exons in total). It is known that G4s can induce premature transcription termination [51]. However, the identified G4 locus is located within exon 6, which is relatively far away from the transcription termination site, so its possible role in this particular event is rather speculative. Nonetheless, various minor non-canonical splice site combinations were recently detected [52].

Next, we inspected the G4-forming potential of selected sequences via circular dichroism (CD) measurements (Figure 2a). All inspected homologous sequences showed clear G4 signatures in differential CD spectra with the characteristic positive peaks at specific wavelength ranges depicted (Figure 2a) by grey vertical dashed lines, whereas the negative control had no significant positive differential CD signal in this spectral region. Interestingly, we found sequence-dependent differences in molar ellipticity across tested species. Such variability might be caused by the different composition of tested sequences, resulting in different folding motifs and, thus, structure. When compared with G4-forming potential (Table 1), only a single discrepancy was identified in the *Cyanidioschyzon merolae* sequence. More specifically, the *RPB1* locus of this unicellular organism has low theoretical potential to form a G4 structure, and, therefore, the obtained CD signal was unexpected. This may be due to the involvement of other nucleotide residues in G4 tetrads (so-called mixed tetrads) that comprise, for example, cytosine residue(s) [53]. Unfortunately, no tool is currently available to determine the G4s formed by other nucleotides than guanine residues. However, existence of such G4s in vitro has been documented [54,55].

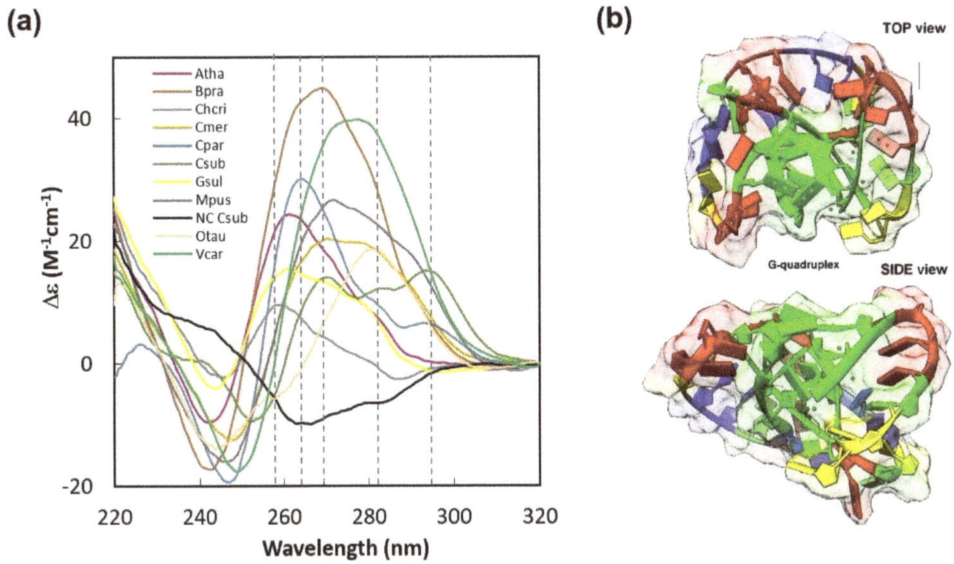

Figure 2. Conformational characterization of conserved G4 locus in the *RPB1* gene. (**a**) The differential CD spectra of the putative G4s in the *RPB1* gene were calculated as the difference of the CD spectra measured at 20 °C and 90 °C. It can be seen that the negative control is different from all other samples analyzed. It also shows different G4 conformations (4–5 CD bands at approx. 258, 264, 270, 282, and 295 nm—grey vertical dashed lines). Atha—*Arabidopsis thaliana*; Bpra—*Bathycoccus prasinos*; Chcri—*Chondrus crispus*; Cmer—*Cyanidioschyzon merolae*; Cpar—*Cyanophora paradoxa*; Csub—*Coccomyxa subellipsoidea*; Gsul—*Galdieria sulphuraria*; Mpus—*Micromonas pusilla*; Otau—*Ostreococcus tauri*; Vcar—*Volvox carteri*; NC—negative control. (**b**) Structural modeling of parallel G4 in *Bathycoccus prasinos*. The coloring of the nucleotides is default NDB (green for guanines, red for adenines, yellow for cytosines, and blue for thymines). The resulting structure in PDB format can be found in Supplementary Material S4.

To better visualize the structure of a parallel G4 in the *RPB1* gene, we selected one representative sequence from *Bathycoccus prasinos* and modeled its parallel G4 structure in silico. The model is based on information obtained by CD measurement, and it mimics parameters of existing PDB structures using the 3DNus algorithm [56] (Figure 2b).

To further validate temperature stability and the reversibility of G4 folding, we performed thermal denaturation followed by the subsequent renaturation and a CD measurement at all three points (Figure 3a–j). Temperatures above 80 °C are generally considered to be enough to melt all common G4 structures [57], and our plots clearly show a decreasing G4 signature in the CD spectrum at 90 °C. After cooling and a short incubation period at 20 °C, the G4 structures renatured, serving as direct evidence of G4 formation. This phenomenon was not observed in the NC sample (Figure 3k).

In natural conditions, plants are often exposed to stress factors that may cause substantial DNA damage, such as high soil salinity, drought, or high irradiation. Plants need light for their growth; however, UV light of all wavelengths (UVC, UVB, and even UVA) induces DNA damage, mainly in the form of cyclobutane pyrimidine dimers [58]. Recently, it was found that low-energy UV radiation (266 nm) can photo-ionize human telomeric G-quadruplexes (GGG(TTAGGG)3) in the presence of K^+ ions in vitro [59]. Here, for the first time, we propose a hypothesis that G4s might function as additional UV sensors, allowing plants to rapidly regulate the rate of DNA replication, gene expression, and protein binding (Figure 4).

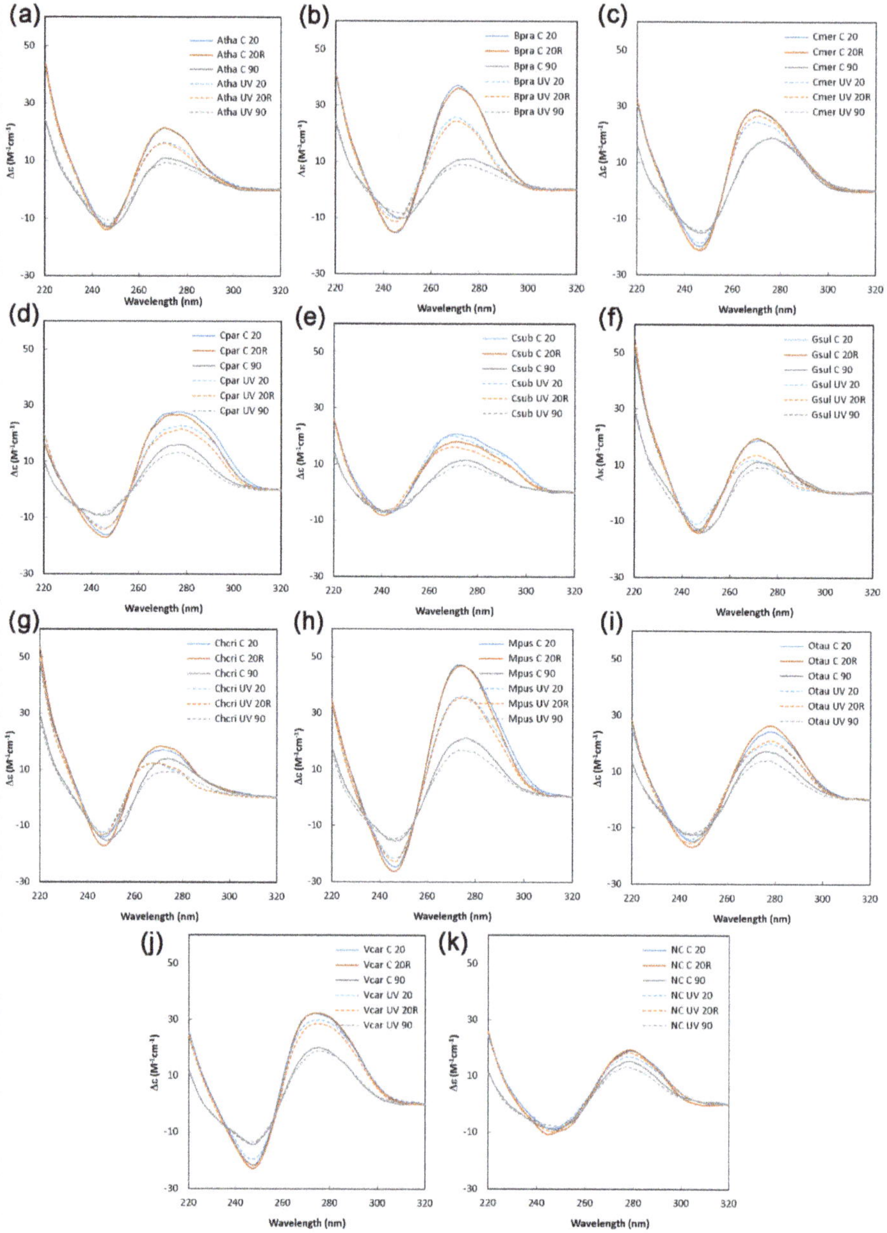

Figure 3. CD spectra of putative G4s in the *RPB1* gene in selected plant species. (**a**) Atha—*Arabidopsis thaliana;* (**b**) Bpra—*Bathycoccus prasinos;* (**c**) Cmer—*Cyanidioschyzon merolae;* (**d**) Cpar—*Cyanophora paradoxa;* (**e**) Csub—*Coccomyxa subellipsoidea;* (**f**) Gsul—*Galdieria sulphuraria;* (**g**) Chcri—*Chondrus crispus;* (**h**) Mpus—*Micromonas pusilla;* (**i**) Otau—*Ostreococcus tauri;* (**j**) Vcar—*Volvox carteri,* (**k**) including negative control (NC). CD spectra were measured at 20 °C (20; blue line), after denaturation (90; grey line) and renaturation (20R; red line) without (C; solid line) or with previous UV irradiations (UV; dashed line). The total exposure was for one hour at 4.1 W/m^2 of UV-A and 4.1 W/m^2 of UV-B.

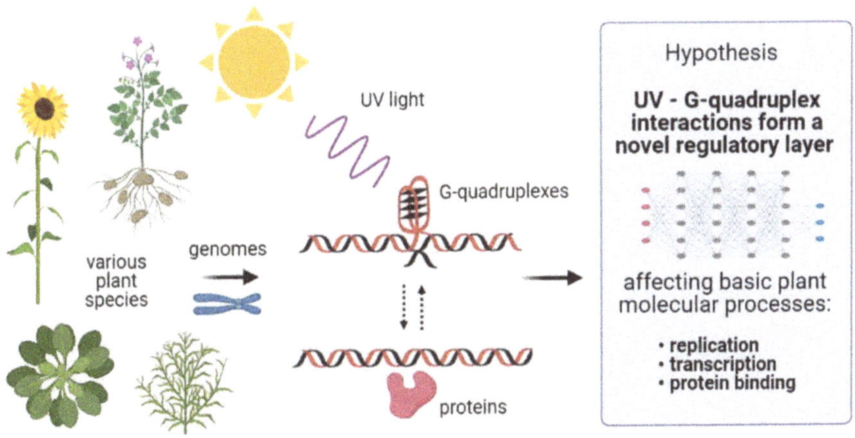

Figure 4. Schematic of UV interacting with G4s. Solar UV radiation penetrates the cell wall, cytoplasmic membrane, and nuclear membrane, and it can directly interact with genomic DNA. We hypothesize that G4s are exceptionally sensitive to UV due to their central metallic K^+ stabilizing ions and Hoogsteen base pairs forming stacked G4 tetrads. Generally, we propose that the interaction of G4s with UV leads to partial destabilization of the G4 structure and, thus, allows relatively rapid and finely tuned changes of molecular process rates, which affects signaling pathways and plant responses to UV irradiation.

To explore our hypothesis of G4 structures being a regulatory element of gene expression in plant cells, we exposed induced G4s to UV for one hour. Interestingly, we found that UV irradiation has a partial inhibitory effect on G4 folding, which is depicted in Figure 3a–j by the dashed lines. It is noteworthy that the decrease in molar ellipticity caused by UV varies between G4-forming oligonucleotides from different plant species. For example, *Cyanidioschyzon merolae* showed a mild decrease, and *Arabidopsis thaliana* showed a medium decrease. In contrast, *Bathycoccus prasinos* or *Micromonas pusilla* displayed a highly pronounced decrease in molar ellipticity associated with G4 presence (Figure 5a). The described variability between plant species is obviously caused by a different nucleotide composition, and, thus, different folding substructures lead to variable G4 sensitivity to UV light. We also confirmed that there were no strand breaks in the oligonucleotides by polyacrylamide gel electrophoresis (PAGE) and that G4s were preserved before and after UV treatment, as verified by thioflavin T (ThT) staining (Supplementary Material S5), which is in accordance with the CD spectroscopy measurements. Figure 5b schematically depicts G4 with adjacent thymines in the loop resulting in thymine dimer formation and G4 structure loosening. Cyclobutane pyrimidine dimers can later be repaired by direct photoreactivation and/or excision repair [60–62].

In vivo evidence of G4s has been studied in connection with cancer [63]; genomic instability [64,65]; telomere formation [19]; and the general ability to regulate transcription [66,67], translation [68], and replication [69,70]. It has been shown that chromatin remodeling, which affects G4 formation, can lead to parental loss of chromatin marks [71], showing the important role of epigenetic modifications. Recently, a single-molecule fluorescent probe, which allows visualization of formed G4s in single DNA molecules in living cells, has been developed [72]. Unfortunately, none of these in vivo experiments were, to the best of our knowledge, performed in plants. However, as there is evidence of in vivo G4 formation in different model organisms, we expect that even in in vivo chromatin G4s can form in plants. It has been well documented for several decades that UV-A can induce thymine dimer formation in vivo even in algae [73]; thus, G4s could serve as a sensor for UV radiation. Therefore, their partial disruption could lead to the initiation of specific processes, possibly resulting in the modulation of gene expression.

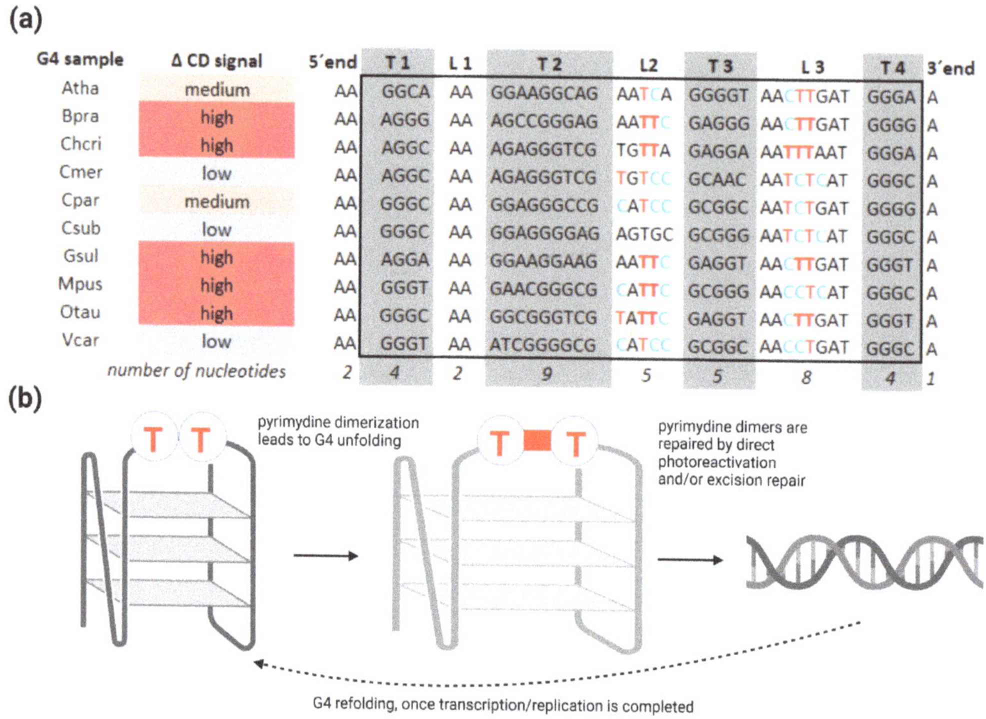

Figure 5. Plants G4s interacts with UVB/UVA light. (**a**) Decrease in molar ellipticity in UV-irradiated G4 samples expressed on the categorical scale (low, moderate, and high decrease). Guanine tracks are highlighted in grey color and designated T1–T4. Loop regions are designated L1–L3. Pyrimidines with the ability to form cyclobutane pyrimidine dimers are depicted in red (thymines) and blue (cytosines). Adjacent thymines with the highest probability to form thymine dimers are in bold. (**b**) Adjacent or opposite pyrimidines in the G4 loops can form cyclobutane pyrimidine dimers [22], which lead to conformational change and/or unfolding of G4 structure. Pyrimidine dimers are then repaired by photoreactivation and/or excision repair [60–62], and G4s can then reform via refolding. Concurrently, important molecular processes (DNA replication and transcription) can take place.

3. Materials and Methods

3.1. Bioinformatics and Structural Modeling

RPB1 coding regions (CDSs) from 40 model plant species that are evolutionarily distant from one another (Supplementary Material S6) were chosen for the bioinformatic analysis. The MUSCLE algorithm [34] running via UGENE workflow [35] was employed to construct multiple alignments of *RPB1* coding regions (Supplementary Material S7). Analyzed *RPB1* paralogs (FASTA sequences) in *Arabidopsis thaliana* are enclosed in Supplementary Material S8.

The potential to form G4s was predicted via the QGRS mapper [37] and G4screener web server [39], and the resulting scores for the inspected putative G4 sites (obtained by four independent approaches) are enclosed in the supporting data for this article (Supplementary Material S1). The taxonomic tree with the time of branching estimations was constructed using the TimeTree tool [36]. G4 from *Bathycoccus prasinos* was modeled in a 3DNus environment [56] using a supervised approach based on a typical parallel conformation measured by CD assessment. The resulting structure was visualized using UCSF Chimera [74].

3.2. Circular Dichroism Measurement

All G4-forming oligonucleotides were purchased in HPLC purity from Elisabeth Pharmacon (Czech Republic) and inducted as reported earlier [75]. CD spectra were recorded in the range of 200–350 nm with a J-815 spectropolarimeter (Jasco, Tokyo, Japan). Spectra were recorded in steps of 0.5 nm with an integration time of 1 s, a bandwidth of 2 nm, and a scanning speed of 50 nm·min^{-1} with 3 accumulations. For all CD analyses, a final concentration of 50 mM KCl was used. To denature the G4 structures, a heating rate of 10 °C·min^{-1} was maintained using a programmable Peltier thermostat up to 90 °C followed by cooling to 20 °C for the CD spectra measurement of renatured G4 structures. A quartz glass cell with a 10 mm path length was used for all CD measurements. The sequence of negative control (NC) was as follows: AAGGGCAAGGAGTGGAGAGTGCGCGTGAATCTCATGTGCAA (designed using the G4Killer tool) [76]. To determine whether the prepared G4 structures have the potential to be a regulatory element, the oligonucleotides were illuminated by a lamp (Philips, TL 20W/12RS UV-B medical, Made in Holland) in a quartz glass cuvette for one hour at 4.1 W/m^2 UV-A and 4.1 W/m^2 UV-B radiation. The control and UV-irradiated samples were compared with respect to height of the CD peak (decrease in molar ellipticity for approximately half was judged as high). The decrease in molar ellipticity was computed, and, for later purposes, it was expressed on the categorical scale (low, moderate, and high decrease in molar ellipticity) using the highest and lowest decreases as borders and then evenly divided into these categories. For the detailed spectrum of the UV lamp used in this study, see the Supplementary Material S9. Differential CD spectra are enclosed in Supplementary Material S10.

3.3. Gel Electrophoresis and Thioflavin T Staining

Gel electrophoresis of the selected G4 samples was performed on a nondenaturing 15% acrylamide gel supplemented with 10 mM KCl. The gel was electrophoresed at room temperature (20 °C). After electrophoresis, the gel was stained in a bath of 0.5 μM ThT (which is a widely used fluorescent light-up probe for G4 formation [77]) for 15 min under agitation and then destained for 15 min in an electrophoresis buffer. Gel images were taken on the BioRad ChemiDoc system (Bio-Rad, Hercules, CA, USA) with an automatically optimized exposure time (Supplementary Material S5).

Supplementary Materials: The following are available online at https://zenodo.org/record/4973514#.YMtNZqj7RPY (accessed on 7 July 2021). Supplementary Material S1: Analyzed *RPB1* sequences in 40 plant species together with detailed characteristics and G4 prediction using four different computational approaches. Supplementary Material S2: G4 locus is the most conserved within the *RPB1* gene. Supplementary Material S3: Multiple alignment of G4 locus in *RPB1* paralogs (*Arabidopsis thaliana*). Supplementary Material S4: Modelled 3D structure of G4 from *Bathycoccus prasinos*. Supplementary Material S5: Gel electrophoresis and ThT staining of the selected G4-forming sequences. Supplementary Material S6: All analyzed *RPB1* sequences in FASTA format. Supplementary Material S7: Aligned *RPB1* sequences in FASTA format. Supplementary Material S8: *RPB1* paralogs in *Arabidopsis thaliana*. Supplementary Material S9. Spectral composition of light used in the UV experiment. Analysis of emitted light was performed using an Ocean Optics (HR4000CG-UV-NIR, USA) device. Supplementary Material S10: Differential CD spectra—comparison without and with previous UV treatment.

Author Contributions: A.V., V.K. and J.N. conceived the wet-lab experiments. M.B., P.P. and K.K. performed bioinformatics and 3D structural modeling. M.B., J.Č. and A.V. prepared the manuscript. J.Č. and V.Š. managed and supervised work. All authors discussed the results and contributed to the final manuscript. All authors have read and agreed to the published version of the manuscript.

Funding: This research was funded by the University of Ostrava (project numbers SGS02/PrF/2021; SGS11/PrF/2021), the Czech Science Foundation (GA ČR; project ID: 18-23702S), and the National Agency for Agricultural Research (NAZV) of the Czech Republic grant no. QK1810391 "Utilization of genomic and transcriptomic approaches to create genetic resources and breeding materials of poppy with specific traits". Participation of V.K. and V.Š. was also supported by the project

"SustES—Adaptation strategies for sustainable ecosystem services and food security under adverse environmental conditions" (CZ.02.1.01/0.0/0.0/16_019/0000797).

Institutional Review Board Statement: Not applicable.

Informed Consent Statement: Not applicable.

Data Availability Statement: All supporting data generated for this manuscript are freely available here together with a short commentary: https://zenodo.org/record/4541573#.YCrckWj0lPY (accessed on 7 July 2021).

Acknowledgments: Our acknowledgment belongs to the University of Ostrava for friendly and supportive working conditions, our collaborators who often contributed with good advice, and our colleagues from the Institute of Biophysics in Brno for their valuable comments and recommendations.

Conflicts of Interest: The authors declare no conflict of interest.

References

1. Watson, J.D.; Crick, F.H. Molecular Structure of Nucleic Acids: A Structure for Deoxyribose Nucleic Acid. *Nature* **1953**, *171*, 737–738. [CrossRef]
2. Klug, A. Rosalind Franklin and the Discovery of the Structure of DNA. *Nature* **1968**, *219*, 808–810. [CrossRef]
3. Gellert, M.; Lipsett, M.N.; Davies, D.R. Helix Formation by Guanylic Acid. *Proc. Natl. Acad. Sci. USA* **1962**, *48*, 2013. [CrossRef] [PubMed]
4. Yang, X.; Cheema, J.; Zhang, Y.; Deng, H.; Duncan, S.; Umar, M.I.; Zhao, J.; Liu, Q.; Cao, X.; Kwok, C.K. RNA G-Quadruplex Structures Exist and Function in Vivo in Plants. *Genome Biol.* **2020**, *21*, 1–23. [CrossRef] [PubMed]
5. Li, X.; Sánchez-Ferrer, A.; Bagnani, M.; Adamcik, J.; Azzari, P.; Hao, J.; Song, A.; Liu, H.; Mezzenga, R. Metal Ions Confinement Defines the Architecture of G-Quartet, G-Quadruplex Fibrils and Their Assembly into Nematic Tactoids. *Proc. Natl. Acad. Sci. USA* **2020**, *117*, 9832–9839. [CrossRef]
6. Garg, R.; Aggarwal, J.; Thakkar, B. Genome-Wide Discovery of G-Quadruplex Forming Sequences and Their Functional Relevance in Plants. *Sci. Rep.* **2016**, *6*, 1–13. [CrossRef]
7. Yang, D. G-Quadruplex DNA and RNA. In *G-Quadruplex Nucleic Acids: Methods and Protocols*; Yang, D., Lin, C., Eds.; Methods in Molecular Biology; Springer: New York, NY, USA, 2019; pp. 1–24. ISBN 978-1-4939-9666-7.
8. Warner, E.F.; Bohálová, N.; Brázda, V.; Waller, Z.A.E.; Bidula, S. Analysis of Putative Quadruplex-Forming Sequences in Fungal Genomes: Novel Antifungal Targets? *Microb. Genom.* **2021**, *7*, 000570. [CrossRef]
9. Bartas, M.; Čutová, M.; Brázda, V.; Kaura, P.; Šťastný, J.; Kolomazník, J.; Coufal, J.; Goswami, P.; Červeň, J.; Pečinka, P. The Presence and Localization of G-Quadruplex Forming Sequences in the Domain of Bacteria. *Molecules* **2019**, *24*, 1711. [CrossRef] [PubMed]
10. Brázda, V.; Luo, Y.; Bartas, M.; Kaura, P.; Porubiaková, O.; Šťastný, J.; Pečinka, P.; Verga, D.; Da Cunha, V.; Takahashi, T.S. G-Quadruplexes in the Archaea Domain. *Biomolecules* **2020**, *10*, 1349. [CrossRef]
11. Saranathan, N.; Vivekanandan, P. G-Quadruplexes: More than Just a Kink in Microbial Genomes. *Trends Microbiol.* **2019**, *27*, 148–163. [CrossRef]
12. Bartas, M.; Brázda, V.; Bohálová, N.; Cantara, A.; Volná, A.; Stachurová, T.; Malachová, K.; Jagelská, E.B.; Porubiaková, O.; Červeň, J. In-Depth Bioinformatic Analyses of Nidovirales Including Human SARS-CoV-2, SARS-CoV, MERS-CoV Viruses Suggest Important Roles of Non-Canonical Nucleic Acid Structures in Their Lifecycles. *Front. Microbiol.* **2020**, *11*, 1583. [CrossRef]
13. Bohálová, N.; Cantara, A.; Bartas, M.; Kaura, P.; Šťastný, J.; Pečinka, P.; Fojta, M.; Mergny, J.-L.; Brázda, V. Analyses of Viral Genomes for G-Quadruplex Forming Sequences Reveal Their Correlation with the Type of Infection. *Biochimie* **2021**, *186*, 13–27. [CrossRef]
14. Métifiot, M.; Amrane, S.; Litvak, S.; Andreola, M.-L. G-Quadruplexes in Viruses: Function and Potential Therapeutic Applications. *Nucleic Acids Res.* **2014**, *42*, 12352–12366. [CrossRef] [PubMed]
15. Ruggiero, E.; Richter, S.N. Viral G-Quadruplexes: New Frontiers in Virus Pathogenesis and Antiviral Therapy. *Annu. Rep. Med. Chem.* **2020**, *54*, 101–131. [CrossRef] [PubMed]
16. Lopes, J.; Piazza, A.; Bermejo, R.; Kriegsman, B.; Colosio, A.; Teulade-Fichou, M.-P.; Foiani, M.; Nicolas, A. G-Quadruplex-Induced Instability during Leading-Strand Replication. *EMBO J.* **2011**, *30*, 4033–4046. [CrossRef] [PubMed]
17. Lerner, L.K.; Sale, J.E. Replication of G Quadruplex DNA. *Genes* **2019**, *10*, 95. [CrossRef] [PubMed]
18. Broxson, C.; Beckett, J.; Tornaletti, S. Transcription Arrest by a G Quadruplex Forming-Trinucleotide Repeat Sequence from the Human c-Myb Gene. *Biochemistry* **2011**, *50*, 4162–4172. [CrossRef]
19. Carvalho, J.; Mergny, J.-L.; Salgado, G.F.; Queiroz, J.A.; Cruz, C. G-Quadruplex, Friend or Foe: The Role of the G-Quartet in Anticancer Strategies. *Trends Mol. Med.* **2020**, *26*, 848–861. [CrossRef]
20. Szlachta, K.; Thys, R.G.; Atkin, N.D.; Pierce, L.C.T.; Bekiranov, S.; Wang, Y.-H. Alternative DNA Secondary Structure Formation Affects RNA Polymerase II Promoter-Proximal Pausing in Human. *Genome Biol.* **2018**, *19*, 89. [CrossRef]

21. Endoh, T.; Sugimoto, N. Conformational Dynamics of the RNA G-Quadruplex and Its Effect on Translation Efficiency. *Molecules* **2019**, *24*, 1613. [CrossRef]
22. Wolna, A.H.; Fleming, A.M.; Burrows, C.J. Single-Molecule Analysis of Thymine Dimer-Containing G-Quadruplexes Formed from the Human Telomere Sequence. *Biochemistry* **2014**, *53*, 7484–7493. [CrossRef] [PubMed]
23. Su, D.G.T.; Fang, H.; Gross, M.L.; Taylor, J.-S.A. Photocrosslinking of Human Telomeric G-Quadruplex Loops by Anti Cyclobutane Thymine Dimer Formation. *Proc. Natl. Acad. Sci. USA* **2009**, *106*, 12861–12866. [CrossRef] [PubMed]
24. Griffin, B.D.; Bass, H.W. Plant G-Quadruplex (G4) Motifs in DNA and RNA.; Abundant, Intriguing Sequences of Unknown Function. *Plant Sci.* **2018**, *269*, 143–147. [CrossRef]
25. Ream, T.S.; Haag, J.R.; Pontvianne, F.; Nicora, C.D.; Norbeck, A.D.; Paša-Tolić, L.; Pikaard, C.S. Subunit Compositions of Arabidopsis RNA Polymerases I and III Reveal Pol I- and Pol III-Specific Forms of the AC40 Subunit and Alternative Forms of the C53 Subunit. *Nucleic Acids Res.* **2015**, *43*, 4163–4178. [CrossRef] [PubMed]
26. Wang, Y.; Ma, H. Step-Wise and Lineage-Specific Diversification of Plant RNA Polymerase Genes and Origin of the Largest Plant-Specific Subunits. *New Phytol.* **2015**, *207*, 1198–1212. [CrossRef] [PubMed]
27. Armache, K.-J.; Mitterweger, S.; Meinhart, A.; Cramer, P. Structures of Complete RNA Polymerase II and Its Subcomplex, Rpb4/7. *J. Biol. Chem.* **2005**, *280*, 7131–7134. [CrossRef] [PubMed]
28. Schubert, V.; Weisshart, K. Abundance and Distribution of RNA Polymerase II in Arabidopsis Interphase Nuclei. *J. Exp. Bot.* **2015**, *66*, 1687–1698. [CrossRef]
29. Haag, J.R.; Pikaard, C.S. Multisubunit RNA Polymerases IV and V: Purveyors of Non-Coding RNA for Plant Gene Silencing. *Nat. Rev. Mol. Cell Biol.* **2011**, *12*, 483–492. [CrossRef]
30. Thieme, M.; Lanciano, S.; Balzergue, S.; Daccord, N.; Mirouze, M.; Bucher, E. Inhibition of RNA Polymerase II Allows Controlled Mobilisation of Retrotransposons for Plant Breeding. *Genome Biol.* **2017**, *18*, 134. [CrossRef]
31. Core, L.; Adelman, K. Promoter-Proximal Pausing of RNA Polymerase II: A Nexus of Gene Regulation. *Genes Dev.* **2019**, *33*, 960–982. [CrossRef]
32. Zhu, J.; Liu, M.; Liu, X.; Dong, Z. RNA Polymerase II Activity Revealed by GRO-Seq and PNET-Seq in Arabidopsis. *Nat. Plants* **2018**, *4*, 1112–1123. [CrossRef]
33. Ferrari, C.; Proost, S.; Janowski, M.; Becker, J.; Nikoloski, Z.; Bhattacharya, D.; Price, D.; Tohge, T.; Bar-Even, A.; Fernie, A.; et al. Kingdom-Wide Comparison Reveals the Evolution of Diurnal Gene Expression in Archaeplastida. *Nat. Commun.* **2019**, *10*, 737. [CrossRef]
34. Edgar, R.C. MUSCLE: Multiple Sequence Alignment with High Accuracy and High Throughput. *Nucleic Acids Res.* **2004**, *32*, 1792–1797. [CrossRef]
35. Okonechnikov, K.; Golosova, O.; Fursov, M.; Team, U. Unipro UGENE: A Unified Bioinformatics Toolkit. *Bioinformatics* **2012**, *28*, 1166–1167. [CrossRef] [PubMed]
36. Kumar, S.; Stecher, G.; Suleski, M.; Hedges, S.B. TimeTree: A Resource for Timelines, Timetrees, and Divergence Times. *Mol. Biol. Evol.* **2017**, *34*, 1812–1819. [CrossRef] [PubMed]
37. Kikin, O.; D'Antonio, L.; Bagga, P.S. QGRS Mapper: A Web-Based Server for Predicting G-Quadruplexes in Nucleotide Sequences. *Nucleic Acids Res.* **2006**, *34*, W676–W682. [CrossRef]
38. Brázda, V.; Kolomazník, J.; Lýsek, J.; Bartas, M.; Fojta, M.; Šťastný, J.; Mergny, J.-L. G4Hunter Web Application: A Web Server for G-Quadruplex Prediction. *Bioinformatics* **2019**, *35*, 3493–3495. [CrossRef]
39. Garant, J.-M.; Perreault, J.-P.; Scott, M.S. G4RNA Screener Web Server: User Focused Interface for RNA G-Quadruplex Prediction. *Biochimie* **2018**, *151*, 115–118. [CrossRef] [PubMed]
40. Beaudoin, J.-D.; Jodoin, R.; Perreault, J.-P. New Scoring System to Identify RNA G-Quadruplex Folding. *Nucleic Acids Res.* **2014**, *42*, 1209–1223. [CrossRef]
41. Garant, J.-M.; Perreault, J.-P.; Scott, M.S. Motif Independent Identification of Potential RNA G-Quadruplexes by G4RNA Screener. *Bioinformatics* **2017**, *33*, 3532–3537. [CrossRef]
42. Sabouri, N.; Capra, J.A.; Zakian, V.A. The Essential Schizosaccharomyces Pombe Pfh1 DNA Helicase Promotes Fork Movement Past G-Quadruplex Motifs to Prevent DNA Damage. *BMC Biol.* **2014**, *12*, 1–14. [CrossRef] [PubMed]
43. Puig Lombardi, E.; Holmes, A.; Verga, D.; Teulade-Fichou, M.-P.; Nicolas, A.; Londoño-Vallejo, A. Thermodynamically Stable and Genetically Unstable G-Quadruplexes Are Depleted in Genomes across Species. *Nucleic Acids Res.* **2019**, *47*, 6098–6113. [CrossRef] [PubMed]
44. Cramer, P.; Bushnell, D.A.; Fu, J.; Gnatt, A.L.; Maier-Davis, B.; Thompson, N.E.; Burgess, R.R.; Edwards, A.M.; David, P.R.; Kornberg, R.D. Architecture of RNA Polymerase II and Implications for the Transcription Mechanism. *Science* **2000**, *288*, 640–649. [CrossRef] [PubMed]
45. Siddiqui-Jain, A.; Grand, C.L.; Bearss, D.J.; Hurley, L.H. Direct Evidence for a G-Quadruplex in a Promoter Region and Its Targeting with a Small Molecule to Repress c-MYC Transcription. *Proc. Natl. Acad. Sci. USA* **2002**, *99*, 11593–11598. [CrossRef] [PubMed]
46. Cogoi, S.; Xodo, L.E. G-Quadruplex Formation within the Promoter of the KRAS Proto-Oncogene and Its Effect on Transcription. *Nucleic Acids Res.* **2006**, *34*, 2536–2549. [CrossRef]
47. Yadav, V.; Kim, N.; Tuteja, N.; Yadav, P. G Quadruplex in Plants: A Ubiquitous Regulatory Element and Its Biological Relevance. *Front. Plant Sci.* **2017**, *8*, 1163. [CrossRef] [PubMed]

48. Marsico, G.; Chambers, V.S.; Sahakyan, A.B.; McCauley, P.; Boutell, J.M.; Antonio, M.D.; Balasubramanian, S. Whole Genome Experimental Maps of DNA G-Quadruplexes in Multiple Species. *Nucleic Acids Res.* **2019**, *47*, 3862–3874. [CrossRef]
49. Bedrat, A.; Lacroix, L.; Mergny, J.-L. Re-Evaluation of G-Quadruplex Propensity with G4Hunter. *Nucleic Acids Res.* **2016**, *44*, 1746–1759. [CrossRef]
50. Thomas, Q.A.; Ard, R.; Liu, J.; Li, B.; Wang, J.; Pelechano, V.; Marquardt, S. Transcript Isoform Sequencing Reveals Widespread Promoter-Proximal Transcriptional Termination in Arabidopsis. *Nat. Commun.* **2020**, *11*, 2589. [CrossRef]
51. Zhang, J.; Zheng, K.; Xiao, S.; Hao, Y.; Tan, Z. Mechanism and Manipulation of DNA: RNA Hybrid G-Quadruplex Formation in Transcription of G-Rich DNA. *J. Am. Chem. Soc.* **2014**, *136*, 1381–1390. [CrossRef]
52. Frey, K.; Pucker, B. Animal, Fungi, and Plant Genome Sequences Harbor Different Non-Canonical Splice Sites. *Cells* **2020**, *9*, 458. [CrossRef]
53. Vinnarasi, S.; Radhika, R.; Vijayakumar, S.; Shankar, R. Structural Insights into the Anti-Cancer Activity of Quercetin on G-Tetrad, Mixed G-Tetrad, and G-Quadruplex DNA Using Quantum Chemical and Molecular Dynamics Simulations. *J. Biomol. Struct. Dyn.* **2020**, *38*, 317–339. [CrossRef]
54. Brown, R.V.; Wang, T.; Chapetta, V.R.; Wu, G.; Onel, B.; Chawla, R.; Quijada, H.; Camp, S.M.; Chiang, E.T.; Lassiter, Q.R.; et al. The Consequences of Overlapping G-Quadruplexes and i-Motifs in the Platelet-Derived Growth Factor Receptor β Core Promoter Nuclease Hypersensitive Element Can Explain the Unexpected Effects of Mutations and Provide Opportunities for Selective Targeting of Both Structures by Small Molecules To Downregulate Gene Expression. *J. Am. Chem. Soc.* **2017**, *2017*, 7456–7475. [CrossRef]
55. Megger, D.A.; Lax, P.M.; Paauwe, J.; Fonseca Guerra, C.; Lippert, B. Mixed Guanine, Adenine Base Quartets: Possible Roles of Protons and Metal Ions in Their Stabilization. *J. Biol. Inorg. Chem.* **2018**, *23*, 41–49. [CrossRef]
56. Patro, L.P.P.; Kumar, A.; Kolimi, N.; Rathinavelan, T. 3D-NuS: A Web Server for Automated Modeling and Visualization of Non-Canonical 3-Dimensional Nucleic Acid Structures. *J. Mol. Biol.* **2017**, *429*, 2438–2448. [CrossRef]
57. Guedin, A.; Gros, J.; Alberti, P.; Mergny, J.-L. How Long Is Too Long? Effects of Loop Size on G-Quadruplex Stability. *Nucleic Acids Res.* **2010**, *38*, 7858–7868. [CrossRef] [PubMed]
58. Nisa, M.-U.; Huang, Y.; Benhamed, M.; Raynaud, C. The Plant DNA Damage Response: Signaling Pathways Leading to Growth Inhibition and Putative Role in Response to Stress Conditions. *Front. Plant Sci.* **2019**, *10*, 653. [CrossRef] [PubMed]
59. Balanikas, E.; Banyasz, A.; Baldacchino, G.; Markovitsi, D. Guanine Radicals Generated in Telomeric G-Quadruplexes by Direct Absorption of Low-Energy UV Photons: Effect of Potassium Ions. *Molecules* **2020**, *25*, 2094. [CrossRef] [PubMed]
60. Cooper, G.M. DNA Repair. In *The Cell: A Molecular Approach*, 2nd ed.; Sinauer Associates: Sunderland, MA, USA, 2000; ISBN 0-87893-106-6.
61. Kimura, S.; Tahira, Y.; Ishibashi, T.; Mori, Y.; Mori, T.; Hashimoto, J.; Sakaguchi, K. DNA Repair in Higher Plants; Photoreactivation Is the Major DNA Repair Pathway in Non-Proliferating Cells While Excision Repair (Nucleotide Excision Repair and Base Excision Repair) Is Active in Proliferating Cells. *Nucleic Acids Res.* **2004**, *32*, 2760–2767. [CrossRef] [PubMed]
62. Spampinato, C.P. Protecting DNA from Errors and Damage: An Overview of DNA Repair Mechanisms in Plants Compared to Mammals. *Cell. Mol. Life Sci.* **2017**, *74*, 1693–1709. [CrossRef]
63. Marquevielle, J.; Robert, C.; Lagrabette, O.; Wahid, M.; Bourdoncle, A.; Xodo, L.E.; Mergny, J.-L.; Salgado, G.F. Structure of Two G-Quadruplexes in Equilibrium in the KRAS Promoter. *Nucleic Acids Res.* **2020**, *48*, 9336–9345. [CrossRef]
64. Wang, Y.; Yang, J.; Wild, A.T.; Wu, W.H.; Shah, R.; Danussi, C.; Riggins, G.J.; Kannan, K.; Sulman, E.P.; Chan, T.A. G-Quadruplex DNA Drives Genomic Instability and Represents a Targetable Molecular Abnormality in ATRX-Deficient Malignant Glioma. *Nat. Commun.* **2019**, *10*, 1–14. [CrossRef] [PubMed]
65. Wu, Y.; Shin-Ya, K.; Brosh, R.M., Jr. FANCJ Helicase Defective in Fanconia Anemia and Breast Cancer Unwinds G-Quadruplex DNA to Defend Genomic Stability. *Mol. Cell. Biol.* **2008**, *28*, 4116–4128. [CrossRef]
66. Armas, P.; David, A.; Calcaterra, N.B. Transcriptional Control by G-Quadruplexes: In Vivo Roles and Perspectives for Specific Intervention. *Transcription* **2017**, *8*, 21–25. [CrossRef] [PubMed]
67. Shen, J.; Varshney, D.; Simeone, A.; Zhang, X.; Adhikari, S.; Tannahill, D.; Balasubramanian, S. Promoter G-Quadruplex Folding Precedes Transcription and Is Controlled by Chromatin. *Genome Biol.* **2021**, *22*, 143. [CrossRef]
68. Bolduc, F.; Garant, J.-M.; Allard, F.; Perreault, J.-P. Irregular G-Quadruplexes Found in the Untranslated Regions of Human MRNAs Influence Translation. *J. Biol. Chem.* **2016**, *291*, 21751–21760. [CrossRef]
69. Kanoh, Y.; Matsumoto, S.; Fukatsu, R.; Kakusho, N.; Kono, N.; Renard-Guillet, C.; Masuda, K.; Iida, K.; Nagasawa, K.; Shirahige, K.; et al. Rif1 Binds to G Quadruplexes and Suppresses Replication over Long Distances. *Nat. Struct. Mol. Biol.* **2015**, *22*, 889–897. [CrossRef] [PubMed]
70. Poggi, L.; Richard, G.-F. Alternative DNA Structures In Vivo: Molecular Evidence and Remaining Questions. *Microbiol. Mol. Biol. Rev.* **2020**, *85*, e00110-20. [CrossRef] [PubMed]
71. Schiavone, D.; Guilbaud, G.; Murat, P.; Papadopoulou, C.; Sarkies, P.; Prioleau, M.-N.; Balasubramanian, S.; Sale, J.E. Determinants of G Quadruplex-Induced Epigenetic Instability in REV1-Deficient Cells. *EMBO J.* **2014**, *33*, 2507–2520. [CrossRef]
72. Di Antonio, M.; Ponjavic, A.; Radzevičius, A.; Ranasinghe, R.T.; Catalano, M.; Zhang, X.; Shen, J.; Needham, L.-M.; Lee, S.F.; Klenerman, D.; et al. Single-Molecule Visualization of DNA G-Quadruplex Formation in Live Cells. *Nat. Chem.* **2020**, *12*, 832–837. [CrossRef]

73. Buma, A.G.J.; Engelen, A.H.; Gieskes, W.W.C. Wavelength-Dependent Induction of Thymine Dimers and Growth Rate Reduction in the Marine Diatom Cyclotella Sp. Exposed to Ultraviolet Radiation. *Mar. Ecol. Prog. Ser.* **1997**, *153*, 91–97. [CrossRef]
74. Pettersen, E.F.; Goddard, T.D.; Huang, C.C.; Couch, G.S.; Greenblatt, D.M.; Meng, E.C.; Ferrin, T.E. UCSF Chimera—A Visualization System for Exploratory Research and Analysis. *J. Comput. Chem.* **2004**, *25*, 1605–1612. [CrossRef]
75. Bartas, M.; Brázda, V.; Karlický, V.; Červeň, J.; Pečinka, P. Bioinformatics Analyses and in Vitro Evidence for Five and Six Stacked G-Quadruplex Forming Sequences. *Biochimie* **2018**, *150*, 70–75. [CrossRef] [PubMed]
76. Brazda, V.; Kolomaznik, J.; Mergny, J.-L.; Stastny, J. G4Killer Web Application: A Tool to Design G-Quadruplex Mutations. *Bioinformatics* **2020**, *36*, 3246–3247. [CrossRef] [PubMed]
77. Renaud de la Faverie, A.; Guédin, A.; Bedrat, A.; Yatsunyk, L.A.; Mergny, J.-L. Thioflavin T as a Fluorescence Light-up Probe for G4 Formation. *Nucleic Acids Res.* **2014**, *42*, e65. [CrossRef]

International Journal of
Molecular Sciences

MDPI

Article

The Changes in the p53 Protein across the Animal Kingdom Point to Its Involvement in Longevity

Martin Bartas [1], Václav Brázda [2], Adriana Volná [3], Jiří Červeň [1], Petr Pečinka [1] and Joanna E. Zawacka-Pankau [4,5,*]

1 Department of Biology and Ecology, Faculty of Science, University of Ostrava, 71000 Ostrava, Czech Republic; martin.bartas@osu.cz (M.B.); jiri.cerven@osu.cz (J.Č.); petr.pecinka@osu.cz (P.P.)
2 Institute of Biophysics of the Czech Academy of Sciences, 61265 Brno, Czech Republic; vaclav@ibp.cz
3 Department of Physics, Faculty of Science, University of Ostrava, Chittussiho 10, 71000 Ostrava, Czech Republic; adriana.volna@osu.cz
4 Faculty of Chemistry, University of Warsaw, Pasteura 1, 02-093 Warsaw, Poland
5 Center for Hematology and Regenerative Medicine, Department of Medicine, Huddinge, Karolinska Institute, SE 171-74 Stockholm, Sweden
* Correspondence: joanna.zawacka-pankau@ki.se

Abstract: Recently, the quest for the mythical fountain of youth has produced extensive research programs that aim to extend the healthy lifespan of humans. Despite advances in our understanding of the aging process, the surprisingly extended lifespan and cancer resistance of some animal species remain unexplained. The p53 protein plays a crucial role in tumor suppression, tissue homeostasis, and aging. Long-lived, cancer-free African elephants have 20 copies of the *TP*53 gene, including 19 retrogenes (38 alleles), which are partially active, whereas humans possess only one copy of *TP*53 and have an estimated cancer mortality rate of 11–25%. The mechanism through which p53 contributes to the resolution of the Peto's paradox in Animalia remains vague. Thus, in this work, we took advantage of the available datasets and inspected the p53 amino acid sequence of phylogenetically related organisms that show variations in their lifespans. We discovered new correlations between specific amino acid deviations in p53 and the lifespans across different animal species. We found that species with extended lifespans have certain characteristic amino acid substitutions in the p53 DNA-binding domain that alter its function, as depicted from the Phenotypic Annotation of p53 Mutations, using the PROVEAN tool or SWISS-MODEL workflow. In addition, the loop 2 region of the human p53 DNA-binding domain was identified as the longest region that was associated with longevity. The 3D model revealed variations in the loop 2 structure in long-lived species when compared with human p53. Our findings show a direct association between specific amino acid residues in p53 protein, changes in p53 functionality, and the extended animal lifespan, and further highlight the importance of p53 protein in aging.

Keywords: p53; aging; longevity; comparative analysis; protein sequence

Citation: Bartas, M.; Brázda, V.; Volná, A.; Červeň, J.; Pečinka, P.; Zawacka-Pankau, J.E. The Changes in the p53 Protein across the Animal Kingdom Point to Its Involvement in Longevity. *Int. J. Mol. Sci.* **2021**, *22*, 8512. https://doi.org/10.3390/ijms22168512

Academic Editors: Razmik Mirzayans and Paola Perego

Received: 2 July 2021
Accepted: 30 July 2021
Published: 7 August 2021

1. Introduction

The promise of eternal life has inspired research into this topic across many civilizations and through the millennia, dating back to Herodotus and his writings 2500 years ago. Although the average human lifespan is increasing, our health span appears to be lagging. Several studies argue that the human lifespan is physiologically and genetically limited [1], yet recent contributions have proposed a future with a potentially unlimited increase in human lifespan [2]. The demographical data show that the death risk increases exponentially up to about age 80, then decelerates and plateaus after age 105 [3]. There are two major theories of aging, senescence theory and programmed theory of aging [4]. The senescence theory converges on the accumulation of cellular damage that cannot be repaired, leading first to permanent cell cycle arrest and, in the end, the loss of organismal fitness. The free radical theory can be classified as a subtype of senescence theory and postulates that organisms age because of the accumulation of the damage inflicted by reactive oxygen species [5,6]. There is also a common agreement that the preservation in the fidelity of the DNA repair process involving the p53 pathway favors longevity [7]. The programmed theory of aging states that aging is tightly controlled and includes the Hayflick limit theory and the central aging clock theory. At the molecular level, biological aging is a complex process that involves genetic factors, mitochondria damage mechanisms, cellular senescence, proteostasis and autophagy, telomere attrition, epigenetics, inflammation, and metabolic switches. Thus, the lifespan is a multi-nodal characteristic [8]. To date, several factors have been found to play important roles in human aging, including mammalian target of rapamycin (mTOR), 5′ AMP-activated protein kinase (AMPK), sirtuin 1 (SIRT1), peroxisome proliferator-activated receptor gamma coactivator 1-alpha (PGC-1α), apolipoprotein E (APOE), lipoprotein (A) (LPA), CDKN2B antisense RNA 1 (CDKN2B-AS1), and p53. Among those, p53 emerges as a central node, linking several pathways together. The p53 is a tumor suppressor that is coded by the most often mutated gene in human cancers [9–12], and the loss of wild-type p53 function is associated with fatal outcomes in cancer patients. p53 is a critical sensor of cellular stress and thus, the dictator of cell fates. Depending on the types of stress, which include DNA damage, oncogene activation, nutrient deprivation, reactive oxygen species accumulation, and telomere shortening, p53 either (1) transiently stops cell proliferation, initiates the DNA repair machinery, and induces cell death when the damage cannot be repaired, or (2) pushes cells to replicative senescence, which is a permanent proliferation arrest.

Given the high cancer susceptibility in humans and the role of p53 in regulating cell fate, p53 is regarded as the key regulator of humans' healthy lifespan [13,14]. When we consider the "lifespan" of tumor cells, it is apparent that cancer cells often gain new functions, including "immortality," which is at least partially attributed to the inactivating mutations in the *TP53* gene and/or in its regulatory pathways [15]. As reviewed by Stiewe and Haran [16], cancer-associated mutations alter p53 in three ways: they promote the loss of wild-type (wt) p53 DNA binding, trigger dominant-negative inhibition of wtp53 by the mutant p53 in the monoallelic mutation setting, or induce the gain of new functions by mutant p53 through new protein–protein–DNA interactions. The loss of binding to the canonical target sequence by mutant p53 can be partial or complete. Different mutant p53 proteins show a variable degree of loss of the DNA-binding capacity. This results in attenuated or target-selective DNA-binding patterns [16]. Multiple functions of p53 have been described and extensively reviewed [17–19]. For example, the p53 protein plays roles in metabolism [20], cell cycle arrest [21,22], apoptosis [23], ferroptosis, angiogenesis [24], DNA repair [25], embryonic development, and cell senescence [18,26]. In the majority of cellular processes, p53 functions as a transcription factor and recognizes and binds to multiple target genes through a recognition sequence (5′-PuPuPuC(A/T)(T/A)GPyPyPy-3′) [27–29]. Owing to its crucial role in protection against the accumulation of DNA damage, p53 is called "the guardian of the genome" [30,31].

From the evolutionary point of view, the *TP53* gene is specific for the Holozoa branch, where its ancestral p63/p73-like genes emerged approximately one billion years ago [32,33]. The p53/p63/p73 protein family plays key roles in several major molecular and biological processes, including tumor suppression, fertility, mammalian embryonic development, and aging [20]. Unlike *TP53*, *TP73* and *TP63* genes are rarely mutated in cancers. Yet, the tumor suppressor function of p73 (tp73) is often attenuated in human cancers. The mechanism of suppression is via the hypermethylation of CpG islands at promoter 1, the binding to the overexpressed dominant-negative p73 isoform, dNp73 [34], or to E3 ubiquitin-protein ligase Mdm2 or p53-binding protein Mdm4. The pharmacological inhibition of protein–protein interactions is currently being explored for improved cancer therapy [35]. Notably, it was demonstrated that all p53 family members take part in regulating aging through the activation of senescence and regulating DNA repair [18,25].

p53 is a major factor that regulates cellular senescence, and the mechanism is via the activation of cyclin-dependent kinase inhibitor 1 *CDKN1*A (p21) and promyelocytic leukemia protein, PML. The study by Tyner et al. showed that heterozygous mice having one *TP53* allele with the deletion of the first six exons ($Tp53^{+/m}$, Δ exon 1–6) aged prematurely. These mutant mice exhibited enhanced resistance to spontaneous tumors, yet displayed accelerated aging compared to $Tp53^{+/+}$ mice [36]. A study by the same group showed that truncated p53 protein stabilized wild-type p53 in non-stressed cells promoted its nuclear accumulation, and induced the hyperstability of wild-type p53 upon irradiation [37]. Based on this observation, the conclusion was that the constitutive expression of p53 accelerates aging. This was not confirmed in a follow-up study [38], as the pro-aging phenotype was not seen in the p53 "super-mice" expressing additional copies of the *TP53* gene. Thus, over-activated p53 *per se* might not be a critical driver of accelerated aging. Yet, the role of p53′s hyperactivity in aging appears to be conflicting. Fibroblasts derived from hereditary segmental progeroid syndrome patients with the homozygous antiterminating mutation c.1492T > C in the *MDM2* gene showed p53 hyperstability and accelerated aging [39]. This study postulated that the hyperstability of p53 due to an aberrant MDM2–p53 axis and the exposure to chronic stress induces the aging phenotype through the induction of chronic senescence. MDM2, the best-described negative regulator of p53, binds to the N terminal domain of p53 via its N-terminus. The knockout $Mdm2^{-/-}$ mice show embryonic lethality in a wtp53 background, which indicates that p53 regulation by MDM2 is critical for development. Yet, the conditional deletion of *Mdm2* in the epidermis induced p53-mediated senescence and accelerated aging [40]. Thus, a deregulated MDM2–p53 axis might play a role in the aging phenotype.

Gradual DNA damage and mitochondrial decline are hallmarks of physiological aging. DNA damage that is activated by telomere attrition in an aging cell induces p53 and mitochondrial dysfunction through the repression of the PPARγ co-activator 1α (PGC1α). This induces senescence [41]. Furthermore, a study on the hereditary segmental progeroid syndrome clearly highlighted the role of Mdm2 inactivation and p53 hyperactivity in the aging phenotype [39]. Despite the emerging evidence, the exact molecular mechanisms underlying the p53-mediated aging phenotype need to be elucidated. For example, it was demonstrated that replicative senescence is facilitated by p53, mainly through the activation of *CDKN1*A. Yet, several other factors contribute to aging, such as the activation of E2F and mTOR, as described elsewhere [18]. In principle, it can be concluded that p53 prevents cancer and protects from aging under physiological conditions; however, chronic-stress-amplified p53 has a detrimental effect on healthy aging despite retaining its tumor suppression function. Hence, p53 can either be a pro-aging or a pro-longevity factor, depending on the physiological context [42].

In addition to full-length p53, p53 isoforms may also play an important role in the modulation of longevity. The expression of certain short and long forms of p53 protein might contribute to a balance between tumor suppression and tissue regeneration [43]. For example, the p53β isoform, which is generated through the alternative splicing of intron 9, is upregulated in normal human senescent fibroblasts and interacts with full-length p53 to induce *CDKN1*A [44].

Considering the critical role of p53 in maintaining tissue homeostasis, high frequency of gain of function (GOF) mutations in cancer, and the limited and conflicting information on p53 role in organismal aging in Animalia, in the present work, we employed currently available datasets and tools to analyze p53 protein sequences in species possessing an extended lifespan. Our thorough analysis depicted a surprising correlation between the changes in the p53 protein sequence and the organismal lifespan, both in short- and long-lived species. Many of the identified changes occurred in the DNA-binding domain and might have a detrimental effect on p53 DNA-binding activity. Overall, we found that, when compared to the majority of closely related organisms within their phylogenetic groups, animals with unusually long lifespans share atypical p53 protein sequence features when compared to human p53 in the position corresponding to the human 180–192 p53 region, which points to the important contribution of loop 2 in the p53 core domain regarding life expectancy.

2. Results

The growing evidence implies that p53 activity might play a pivotal role in aging in humans and little is known about the molecular signatures of extended lifespan in animals including humans; thus, we inspected all currently available sequence data of long-lived animals to explore a link between longevity (maximal lifespan) and p53 protein sequences. For this, we used the longevity data from the AnAge Database [45]. We merged all available p53 protein sequences from the RefSeq database with the AnAge Database (for more detail, refer to the Materials and Methods section). The p53 sequence from 118 species and their lifespan data were cataloged and sorted according to their phylogenetic group (Supplementary Materials File S1).

The longest living animal in our dataset was the bowhead whale (*Balaena mysticetus*) from Artiodactyla (subgroup Cetacea), with a maximal lifespan of 211 ± 35 years [46]. Bowhead whales had a significantly longer lifespan (about four times longer) compared with other whales. A thorough comparison of p53 protein sequences showed that, in contrast to other Cetacea, *Balaena mysticetus* had a unique leucine substitution in the proline-rich region, corresponding to amino acid residue 77 in human p53 (Figure 1). Even though the change in the amino acids was predicted to be neutral according to the Protein Variation Effect Analyzer (PROVEAN) score of −0.993, the substitution still might change the activity of p53. Yet, this could only be addressed by extended functional studies. All other accessible p53 sequences of whales had an identical amino acid residue in this position to human p53.

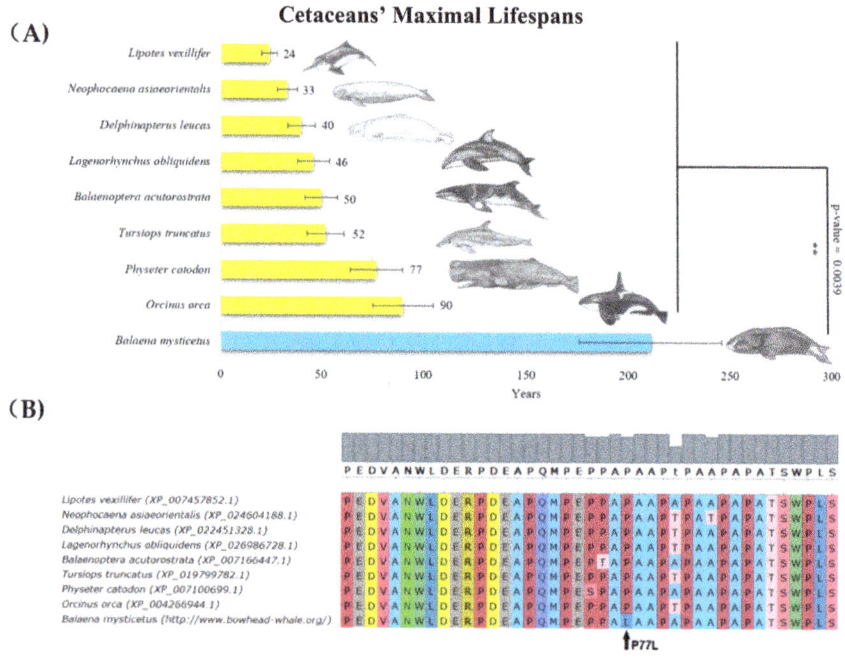

Figure 1. Lifespan of species in the Cetacea order and the corresponding p53 sequence changes. (**A**) Comparison of cetaceans' maximal lifespans in years. The bowhead whale's (*Baleana mysticetus*'s) maximal lifespan was more than twice the maximal lifespan of the rest of Cetacea (Wilcoxon one-sided signed-rank test was used, ** *p*-value < 0.01). (**B**) Multiple sequence protein alignments of p53 proline-rich region, performed in MUSCLE with default parameters [47], colors in "UGENE" style.

Most amphibian species live for less than 30 years [45]; however, the olm (*Proteus anguinus*, Batrachia, Amphibians), which is the only exclusively cave-dwelling chordate, has a maximal documented lifespan of 102 years. A comparison of the p53 protein sequences in amphibians showed a previously unrecognized insertion in *Proteus anguinus*. The p53 protein from this species had additional serine and arginine residues in the core domain (corresponding to an insertion after amino acid L188 in human p53), which had a deleterious effect on p53 functionality according to the PROVEAN tool (Figure 2, Table 1).

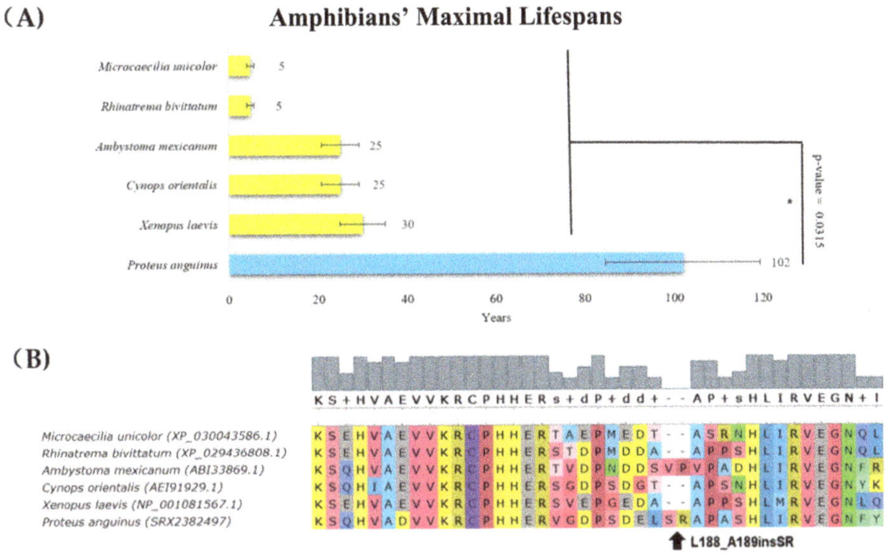

Figure 2. Lifespan of species in amphibians and the corresponding p53 sequence changes. (**A**) Comparison of amphibians' maximal lifespans in years. The olm's (*Proteus anguinus*'s) maximal lifespan was more than three times higher than the maximal lifespan of other amphibians (Wilcoxon one-sided signed-rank test, * p-value < 0.05). (**B**) Multiple protein alignments of the p53 dimerization region. The olm (*Proteus anguinus*) had an insertion that is two amino acid residues long following amino acid residue 188 (related to human p53 canonical sequence). The sequence of the p53 homolog from *Proteus anguinus* was determined using transcriptomic data from the SRA Archive (SRX2382497). The methods and color schemes are the same as in Figure 1B.

The kakapo (*Strigops habroptila*, Aves) is a long-lived, large, flightless, nocturnal, ground-dwelling parrot that is endemic to New Zealand with a lifespan of around 95 years (Figure 3A, blue bar). A comparison of its p53 protein sequence with other related species showed a change at positions 128 and 131, corresponding to the following changes in human p53: P128V and N131H (Figure 3B). Interestingly, N131H mutations in human p53 are found in pancreatic and colon cancers [48,49]. This mutation most probably changes the structure of the p53 core domain and decreases the ability of p53 to bind to a canonical DNA sequence. Relevantly, according to the Phenotypic Annotation of *TP53* Mutations (PHANTM) classifier, the N131H mutation decreases p53 transcriptional activity by 47.19% [50]. In addition, according to the PROVEAN tool, substitutions at position 128 were deleterious with a score of −4.45 (Table 1). These findings support the hypothesis that the change in p53 in the kakapo is linked to the loss of function. We speculate that the lack of exposure to sunlight, thus low incidence of UV-induced DNA damage, might render p53 inactive in this species.

Table 1. Comparison of animals that were characterized by extreme longevity and their atypical p53 features, where the significance of particular changes was predicted. The default PROVEAN threshold of −2.5 was used, insertions and deletions were submitted relative to the human canonical protein sequence (NP_001119584.1). "*" indicates significant PROVEAN values (<−2.5).

Organism Classification	Maximal Lifespan (y)	Adult Weight (kg)	p53 Oddities	Effect Predicted by PROVEAN
Balaena mysticetus Mammalia, Cetacea	211	100,000	Unique substitution in proline rich region	Neutral (P77L, score = −0.993)
Myotis brandtii, Myotis lucifugus Mammalia, Chiroptera	41	0.007	Insertion in DNA-binding domain	Deleterious (P295_H296insPKQPPGS, score = −2.526) *
Strigops habroptila Aves, Psittaciformes	95	1.75	Substitution in core domain	Deleterious (N131H, score = −3.162) *
Proteus anguinus Amphibia, Urodela	102	0.017	Insertion nearby dimerization region	Deleterious (L188_A189insSR, score = −3.357) *
Turritopsis sp. Cnidaria	∞ rejuvenation	0.001	No p53/63/73 protein expressed (unprecedented phenomenon in the whole animal kingdom)	Not applicable

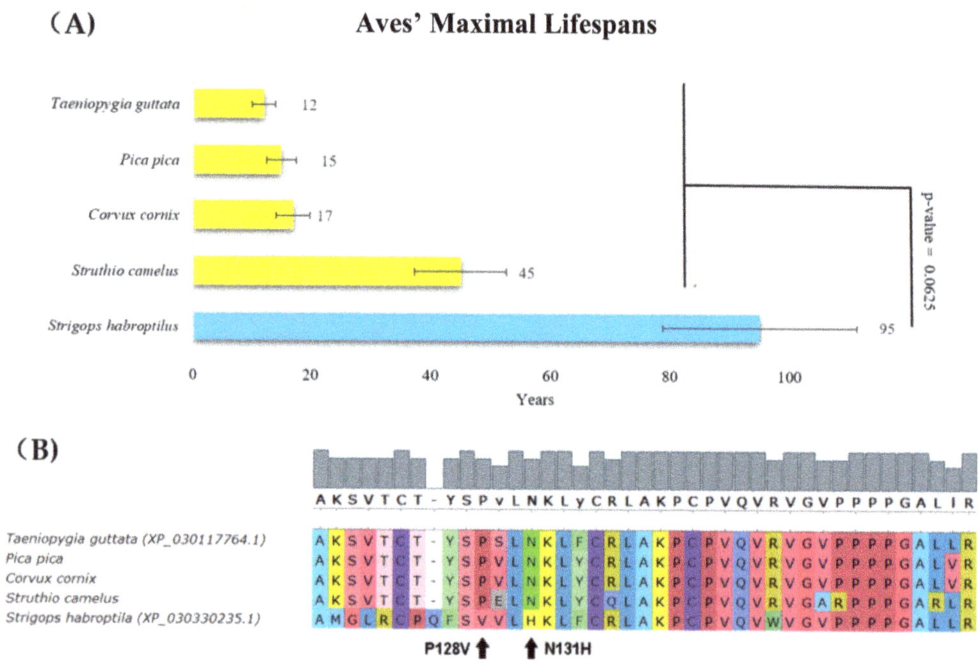

Figure 3. Lifespans of species in the Aves order and the corresponding p53 sequence changes. (**A**) Comparison of Aves' maximal lifespans in years. The kakapo's (*Strigops habroptila*'s) maximal lifespan was more than twice the maximal lifespan of other Aves (Wilcoxon one-sided signed-rank test). (**B**) Multiple protein alignments representing the partial p53 core domain of the accessible Aves sequences. Sequences of all avian p53 homologs were determined using transcriptomic data from the SRA Archive, except for *Strigops habroptila*, where the p53 sequence was known (XP_030330235.1). The methods and color schemes are the same as in Figure 1B.

Next, our analysis identified alterations in p53 in species having a long lifespan in the Chiroptera order. The Brandt's bat (*Myotis brandtii*) is an extremely long-lived bat with a documented lifespan of 41 years [51]. Together with its close relative *Myotis lucifugus*, they had significantly longer lifespans than other bats (Figure 4A, blue bars). These two species share a unique arrangement in the p53 DNA-binding region, with the insertion of seven amino acid residues in the central DNA-binding region (following amino acid 295 in the human p53 canonical sequence) (Figure 4B). To assess how this rearrangement in the DNA-binding region changes the interaction of p53 with DNA, we next modeled the p53 tetramer using the SWISS-MODEL workflow. The insertion in the DNA-binding domain of bats with a long lifespan occurred in the DNA interaction cavity, suggesting the decreased affinity of p53 for binding to DNA (Supplementary Materials File S2). *Myotis brandtii* and *Myotis lucifugus* are very small bats (max 8 g body weight) and provide a significant exception from Max Kleiber's law (mouse-to-elephant curve) since their lifespan is extremely long in relation to their small body size [52].

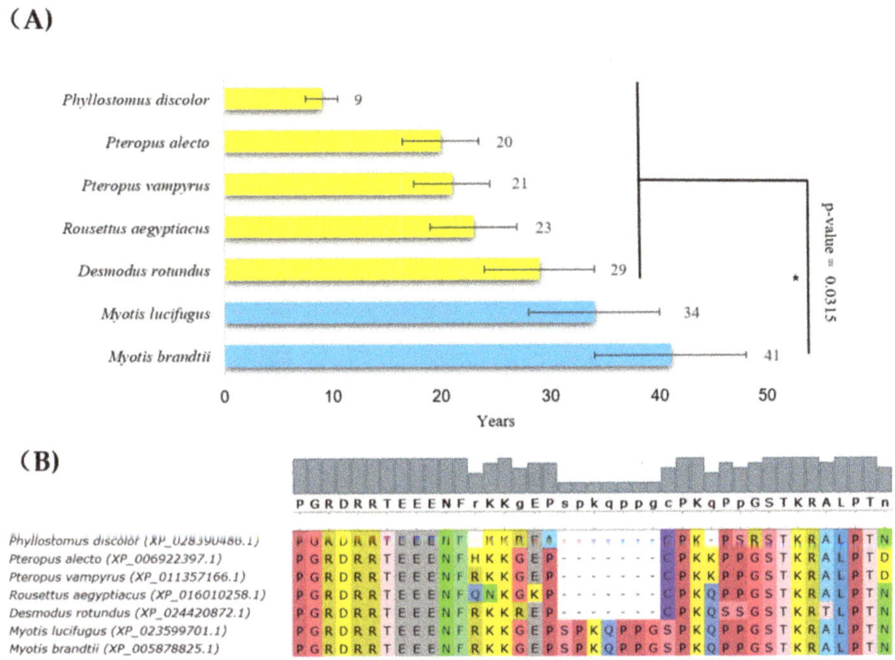

Figure 4. Lifespans of species in the Chiroptera order and the corresponding p53 sequence changes. (**A**) Comparison of Chiropteras' maximal lifespans in years. The bats', *Myotis brandtii*'s and *Myotis lucifugus*'s maximal lifespans were significantly longer compared with other sequenced bats (Wilcoxon one-sided signed-rank test, * p-value < 0.05). (**B**) Multiple protein alignments of the C-terminal part of the p53 core domain of accessible Chiroptera sequences. Methods and color schemes are the same as in Figure 1B.

The abovementioned analysis of long-lived organisms in various animal groups led us to conclude that the amino acid sequence of p53 was associated with organismal lifespan. Therefore, we continued the analysis by further correlating the p53 amino acid sequences with the lifespan across the animal kingdom. Due to the low similarity between the p53 N-terminal and C-terminal domains across species and the significant role of mutations in the p53 DNA-binding domain in cancer, we focused on the most conserved core domain of p53 and constructed the p53-based tree (Figure 5, left panel) [32]. We then compared the contemporary phylogenetic tree with the tree based on the p53 protein sequence (Figure 5). Then, the dataset with p53 sequences and animal lifespans were divided into 12 groups based on their phylogenetic relationships. Interestingly, some p53 sequences were not closely associated with the phylogenetic tree, indicating several parallel evolutionary processes leading to modified p53 activity. Even closely related species in various groups had significantly different lifespans (Supplementary Materials File S1), and therefore, were suitable for correlation analyses according to the method introduced by Jensen and colleagues [53].

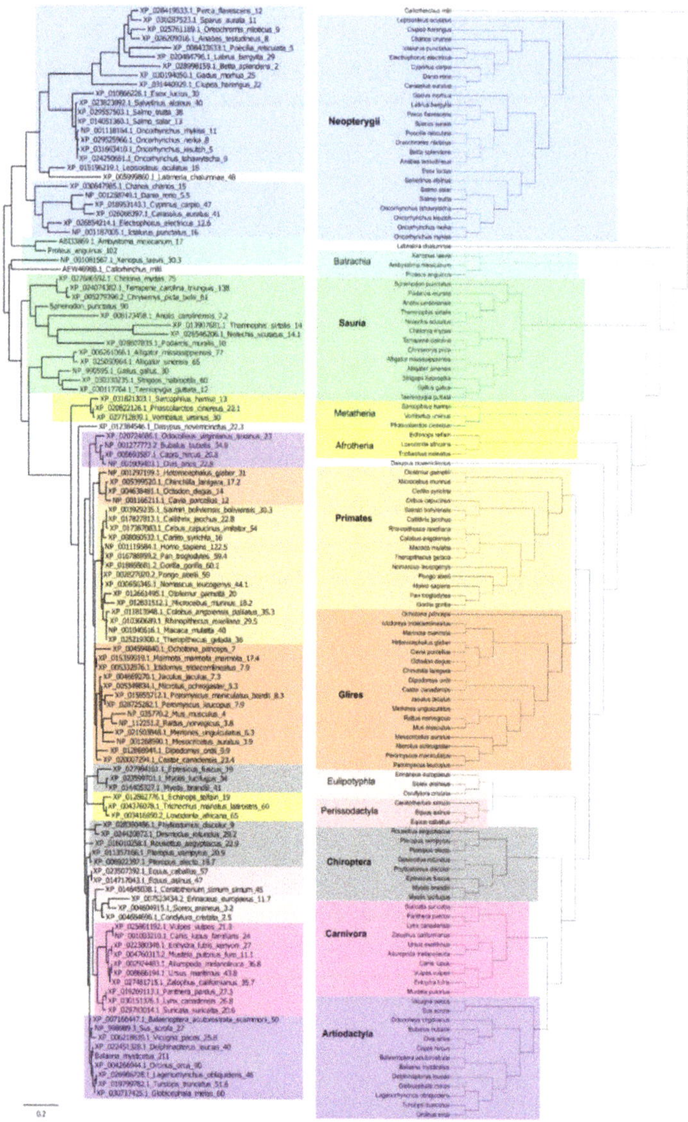

Figure 5. The p53-based and contemporary phylogenetic trees. Comparison of the p53 protein tree (**left**) and the real phylogenetic tree (**right**). The protein tree was built using the Phylogeny.fr platform. Organismal phylogeny was reconstructed using PhyloT and visualized in iTOL (see the Materials and Methods section for details). The same color backgrounds represent the same phylogenetic groups.

Figure 6 summarizes the lifespan data and the total number of analyzed animals for each group, with minimal and maximal values (shown in Supplementary Materials File S3). Only datasets with more than five members in the group were used in the correlation analyses.

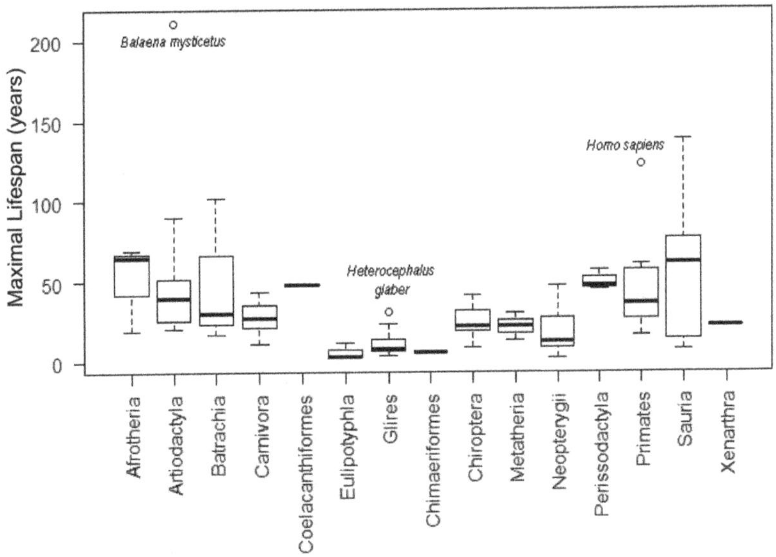

Figure 6. Representation of lifespans for all tested phylogenetic groups.

The organisms with the longest lifespan in the Neopterygii dataset were the carp (*Cyprinus carpio* (47 years)), followed by the goldfish (*Carassius auratus* (41 years)). The Siamese fighting fish (*Betta splendens)* had the shortest lifespan in the group (2 years). The correlation analyses show that fifteen amino acid residues in the p53 core domain were significantly associated with a prolonged lifespan (Figure 7A). We found that the most common variation in the long-lived Neopterygii was the presence of a serine (S) at positions corresponding to 98, 128, and 211 of human p53, and the presence of valine (V) at positions 128, 150, 217, and 232. On the other hand, in the short-lived organisms in Neoropterygii, we identified threonine (T) at positions 98, 100, 141, 217, and 260; glutamic acid (E) at positions 110, 128, 150, and 291; and serine at positions 141, 203, and 235. We reasoned that the abundance of glutamic acid could result in the decreased affinity of p53 to DNA due to the local change in the ionic charge at the site of the amino acid p53 variant. Indeed, the PROVEAN tool predicted a deleterious effect on p53 function for glutamic acid at position 128. In addition, according to the PHANTM classifier, C141S substitution led to a decrease in p53 transcription activity by 41.08% as compared to wtp53.

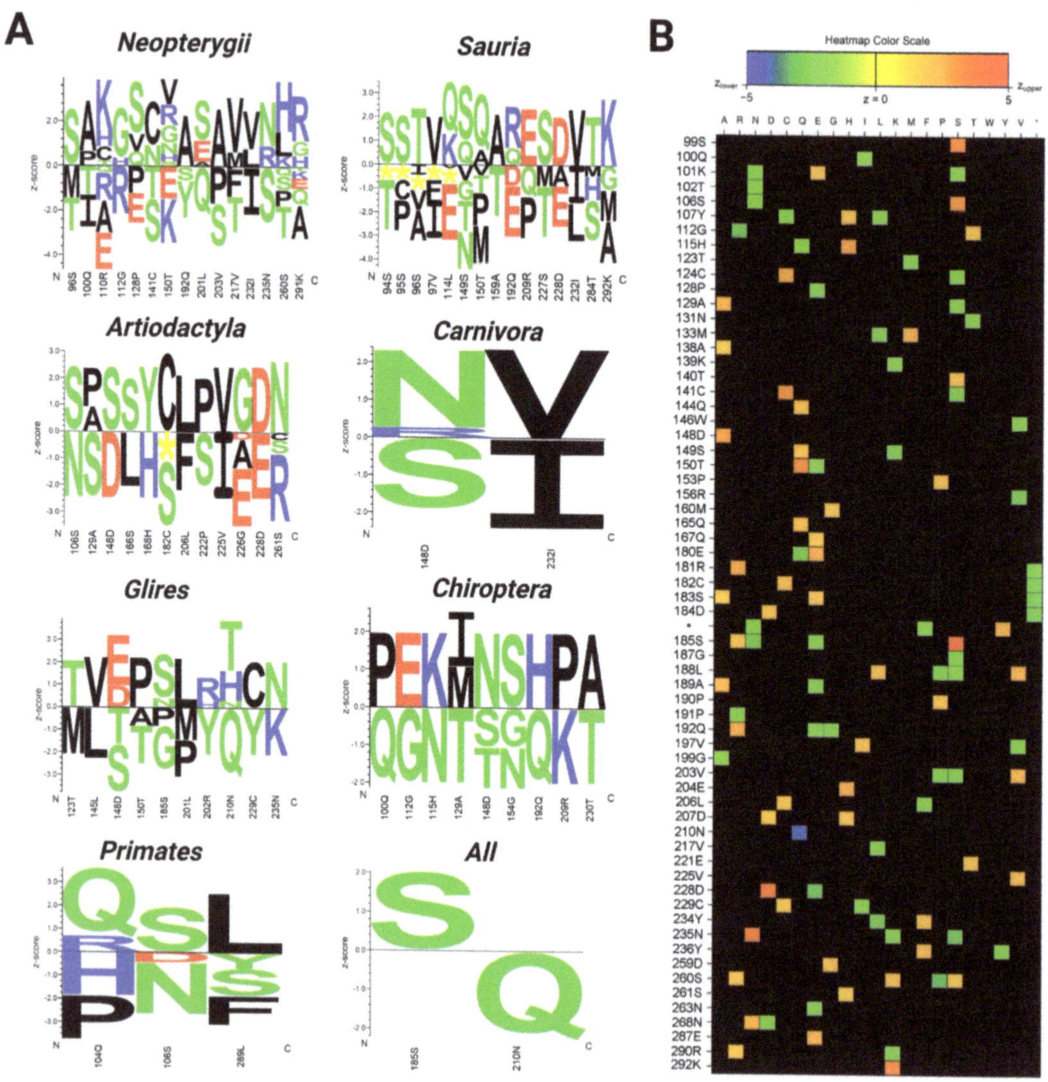

Figure 7. Correlation of the most commonly altered p53 amino acid residues with the maximal lifespans of the analyzed species. (**A**) Logos quantifying the strength of the p53 core domain residue association (related to the human aa 94–293 according to the p53 canonical sequence) with the maximal lifespan in years in the analyzed subgroups of animals. Amino acid residues on the positive y-axis were significantly associated with the prolonged lifespan phenotype and residues on the negative y-axis were significantly associated with the shorter lifespan phenotype (significance threshold *p*-value ≤ 0.05). The height of each letter representing the strength of the statistical association between the residue and the data set phenotype. The amino acids are colored according to their chemical properties as follows: acidic (DE): red, basic (HKR): blue, hydrophobic (ACFILMPVW): black, and neutral (GNQSTY): green. (**B**) Heatmap visualization of the strength of the residue association (without a Bonferroni correction). The color scale ranges from blue ($z < -5$) to red ($z > 5$). Each column corresponds to one of the 20 proteinogenic amino acids and each row to a position in the submitted multiple sequence alignment (Supplementary Materials File S4). * Indicates site of the amino acid insertion.

The lifespans of species in Sauria were significantly variable. The organisms with the longest lifespan in this group were the three-toed box turtle (*Terrapene carolina triunguis* (138 years)) and the kakapo (*Strigops habroptila* (95 years)). The green anole (*Anolis caroli-*

nensis) had the shortest lifespan in the group (7.2 years). The correlation analyses showed that, similar to Neopterygii, a specific fifteen-amino-acid-residue fragment in the p53 core domain was associated with a prolonged lifespan (Figure 7A). The most common p53 variation in long-lived Sauria was similar to Neopterygii and it was the higher abundance of serine (corresponding to positions 94, 95, 149, and 227 in human p53) and the presence of valine (at positions 97 and 232, identical to Neopterygii). When compared to human p53, in short-lived organisms, we identified threonine at positions 94, 149, 159, and 227, and glutamic acid at positions 114, 192, and 228. In addition, deletions in the p53 sequence were found at positions 94–97 and 114 (Figure 7A). Similar to Neopterygii, the most common p53 variation in short-lived Sauria was the presence of threonines and glutamic acid residues. However, more studies are needed to elucidate the functionality of these p53 sequences.

The organisms with the longest lifespans in the Primates group were humans (*Homo sapiens* (122 years)) and the western gorilla (*Gorilla gorilla* (60 years)). Tarsier (*Carlito syrichta)* had the shortest lifespan in the group (16 years). The correlation analyses showed that the specific amino acid triad—Q^{104}, S^{106}, L^{289}—was significantly associated with a prolonged lifespan (Figure 7A). Besides serine at position 106, two others—glutamine (Q) at position 104 and leucine (L) at position 289—are both hydrophobic and might impact the structure of the DNA-binding domain.

In contrast, proline (P) or histidine (H) at position 104, asparagine (N) at position 106, and phenylalanine (F), serine (S), or tyrosine (Y) at position 289 were associated with short-living primates. While studying human longevity, one needs to consider that the prolonged lifespan of *Homo sapiens* is associated with cultural and socio-economical advantages. Therefore, we performed additional analyses after excluding *Homo sapiens* from the dataset. The same variations were observed in the correlation analyses. Taken together, our results demonstrate that the amino acid variations shown in Figure 7A were conserved in the following closely related species: *Homo sapiens*, *Pan troglodytes*, and *Gorilla gorilla*.

The dataset of Glires contained seventeen species with lifespans ranging from 3.8 to 31 years. The organisms with the longest lifespan in this group were *Heterocephalus glaber* (31 years) and *Castor canadensis* (23). The shortest lifespan in the group was *Rattus norvegicus* (3.8 years). The correlation analyses showed that ten amino acid residues were significantly associated with a prolonged lifespan (Figure 7A). Two threonine residue variations (positions 123 and 210) were present in long-lived Glires. Other amino acid changes occurred only once. Interestingly, in short-lived Glires, there was also a significant presence of threonine at two other locations (positions 148 and 150). Similar variations were also observed in the methionine residues (at positions 123 and 201), tyrosine (positions 202 and 229), and proline (positions 185 and 201).

The organisms with the longest lifespan in the dataset of Chiroptera were the brandt bat (*Myotis brandtii* (41 years)) and the little brown bat (*Myotis lucifugus* (29 years)). The pale spear-nosed bat (*Phyllostomus discolour)* had the shortest lifespan (9 years). The correlation analyses showed that nine-amino-acid-long motif was associated with a significantly prolonged lifespan (Figure 7A).

The organisms with the longest lifespans in the Carnivora group were the polar bear (*Ursus maritimus*) and the panda (*Ailuropoda melanoleuca*). The shortest lifespan in the group was that of the ferret (*Mustela putorius furo*). The correlation analyses showed that two amino acids in the p53 core domain, position 148 and 232, were significantly associated with lifespan (Figure 7A). While the presence of asparagine at position 148 and valine at position 232 were associated with a long lifespan, the presence of a serine at position 148 and isoleucine at position 232 was associated with a short lifespan.

The organism with the longest lifespan in the Artiodactyla group was the bowhead whale (*Balaena mysticetus* (211 years)) followed by the orca (*Orcinus orca* (90 years)). Correlation analyses showed that twelve amino acid residues in the p53 core domain were significantly associated with prolonged lifespan (Figure 7A). Similar to Neopterygii and Sauria, the most common variation present in the long-lived organisms were associated with high abundance of serine (corresponding to positions of 106, 148, and 166 in human p53). The variations of serine at positions 129, 182, and 222 were the most common variations for the short-lived Artiodactyla, together with variations in the glutamic acid residues at positions 226 and 228.

Next, we investigated all 118 RefSeq p53 sequences to evaluate the associations between amino acid variations and maximal lifespan (Figure 7A). When applying the Bonferroni correction, only two significantly associated residues were revealed, corresponding to human serine (S) 185 and asparagine (N) 210. Organisms that have serine at position 185 live statistically longer than organisms with another amino acid in this position. Interestingly, p53 S185 variants are rare in humans and only a few variants were found to be cancer-specific, suggesting that S185 might be a conserved amino acid that is critical for organismal longevity [54]. On other hand, organisms that contained glutamine (Q) instead of asparagine (N) at position 210 had a significantly shorter maximal lifespan. Without a Bonferroni correction, from the 200 analyzed positions of the aligned p53 core domains (related to human p53 94–293 aa), 64 positions were significantly associated with lifespan (Figure 7B). Positive correlations with longevity are shown using orange and red colors, green and blue show negative correlations. However, more detailed studies are needed to fully apprehend the functionality of the changed p53, both in the short-lived and in the long-lived organisms.

The changes at the molecular level are often a result of the adaptation of species to environmental forces. To evaluate whether the amino acid residues in the p53 core domains (aa 94—293 of the human p53 canonical sequence) share some relevant features in relation to the convergent evolution, we constructed a sequential circular representation of the multiple sequence alignments and the mutual information it contains (Figure 8A). The figure shows that the amino acid residues that were significantly associated with longevity (extracted from the heatmap (Figure 7B), highlighted in light green) very often coevolved together (represented by connected lines). This observation may provide evidence for the convergent evolution of p53 proteins in organisms with extreme longevity. According to Passow and colleagues, taxa with evidence of positive selection in the *TP53* gene are those with the lowest incidences of cancer reported in amniotes (elephants, snakes, lizards, crocodiles, and turtles) [55].

The longest region in p53 that was associated with longevity spanned amino acids in loop 2 (L2) of human p53 (Figure 8A, green, dashed circle, residues 180–192). L2 is the minor groove binding region of human p53 and the stability of this region is maintained by Zn^{2+}. The loss of Zn^{2+} triggers the aggregation of L2 and the loss of DNA-binding specificity. The 3D structures of DNA binding domain (DBD) in selected long-lived species revealed intrinsic variations in L2 structure when compared to human p53 (Figure 8B). It is possible that intrinsic changes in L2 due to amino acid changes in long-lived species amend p53 DNA-binding specificity and/or the binding of co-factors via an allosteric mechanism and alter the p53-driven senescence program.

Table 1 lists the species characterized by the extreme longevity together with the associated p53 variations identified in our study. Apart from the unique substitutions (*Strigops habroptila, Balaena mysticetus*) and insertions (*Myotis brandtii, Myotis lucifugus, Proteus anguinus*), a complete lack of p53 mRNA expression was found in *Turritopsis* sp.

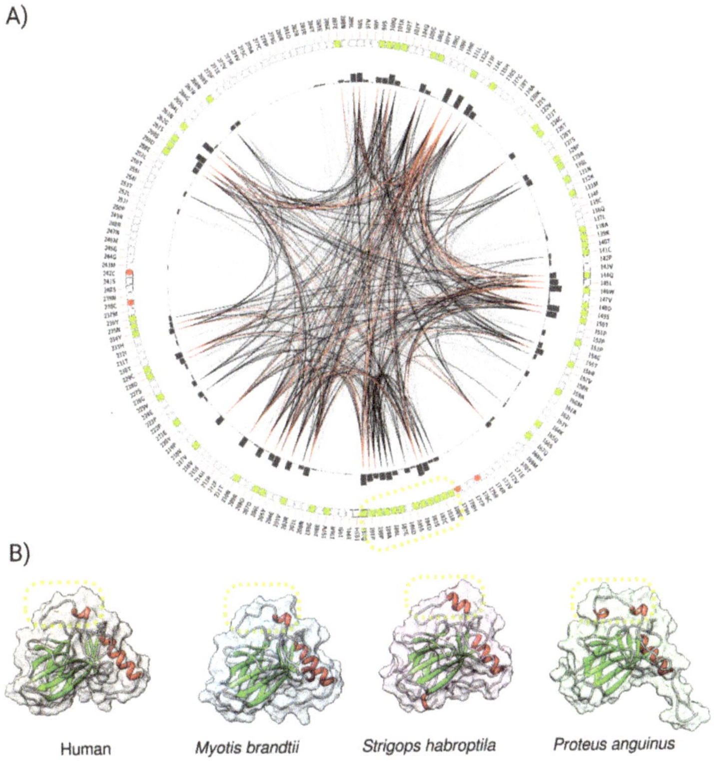

Figure 8. A graphic representation of the positions of the p53 amino acids' linked to longevity in the animal kingdom. (**A**) Mutual information to infer the convergent evolution of p53 core domains. A circos plot is a sequential circular representation of the multiple sequence alignment and the information it contains. Green boxes in the outer circle indicate the positions of the amino acids' changes correlating with longevity. The dashed oval highlights the longest region associated with longevity, which spans the loop 2 (L2, residues 180–192) region of human p53 DBD including S185. Lines connect pairs of positions with mutual information greater than 6.5 [51]. Red edges represent the top 5%, black represents between 70 and 95%, and gray edges account for the remaining 70%. (**B**) p53 core domains of three different, long-lived organisms compared to humans, as modeled by trRosetta.

To gain a better insight into the putative changes in the p53 regulatory pathways in the long-lived species in which the p53 protein remains unaltered, we analyzed the sequence of p53 regulators. SIRT1 deacetylates p53 in an NAD^+-dependent manner and inhibits p53 transcription activity [56]. We found that SIRT1 had an atypical protein sequence in *Cebus imitator*, a model organism for studying extreme longevity in primates, where the amino acid sequence is different from all other primate SIRT1s. 3D modeling revealed that the predicted structure of *Cebus imitator*'s SIRT1 was significantly different from the structures of SIRT1 from *Homo sapiens* and *Sapajus apella* (a close relative of *Cebus imitator*) (Figure 9).

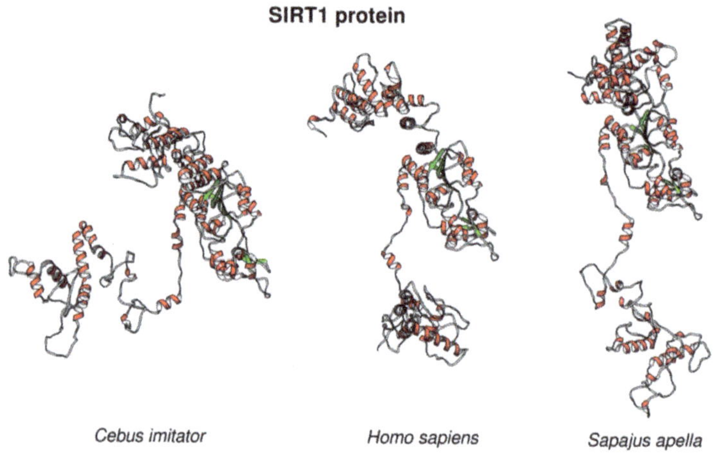

Figure 9. The 3D structures of SIRT1 proteins from long-lived species compared to *Homo sapiens*. SIRT1 structures from *Cebus imitator* (XP_017357564.1), *Homo sapiens* (NP_001135970.1), and *Sapajus apella* SIRT1 (XP_032108492.1) showed differences in the protein structures in the given species.

We hypothesized that SIRT1 from *Cebus imitator* gained new functions, which might result in the decreased activity of p53 when compared to other primates and slow down the aging processes, most likely via the transient inhibition of p53. Yet, it remains to be elucidated which factors might be affecting the altered target recognition by SIRT1.

In addition to SIRT1, we investigated other key factors in the p53 pathway. Surprisingly, we found that *Myotis brandtii* (long-living bat described above), in addition to p53, had two atypical protein sequences, one in UFM1 (Ubiquitin-fold modifier 1, XP_005862786.1) and the other in the p73 (tumor protein 73, XP_014401672).

Recently, it was reported that UFM1 covalently modifies p53; this phenomenon is called UFMylation [57]. UFMylated p53 is stabilized at the protein level, as this covalent modification antagonizes p53 ubiquitination and proteasomal degradation.

In UFM1, we found an approximately 20-amino-acid-long extension of the C-terminal end in *Myotis brandtii* (and also in two other myotis bats—*Myotis lucifugus* and *Myotis myotis*) (see Figure 10). In contrast, in other bats that live much shorter (e.g., the closest myotis bats relative is *Pipistrellus kuhlii*, with a maximal lifespan of only 8 years) and in the rest of mammals, including humans, no such extension occurs. We hypothesize that the extended UFM1 protein might contribute to the extreme longevity in myotis bats through the loss of function and consequent changes in p53 protein degradation patterns. Yet, more experimental evidence is needed to draw a clear conclusion.

Lastly, we found unique changes in the p73 protein sequence in the *Myotis brandtii* bat (XP_014401672). There were multiple large deletions (>10 amino acid residues) in critical p73 regions, which were found exclusively in this extremely long-lived bat. p73 is the transcription factor that undergoes similar cellular regulation as p53 protein and its role in aging is attributed to the induction of senescence through the upregulation of *CDKN*1A [58].

Taken together, our analyses revealed the unexpected correlation between p53 sequence variations and longevity in the animal kingdom. The changes may affect p53 functionality and, thus, influence the activation of replicative senescence, a hallmark of molecular aging. In long-lived species, with no changes in p53, the upstream regulatory proteins, including SIRT1 and UFM1, displayed amino acid changes that may affect their functionality and in consequence alter p53 activity. Yet, further studies are needed to fully comprehend the role of amino acid changes in p53 and its role in the long-lived species described in our work.

Ubiquitin-Fold Modifier 1 in Bats

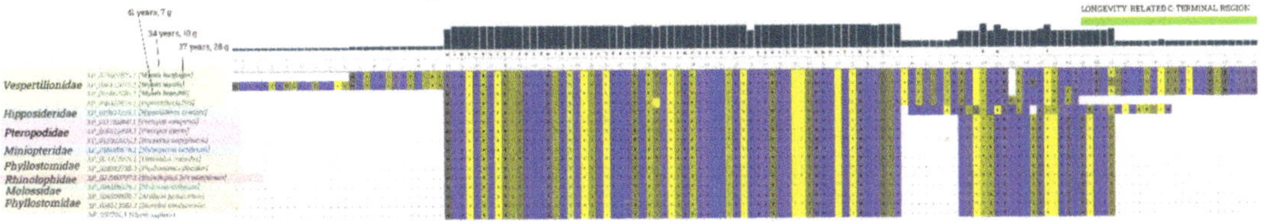

Figure 10. Sequence alignment of UFM1 proteins in bat species and humans. The 20-amino-acid-long extension of the C-terminal end in three long-living bats is depicted in light green. Multiple sequence protein alignments of UFM1 reference protein sequences were performed in MUSCLE with default parameters [47]; the colors express strand propensity.

3. Discussion

The p53 protein is a well-known tumor suppressor and *TP53* is the most often mutated gene in human cancers. On the cellular level, in humans, increased p53 activity protects from cellular stress and enables genome stability, whereas altered mutant p53 protein functionality is essential for cells' immortalization and neoplastic transformation [59]. However, the role of variations in the p53 amino acid sequence at the organism level in other animals has not been studied systematically. Here, we addressed the role of p53 in longevity in the animal kingdom by presenting an in-depth correlation analysis manifesting the dependencies between p53 variations and organismal lifespan. To date, p53 expression has been detected in all sequenced animals from unicellular Holozoans to vertebrates [32]. The seminal work by Kubota provided important evidence demonstrating that immortality is not just a hypothetical phenomenon. He demonstrated that the Cnidarian species *Turritopsis* jellyfish is immortal and can repeatedly rejuvenate, reverse its life cycle, and thus, was the first and only known "immortal" animal on Earth [60]. Here, we inspected recently published data from the whole-transcriptome data of "immortal" *Turritopsis* sp. [61] and surprisingly found no expression of any of the p53 protein family members in the pooled data from all individuals at all developmental stages (polyp, dumpling with a short stolon, dumpling, and medusa). This points to the possibility that the absence of p53 in *Turritopsis* might be directly related to its unique ability of life cycle reversal and "immortality."

Telomere shortening in humans induces replicative senescence, which is a process that is regulated by p53. In the absence of p53, the replicative lifespan of human cells is extended and the concurrent loss of retinoblastoma protein (RB) extends the replicative lifespan to a greater extend (reviewed in [26]).

Intriguingly, our results obtained using Protein Variation Effect Analyzer [62] show that the variability in lifespan among closely related species correlated with specific p53 amino acids' variations. Long-lived organisms were characterized by specific substitutions in the p53 amino acid sequence. It is likely that the amino acid changes imposed on p53 in long-lived species enable p53 to interact with different multiple protein partners to induce gene expression programs that vary from those induced in species with a relatively normal lifespan.

We identified the 180–192 region, corresponding to the loop 2 (L2) region of human p53, as the longest region that is associated with longevity. The 3D model revealed variations in L2 structure in long-lived species when compared to human p53. Loop 2 is responsible for binding to the minor groove and its structure depends on the presence of Zn^{2+}. We speculate that in long-lived species, L2 affects the p53 binding to DNA and/or other transcription factors and, consequently, affects the replicative senescence program. On other hand, in humans, the L1 region is responsible for p53 binding to the major groove and was reported to undergo the most dynamic changes among the DNA contacting loops (L1–L3) when located on a non-target or target DNA sequence [63]. Our findings indicate that the L2 region, but not L1, might play a role in modulating the senescence (or other pro-aging program) in long-lived species. Yet, detailed functional studies are needed to fully comprehend the role of p53 alterations in longevity.

Based on what is known about the processes underlying aging, we anticipate that the altered gene expression programs would enable the following changes (Figure 11): (1) more efficient tissue repair through autophagy, (2) loss of replicative senescence, (3) enhanced clearance of senescent cells by the immune system, (4) enhanced regulation of intracellular ROS levels, (5) improved resistance of mitochondria to ROS-induced damage, or (6) loss of immune senescence that occurs in humans during healthy aging. All of the above processes were previously described as significantly contributing to longevity in humans (reviewed in [18]). Intriguingly, a recent GWAS study on 1 million parent lifespans identified only several variants influencing lifespan at genome-wide significance, including *CDKN2B-AS1* and *IGF2R*. The *TP53* gene was not among the singled-out variants, which, in accordance with our observations, indicates that no changes in human p53 might be attributing to longevity in humans [64]. Our analysis demonstrates that the long-lived organisms might have different mechanisms of protection against cancer that are not directly linked to p53 activity. We speculate that their lifespan is not limited by somatic cells' senescence caused by the chronic stress-induced hyperactive p53 protein, which is the case for other species with shorter lifespans (Figure 11).

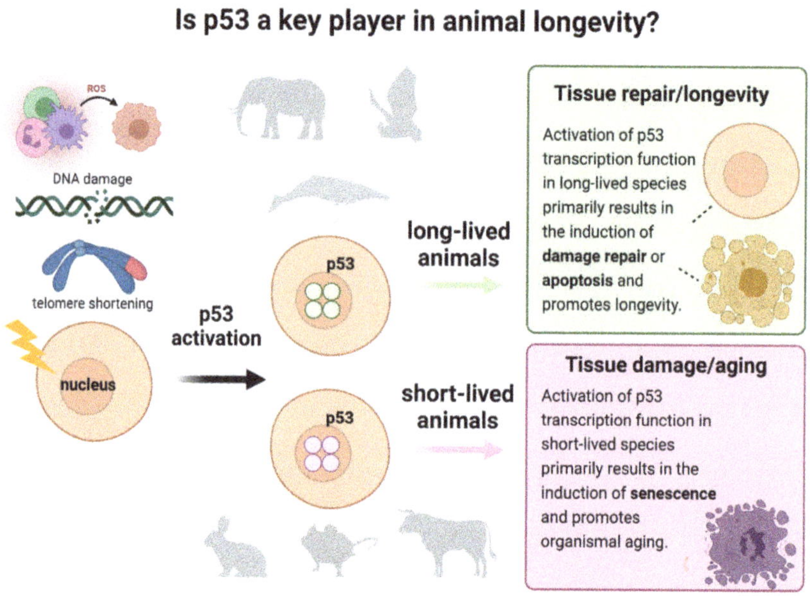

Figure 11. Proposed p53-centric theory of extreme longevity. Cell damage caused by ROS, DNA damage, telomere shortening, or other factors activates p53 to enable DNA repair and/or apoptosis. On the other hand, a high activity of p53 promotes organismal aging, thus shortening the lifespan. We hypothesize that long-lived animals developed the "improved" p53 proteins, which are less active than in their short-lived counterparts but still may sufficiently contribute to DNA damage repair and apoptosis in species that are exposed to environmental genotoxic stresses.

The maximal lifespan according to the AnAge database is attributed to the Greenland shark, with an estimated maximal life span of 300–500 years. Unfortunately, no transcriptomic nor genomic data for the Greenland shark (*Somniosus microcephalus*) are available. Compared to other sharks (with a life expectancy of up to seventy years), its lifespan is exceptional. It will thus be very interesting to know the sequence of their p53 protein. A recent study suggests that certain animals may have evolved to have longer lifespans compared to other species belonging to the same taxa [65]. In addition, the authors found that the outliers among taxa (in terms of maximum age) always had longer lifespans, not shorter. This would support our hypothesis that extreme longevity is a result of adaptive mutational changes in the particular critical gene(s), allowing the organisms to escape the senescence machinery.

Experimental data support our hypothesis that specific p53 variations are associated with longevity. For example, it was found that the reduced expression of the *Caenorhabditis elegans* p53 ortholog, namely, cep-1, results in increased longevity [66]. It was also demonstrated that the neuronal expression of p53 dominant-negative proteins in adult *Drosophila melanogaster* inhibits the function of full-length p53 and extends their lifespan [67]. The same principle is most probably present in humans, where, for example, p53 variants that predispose to cancer are present in healthy centenarians [68] and a meta-analysis showed that the codon 72 polymorphic variant of p53 with proline (compared to arginine) was associated with increased cancer risk and with the increased survival [69]. In a recent study by Zhao et al., polymorphism at position 72 (P72 compared to R72) was reported to have a positive effect on lifespan and to delay the development of aging-related phenotypes in mice, supporting a role of the changed p53 activity in longevity [70]. Another example of a long-lived vertebrate is the elephant, which has 20 copies of the *Tp53* gene [71]. In this species, part of the DNA-binding region of p53 is deleted in all but one of the *TP53* gene copies, which may result in the formation of dysfunctional p53 tetramers, thus presumably modulating p53 transcriptional activity in response to stress [71]. In contrast, a study by Tejada-Martinez et al. [72] in cetaceans did not single out *TP53* as a gene associated with extreme longevity. Yet, the authors provided evidence that natural selection in tumor suppressor genes (including *TP53*) could act on species with an extended lifespan. In the naked mole rat (*Heterocephalus glaber*), which is the longest-living small rodent and weighs only around 35 g, a unique hyperstabilization and nuclear accumulation of the p53 protein were recently reported [73]. The naked mole rats' natural habitat is in the hypoxic environment underground in constant darkness. Despite their extremely long lifespan of up to 30 years, naked mole rats show very little biological decline, neurodegeneration, and senescence [74]. The hyperstability of the naked mole rat's p53 when compared to murine p53, which is independent of genotoxic stress, might be a consequence of the change in the amino acid sequence in the p53 protein. Yet, an in-depth analysis must be performed to decipher the mechanisms leading to the unprecedented stability of p53 in the naked mole rat and to understand the role of hyperstable p53 in the longevity in this species.

Intriguingly, the animals with extreme longevity that we have identified are mostly nocturnal or live in absolute darkness. These include *Strigops habroptila*, *Myotis brandtii*, *Myotis lucifugus*, *Proteus anguinus*, *Balaena mysticetus*, and *Heterocephalus glaber*. It is thus likely that low or no exposure to the UV light promotes the evolutionary changes in the p53 protein structure that alter the p53′s pro-senescence activity. In addition, we speculate that in those species, the endogenous levels of reactive oxygen species might be lower when compared to animals from other, less extreme habitats. This might be a consequence of the changes in the metabolism rates that might affect the overall rate of oxidative phosphorylation and, consequently, slow down the generation of free radical species through the electron transport chain. This hypothesis agrees with the recent study showing that rapamycin, which is a widely studied inhibitor of mTOR, prevents UV-induced skin aging through the inhibition of p53, reversal of UVA-induced cellular senescence, and induction of autophagy [75].

Despite the high complexity of the p53 proteins family, modern methods of comparative genomics provide useful tools for exploring protein variations in closely related species and correlating the extracted molecular information with lifespan [76]. According to Sahin and DePinho, the hyperactivity of p53 in the presence of accumulated DNA damage and ROS is one of the main causes of aging [41]. This observation is in congruence with our hypothesis that organisms with atypical p53 sequences, likely attenuating the wtp53 activity, are extremely long-lived. Of note, even though several p53 amino acid changes were found in various animal groups, some variations developed in convergent evolutions in different groups of species. For example, the presence of threonine and glutamic acid was observed in short-lived organisms of different groups, and the richness of serine residues was typical for long-lived organisms in several groups. Next, a serine residue at position 185 was significantly associated with a prolonged lifespan across all analyzed species. Yet, further mechanistic studies are needed to pin down how the identified p53 changes affect its functionality, how the amino acid changes contribute to longevity, and how this knowledge can be translated for prolonging healthy aging in humans.

4. Materials and Methods

4.1. Searches of Maximal Lifespan

To access data on longevity and maximal lifespan, we used the AnAge Database (https://genomics.senescence.info/species/, accessed on 4 May 2020); AnAge currently contains data on the longevity of more than four thousand animals [45]. We downloaded the whole dataset and selected species that were presented in the NCBI RefSeq database.

4.2. Protein Similarity Searches

For the protein similarity searches, we downloaded all available p53 sequences from the RefSeq database (https://www.ncbi.nlm.nih.gov/refseq/, accessed on 10 October 2020) and merged them with the AnAge Database. We received the p53 sequence information of 118 species with information about their lifespan and sorted them according to their phylogenetic group (Supplementary Materials File S1). For animals with extreme longevity, where the p53 homologs were not present in the NCBI database, local BLAST searches (tblastn) applied to de novo assembled transcriptomes were used together with the default "BLAST+ make database" command and searching parameters within the UGENE standalone program [77].

4.3. Transcriptome Assemblies

Transcriptomic data for the bowhead whale was obtained from http://www.bowhead-whale.org/, accessed on 3 July 2020 [46]. When there were only raw seq reads from the RNA-seq experiments available (deposited in the NCBI SRA), we performed the de novo assembly first using the Trinity tool [78] from the Galaxy webserver (https://usegalaxy.eu/, accessed on 7 September 2020) [79] with default settings. This was done for *Proteus anguinus* (SRX2382497) and *Sphenodon punctatus* (SRX4014663); the resulting assemblies are enclosed in Supplementary Materials File S5.

4.4. p53 Protein Tree and Real Phylogenetic Tree Construction

The protein tree was built using the Phylogeny.fr platform (http://www.phylogeny.fr/alacarte.cgi, accessed on 15 October 2020) [80,81] and comprised the following steps. First, the sequences were aligned with MUSCLE (v3.8.31) [47], which was configured for the highest accuracy (MUSCLE with default settings). After the alignment, ambiguous regions (i.e., containing gaps and/or poorly aligned) were removed with Gblocks (v0.91b) [82] using the following parameters: minimum length of a block after gap cleaning: 10; no gap positions were allowed in the final alignment; all segments with contiguous non-conserved positions longer than 8 were rejected; minimum number of sequences for a flank position: 85%. The phylogenetic tree was constructed using the maximum likelihood method implemented in the PhyML program (v3.1/3.0 aLRT) [83,84]. The JTT substitution model was selected by assuming an estimated proportion of invariant sites (of 0.204) and 4 gamma-distributed rate categories to account for the rate heterogeneity across sites. The gamma shape parameter was estimated directly from the data (gamma = 0.657). Reliability for the internal branch was assessed using the bootstrapping method (100 bootstrap replicates). Graphical representation and editing of the phylogenetic tree were performed with TreeDyn (v198.3) [85]. The real phylogenetic tree was reconstructed using PhyloT (https://phylot.biobyte.de/, accessed on 4 September 2020) and visualized in iTOL (https://itol.embl.de/) [86].

4.5. Prediction and Statistical Evaluation Using PROVEAN

The effect of the p53 variations in long-lived organisms was predicted and statistically evaluated using the Protein Variation Effect Analyzer web-based tool (PROVEAN; http://provean.jcvi.org/index.php, accessed on 20 May 2021) [62,87]. PROVEAN is a software tool that predicts whether an amino acid substitution or in/del has an impact on the biological function of a protein [62]. All inspected p53 variations in selected animals were statistically evaluated and numbered according to the human canonical p53 sequence (NP_000537.3).

4.6. Modeling of 3D Protein Structures

We used the SWISS-MODEL template-based approach (https://www.swissmodel.expasy.org/interactive, accessed on 3 March 2021) [88] to predict the 3D structures using individual FASTA sequences and reference PDB:4mzr as the crystal structure of the p53 tetramer from *Homo sapiens* with bound DNA [89]. The resulting PDB files are enclosed in Supplementary Materials File S6. The predicted p53 structures were visualized in UCSF Chimera 1.12 [90]. Effects of the novel mutation on SIRT tertiary structure were predicted using RaptorX [91].

4.7. Correlation of the Maximal Lifespan and Alterations within the p53 Core Domain in Vertebrates

Residue level genotype/phenotype correlations in p53 multiple sequence alignment were performed using SigniSite 2.1 (http://www.cbs.dtu.dk/services/SigniSite/, accessed on 22 October 2020) [53] with a significance threshold *p*-value of $\leq$0.05. A Bonferroni single-step correction for multiple testing was applied for the global correlation of all sequences, no correction was applied for smaller groups of taxonomically related animals. The manually curated set of 118 high-quality p53 protein sequences obtained from the NCBI (https://www.ncbi.nlm.nih.gov/, accessed on 20 July 2020) was used as the input file. These sequences were taken from the RefSeq database and the canonical isoform corresponding to the human full-length p53 isoform (NP_001119584.1) was manually filtered for each vertebrate species. The resulting set of these 118 p53 sequences was aligned within the UGENE workflow [77] and the MUSCLE algorithm [47] with default parameters. All sequences were then manually trimmed to preserve only the core domain, which corresponded to human 94–293 aa. Then, the numerical values of the maximal lifespan of each organism were added into the resulting FASTA file based on the information in the reference AnAge database (http://genomics.senescence.info/species/, accessed on 5 September 2020) [45].

4.8. Convergent Evolution

Multiple sequence alignments of the p53 core domains from 118 species were uploaded to the MISTIC webserver (http://mistic.leloir.org.ar/index.php, accessed on 18 November 2020), with PDB 2ocj (A) as the reference and using default parameters [92].

4.9. Gene Gain and Losses

TP53 gene gain or losses were inspected using Ensembl Comparative Genomics toolshed [93] via Ensembl web pages and *TP53* gene query ENSG00000141510: https://www.ensembl.org/Homo_sapiens/Gene/SpciesTree?db=core;g=ENSG00000141510;r=17:7661779-7687550, accessed on 8 December 2020.

5. Conclusions

This study revealed a previously overlooked correlation between longevity and changes in p53 function due to the amino acid variations in the animal kingdom. Strikingly, several long-lived species, including *Myotis brandtii*, *Myotis lucifugus*, *Balaena mysticetus*, *Heterocephalus glaber*, *Strigops habroptila*, and *Proteus anguinus* displayed unique p53 protein sequence properties that were not shared with their close relatives that have a shorter lifespan. Altogether, our data show the convergent evolution of p53 sequences supporting a higher insensitivity to p53-mediated senescence under prolonged stress conditions in long-lived vertebrates. Our observations that specific variations of p53 protein are correlated with lifespan provide important grounds for the further exploration of p53 sequences in species displaying extreme longevity. Most importantly, our data implies a general mechanism at work in all vertebrates that leads to extended lifespan, which might be translated to studies on the extension of the health span in humans.

Supplementary Materials: The following are available online at https://www.mdpi.com/article/10.3390/ijms22168512/s1.

Author Contributions: Conceptualization, M.B., V.B. and P.P.; methodology, M.B.; validation, V.B. and J.Č.; formal analysis, M.B., V.B. and A.V.; investigation, M.B. and J.E.Z.-P.; resources, M.B., A.V. and V.B.; data curation, M.B., A.V. and J.Č.; writing—original draft preparation, M.B., J.E.Z.-P. and V.B.; writing—review and editing, M.B., J.Č., P.P. and J.E.Z.-P.; visualization, M.B. and A.V.; supervision, P.P. and J.E.Z.-P.; project administration, J.E.Z.-P.; funding acquisition, P.P. All authors have read and agreed to the published version of the manuscript.

Funding: This work was supported by The Czech Science Foundation (18-15548S), by University of Ostrava (SGS01/PřF/2020; SGS11/PřF/2021), and by the Warsaw University's Integrated Development Programme (ZIP), co-financed by the European Social Fund, European Union Operational Programme Knowledge Education Development for 2014–2020 (POWER), under the program priority axis III: Higher Education for the Economy and Development, action 3.5. This work was implemented on the basis of an agreement between the University of Warsaw and The National Center for Research and Development, which is an implementing agency of the Ministry of Science and Higher Education (to J.E.Z.-P.).

Institutional Review Board Statement: Not applicable.

Informed Consent Statement: Not applicable.

Data Availability Statement: The data presented in this study are available in Supplementary Materials.

Acknowledgments: We thank Jean-Christophe Bourdon for valuable comments and discussion, and Philip Coates for the proofreading and editing of an earlier version of the manuscript. We would also like to express our gratitude to Alena Volná and Milan Bolek for their time spent on the illustration preparation. J.E.Z.-P. would like to thank Klas Wiman from Karolinska Institute for a valuable mentorship and outstanding support.

Conflicts of Interest: The authors declare no conflict of interest. The funders had no role in the design of the study; in the collection, analyses, or interpretation of data; in the writing of the manuscript; or in the decision to publish the results.

References

1. Whittemore, K.; Vera, E.; Martínez-Nevado, E.; Sanpera, C.; Blasco, M.A. Telomere Shortening Rate Predicts Species Life Span. *Proc. Natl. Acad. Sci. USA* **2019**, *116*, 15122–15127. [CrossRef]
2. Hughes, B.G.; Hekimi, S. Many Possible Maximum Lifespan Trajectories. *Nature* **2017**, *546*, E8. [CrossRef] [PubMed]
3. Barbi, E.; Lagona, F.; Marsili, M.; Vaupel, J.W.; Wachter, K.W. The Plateau of Human Mortality: Demography of Longevity Pioneers. *Science* **2018**, *360*, 1459–1461. [CrossRef] [PubMed]
4. Hägg, S.; Jylhävä, J. Sex Differences in Biological Aging with a Focus on Human Studies. *eLife* **2021**, *10*, e63425. [CrossRef] [PubMed]
5. Harman, D. Aging: A Theory Based on Free Radical and Radiation Chemistry. *J. Gerontol.* **1956**, *11*, 298–300. [CrossRef] [PubMed]
6. Sohal, R.S.; Weindruch, R. Oxidative Stress, Caloric Restriction, and Aging. *Science* **1996**, *273*, 59–63. [CrossRef] [PubMed]
7. Storci, G.; Carolis, S.D.; Papi, A.; Bacalini, M.G.; Gensous, N.; Marasco, E.; Tesei, A.; Fabbri, F.; Arienti, C.; Zanoni, M.; et al. Genomic Stability, Anti-Inflammatory Phenotype, and up-Regulation of the RNAseH2 in Cells from Centenarians. *Cell Death Differ.* **2019**, *26*, 1845–1858. [CrossRef]
8. Campisi, J. Senescent Cells, Tumor Suppression, and Organismal Aging: Good Citizens, Bad Neighbors. *Cell* **2005**, *120*, 513–522. [CrossRef] [PubMed]
9. Bennett, W.P.; Hussain, S.P.; Vahakangas, K.H.; Khan, M.A.; Shields, P.G.; Harris, C.C. Molecular Epidemiology of Human Cancer Risk: Gene–Environment Interactions and p53 Mutation Spectrum in Human Lung Cancer. *J. Pathol.* **1999**, *187*, 8–18. [CrossRef]

10. Kandoth, C.; McLellan, M.D.; Vandin, F.; Ye, K.; Niu, B.; Lu, C.; Xie, M.; Zhang, Q.; McMichael, J.F.; Wyczalkowski, M.A. Mutational Landscape and Significance across 12 Major Cancer Types. *Nature* **2013**, *502*, 333–339. [CrossRef]
11. Levine, A.J.; Oren, M. The First 30 Years of p53: Growing Ever More Complex. *Nat. Rev. Cancer* **2009**, *9*, 749–758. [CrossRef]
12. Petitjean, A.; Mathe, E.; Kato, S.; Ishioka, C.; Tavtigian, S.V.; Hainaut, P.; Olivier, M. Impact of Mutant p53 Functional Properties on Tp53 Mutation Patterns and Tumor Phenotype: Lessons from Recent Developments in the IARC Tp53 Database. *Hum. Mutat.* **2007**, *28*, 622–629. [CrossRef] [PubMed]
13. Bray, F.; Ferlay, J.; Soerjomataram, I.; Siegel, R.L.; Torre, L.A.; Jemal, A. Global Cancer Statistics 2018: GLOBOCAN Estimates of Incidence and Mortality Worldwide for 36 Cancers in 185 Countries. *A Cancer J. Clin.* **2018**, *68*, 394–424. [CrossRef] [PubMed]
14. Timmis, A.; Townsend, N.; Gale, C.; Grobbee, R.; Maniadakis, N.; Flather, M.; Wilkins, E.; Wright, L.; Vos, R.; Bax, J. European Society of Cardiology: Cardiovascular Disease Statistics 2017. *Eur. Heart J.* **2018**, *39*, 508–579. [CrossRef]
15. Schmidt-Kastner, P.K.; Jardine, K.; Cormier, M.; McBurney, M.W. Absence of p53-Dependent Cell Cycle Regulation in Pluripotent Mouse Cell Lines. *Oncogene* **1998**, *16*, 3003–3011. [CrossRef] [PubMed]
16. Stiewe, T.; Haran, T.E. How Mutations Shape p53 Interactions with the Genome to Promote Tumorigenesis and Drug Resistance. *Drug Resist. Updates* **2018**, *38*, 27–43. [CrossRef]
17. Levine, A.J. p53, the Cellular Gatekeeper for Growth and Division. *Cell* **1997**, *88*, 323–331. [CrossRef]
18. Rufini, A.; Tucci, P.; Celardo, I.; Melino, G. Senescence and Aging: The Critical Roles of p53. *Oncogene* **2013**, *32*, 5129. [CrossRef]
19. Sabapathy, K.; Lane, D.P. Therapeutic Targeting of p53: All Mutants Are Equal, but Some Mutants Are More Equal than Others. *Nat. Rev. Clin. Oncol.* **2018**, *15*, 13. [CrossRef]
20. Vousden, K.H.; Lane, D.P. p53 in Health and Disease. *Nat. Rev. Mol. Cell Biol.* **2007**, *8*, 275–283. [CrossRef]
21. Chen, J. The Cell-Cycle Arrest and Apoptotic Functions of p53 in Tumor Initiation and Progression. *Cold Spring Harb. Perspect. Med.* **2016**, *6*, a026104. [CrossRef] [PubMed]
22. Hafner, A.; Bulyk, M.L.; Jambhekar, A.; Lahav, G. The Multiple Mechanisms That Regulate p53 Activity and Cell Fate. *Nat. Rev. Mol. Cell Biol.* **2019**, *20*, 199–210. [CrossRef] [PubMed]
23. Aubrey, B.J.; Kelly, G.L.; Janic, A.; Herold, M.J.; Strasser, A. How Does p53 Induce Apoptosis and How Does This Relate to p53-Mediated Tumour Suppression? *Cell Death Differ.* **2018**, *25*, 104–113. [CrossRef]
24. Pfaff, M.J.; Mukhopadhyay, S.; Hoofnagle, M.; Chabasse, C.; Sarkar, R. Tumor Suppressor Protein p53 Negatively Regulates Ischemia-Induced Angiogenesis and Arteriogenesis. *J. Vasc. Surg.* **2018**, *68*, 222S–233S. [CrossRef] [PubMed]
25. Nicolai, S.; Rossi, A.; Di Daniele, N.; Melino, G.; Annicchiarico-Petruzzelli, M.; Raschellà, G. DNA Repair and Aging: The Impact of the p53 Family. *Aging* **2015**, *7*, 1050. [CrossRef]
26. Itahana, K.; Dimri, G.; Campisi, J. Regulation of Cellular Senescence by p53. *Eur. J. Biochem.* **2001**, *268*, 2784–2791. [CrossRef]
27. Brázda, V.; Fojta, M. The Rich World of p53 DNA Binding Targets: The Role of DNA Structure. *Int. J. Mol. Sci.* **2019**, *20*, 5605. [CrossRef]
28. El-Deiry, W.S.; Kern, S.E.; Pietenpol, J.A.; Kinzler, K.W.; Vogelstein, B. Definition of a Consensus Binding Site for p53. *Nat. Genet.* **1992**, *1*, 45–49. [CrossRef]
29. Vyas, P.; Beno, I.; Xi, Z.; Stein, Y.; Golovenko, D.; Kessler, N.; Rotter, V.; Shakked, Z.; Haran, T.E. Diverse p53/DNA Binding Modes Expand the Repertoire of p53 Response Elements. *Proc. Natl. Acad. Sci. USA* **2017**, *114*, 10624–10629. [CrossRef]
30. Lane, D.P. Cancer. p53, Guardian of the Genome. *Nature* **1992**, *358*, 15–16. [CrossRef]
31. Toufektchan, E.; Toledo, F. The Guardian of the Genome Revisited: p53 Downregulates Genes Required for Telomere Maintenance, DNA Repair, and Centromere Structure. *Cancers* **2018**, *10*, 135. [CrossRef]
32. Bartas, M.; Brázda, V.; Červeň, J.; Pečinka, P. Characterization of p53 Family Homologs in Evolutionary Remote Branches of Holozoa. *Int. J. Mol. Sci.* **2020**, *21*, 6. [CrossRef] [PubMed]
33. Belyi, V.A.; Levine, A.J. One Billion Years of p53/P63/P73 Evolution. *Proc. Natl. Acad. Sci. USA* **2009**, *106*, 17609–17610. [CrossRef] [PubMed]
34. Engelmann, D.; Meier, C.; Alla, V.; Pützer, B.M. A Balancing Act: Orchestrating Amino-Truncated and Full-Length P73 Variants as Decisive Factors in Cancer Progression. *Oncogene* **2015**, *34*, 4287–4299. [CrossRef]
35. Jiang, L.; Zawacka-Pankau, J. The p53/MDM2/MDMX-Targeted Therapies—a Clinical Synopsis. *Cell Death Dis.* **2020**, *11*, 1–4. [CrossRef] [PubMed]
36. Tyner, S.D.; Venkatachalam, S.; Choi, J.; Jones, S.; Ghebranious, N.; Igelmann, H.; Lu, X.; Soron, G.; Cooper, B.; Brayton, C.; et al. p53 Mutant Mice That Display Early Ageing-Associated Phenotypes. *Nature* **2002**, *415*, 45–53. [CrossRef]
37. Moore, L.; Lu, X.; Ghebranious, N.; Tyner, S.; Donehower, L.A. Aging-Associated Truncated Form of p53 Interacts with Wild-Type p53 and Alters p53 Stability, Localization, and Activity. *Mech. Ageing Dev.* **2007**, *128*, 717–730. [CrossRef] [PubMed]
38. García-Cao, I.; García-Cao, M.; Martín-Caballero, J.; Criado, L.M.; Klatt, P.; Flores, J.M.; Weill, J.-C.; Blasco, M.A.; Serrano, M. 'Super p53' mice Exhibit Enhanced DNA Damage Response, Are Tumor Resistant and Age Normally. *EMBO J.* **2002**, *21*, 6225–6235. [CrossRef]
39. Lessel, D.; Wu, D.; Trujillo, C.; Ramezani, T.; Lessel, I.; Alwasiyah, M.K.; Saha, B.; Hisama, F.M.; Rading, K.; Goebel, I. Dysfunction of the MDM2/p53 Axis Is Linked to Premature Aging. *J. Clin. Investig.* **2017**, *127*, 3598–3608. [CrossRef]
40. Gannon, H.S.; Donehower, L.A.; Lyle, S.; Jones, S.N. Mdm2–p53 Signaling Regulates Epidermal Stem Cell Senescence and Premature Aging Phenotypes in Mouse Skin. *Dev. Biol.* **2011**, *353*, 1–9. [CrossRef]
41. Sahin, E.; DePinho, R.A. Axis of Ageing: Telomeres, p53 and Mitochondria. *Nat. Rev. Mol. Cell Biol.* **2012**, *13*, 397–404. [CrossRef]

42. De Keizer, P.L.; Laberge, R.-M.; Campisi, J. p53: Pro-Aging or pro-Longevity? *Aging* **2010**, *2*, 377. [CrossRef] [PubMed]
43. Maier, B.; Gluba, W.; Bernier, B.; Turner, T.; Mohammad, K.; Guise, T.; Sutherland, A.; Thorner, M.; Scrable, H. Modulation of Mammalian Life Span by the Short Isoform of p53. *Genes Dev.* **2004**, *18*, 306–319. [CrossRef] [PubMed]
44. Olivares-Illana, V.; Fåhraeus, R. p53 Isoforms Gain Functions. *Oncogene* **2010**, *29*, 5113–5119. [CrossRef]
45. De Magalhaes, J.P.; Costa, J. A Database of Vertebrate Longevity Records and Their Relation to Other Life-History Traits. *J. Evol. Biol.* **2009**, *22*, 1770–1774. [CrossRef]
46. Keane, M.; Semeiks, J.; Webb, A.E.; Li, Y.I.; Quesada, V.; Craig, T.; Madsen, L.B.; van Dam, S.; Brawand, D.; Marques, P.I.; et al. Insights into the Evolution of Longevity from the Bowhead Whale Genome. *Cell Rep.* **2015**, *10*, 112–122. [CrossRef] [PubMed]
47. Edgar, R.C. MUSCLE: Multiple Sequence Alignment with High Accuracy and High Throughput. *Nucleic Acids Res.* **2004**, *32*, 1792–1797. [CrossRef] [PubMed]
48. Deuter, R.; Müller, O. Detection of APC Mutations in Stool DNA of Patients with Colorectal Cancer by HD-PCR. *Hum. Mutat.* **1998**, *11*, 84–89. [CrossRef]
49. Pellegata, N.S.; Sessa, F.; Renault, B.; Bonato, M.; Leone, B.E.; Solcia, E.; Ranzani, G.N. K-Ras and p53 Gene Mutations in Pancreatic Cancer: Ductal and Nonductal Tumors Progress through Different Genetic Lesions. *Cancer Res.* **1994**, *54*, 1556–1560.
50. Giacomelli, A.O.; Yang, X.; Lintner, R.E.; McFarland, J.M.; Duby, M.; Kim, J.; Howard, T.P.; Takeda, D.Y.; Ly, S.H.; Kim, E. Mutational Processes Shape the Landscape of Tp53 Mutations in Human Cancer. *Nat. Genet.* **2018**, *50*, 1381. [CrossRef]
51. Wilkinson, G.S.; South, J.M. Life History, Ecology and Longevity in Bats. *Aging Cell* **2002**, *1*, 124–131. [CrossRef] [PubMed]
52. Kleiber, M. Body Size and Metabolism. *Hilgardia* **1932**, *6*, 315–353. [CrossRef]
53. Jessen, L.E.; Hoof, I.; Lund, O.; Nielsen, M. SigniSite: Identification of Residue-Level Genotype-Phenotype Correlations in Protein Multiple Sequence Alignments. *Nucleic Acids Res.* **2013**, *41*, W286–W291. [CrossRef] [PubMed]
54. Olivier, M.; Eeles, R.; Hollstein, M.; Khan, M.A.; Harris, C.C.; Hainaut, P. The IARC Tp53 Database: New Online Mutation Analysis and Recommendations to Users. *Hum. Mutat.* **2002**, *19*, 607–614. [CrossRef] [PubMed]
55. Passow, C.N.; Bronikowski, A.M.; Blackmon, H.; Parsai, S.; Schwartz, T.S.; McGaugh, S.E. Contrasting Patterns of Rapid Molecular Evolution within the p53 Network across Mammal and Sauropsid Lineages. *Genome Biol. Evol.* **2019**, *11*, 629–643. [CrossRef]
56. Ong, A.L.C.; Ramasamy, T.S. Role of Sirtuin1-p53 Regulatory Axis in Aging, Cancer and Cellular Reprogramming. *Ageing Res. Rev.* **2018**, *43*, 64–80. [CrossRef] [PubMed]
57. Liu, J.; Guan, D.; Dong, M.; Yang, J.; Wei, H.; Liang, Q.; Song, L.; Xu, L.; Bai, J.; Liu, C.; et al. UFMylation Maintains Tumour Suppressor p53 Stability by Antagonizing Its Ubiquitination. *Nat. Cell Biol.* **2020**, *22*, 1056–1063. [CrossRef] [PubMed]
58. Qian, Y.; Chen, X. Senescence Regulation by the p53 Protein Family. *Cell Senescence* **2013**, *9*, 37–61. [CrossRef]
59. Soussi, T.; Wiman, K.G. Tp53: An Oncogene in Disguise. *Cell Death Differ.* **2015**, *22*, 1239–1249. [CrossRef]
60. Kubota, S. Repeating Rejuvenation in Turritopsis, an Immortal Hydrozoan (Cnidaria, Hydrozoa). *Biogeography* **2011**, *12*, 101–103.
61. Hasegawa, Y.; Watanabe, T.; Takazawa, M.; Ohara, O.; Kubota, S. De Novo Assembly of the Transcriptome of Turritopsis, a Jellyfish That Repeatedly Rejuvenates. *Zool. Sci.* **2016**, *33*, 366–372. [CrossRef] [PubMed]
62. Choi, Y.; Chan, A.P. PROVEAN Web Server: A Tool to Predict the Functional Effect of Amino Acid Substitutions and Indels. *Bioinformatics* **2015**, *31*, 2745–2747. [CrossRef] [PubMed]
63. Lukman, S.; Lane, D.P.; Verma, C.S. Mapping the Structural and Dynamical Features of Multiple p53 DNA Binding Domains: Insights into Loop 1 Intrinsic Dynamics. *PLoS ONE* **2013**, *8*, e80221. [CrossRef] [PubMed]
64. Linnér, R.K.; Biroli, P.; Kong, E.; Meddens, S.F.W.; Wedow, R.; Fontana, M.A.; Lebreton, M.; Tino, S.P.; Abdellaoui, A.; Hammerschlag, A.R. Genome-Wide Association Analyses of Risk Tolerance and Risky Behaviors in over 1 Million Individuals Identify Hundreds of Loci and Shared Genetic Influences. *Nat. Genet.* **2019**, *51*, 245–257. [CrossRef]
65. Berkel, C.; Cacan, E. Analysis of Longevity in Chordata Identifies Species with Exceptional Longevity among Taxa and Points to the Evolution of Longer Lifespans. *Biogerontology* **2021**, *22*, 329–343. [CrossRef] [PubMed]
66. Arum, O.; Johnson, T.E. Reduced Expression of the Caenorhabditis Elegans p53 Ortholog Cep-1 Results in Increased Longevity. *J. Gerontol. Ser. Biol. Sci. Med. Sci.* **2007**, *62*, 951–959. [CrossRef] [PubMed]
67. Bauer, J.H.; Poon, P.C.; Glatt-Deeley, H.; Abrams, J.M.; Helfand, S.L. Neuronal Expression of p53 Dominant-Negative Proteins in Adult Drosophila Melanogaster Extends Life Span. *Curr. Biol.* **2005**, *15*, 2063–2068. [CrossRef]
68. Bonafè, M.; Olivieri, F.; Mari, D.; Baggio, G.; Mattace, R.; Sansoni, P.; De Benedictis, G.; De Luca, M.; Bertolini, S.; Barbi, C. p53 Variants Predisposing to Cancer Are Present in Healthy Centenarians. *Am. J. Hum. Genet.* **1999**, *64*, 292. [CrossRef]
69. Van Heemst, D.; Mooijaart, S.P.; Beekman, M.; Schreuder, J.; de Craen, A.J.M.; Brandt, B.W.; Eline Slagboom, P.; Westendorp, R.G.J. Variation in the Human Tp53 Gene Affects Old Age Survival and Cancer Mortality. *Exp. Gerontol.* **2005**, *40*, 11–15. [CrossRef]
70. Zhao, Y.; Wu, L.; Yue, X.; Zhang, C.; Wang, J.; Li, J.; Sun, X.; Zhu, Y.; Feng, Z.; Hu, W. A Polymorphism in the Tumor Suppressor p53 Affects Aging and Longevity in Mouse Models. *Elife* **2018**, *7*, e34701. [CrossRef]
71. Sulak, M.; Fong, L.; Mika, K.; Chigurupati, S.; Yon, L.; Mongan, N.P.; Emes, R.D.; Lynch, V.J. Tp53 Copy Number Expansion Is Associated with the Evolution of Increased Body Size and an Enhanced DNA Damage Response in Elephants. *Elife* **2016**, *5*, e11994. [CrossRef]
72. Tejada-Martinez, D.; de Magalhães, J.P.; Opazo, J.C. Positive Selection and Gene Duplications in Tumour Suppressor Genes Reveal Clues about How Cetaceans Resist Cancer. *Proc. R. Soc. B Biol. Sci.* **2021**, *288*, 20202592. [CrossRef]
73. Deuker, M.M.; Lewis, K.N.; Ingaramo, M.; Kimmel, J.; Buffenstein, R.; Settleman, J. Unprovoked Stabilization and Nuclear Accumulation of the Naked Mole-Rat p53 Protein. *Sci. Rep.* **2020**, *10*, 6966. [CrossRef]

74. Boughey, H.; Jurga, M.; El-Khamisy, S.F. DNA Homeostasis and Senescence: Lessons from the Naked Mole Rat. *Int. J. Mol. Sci.* **2021**, *22*, 6011. [CrossRef] [PubMed]
75. Bai, G.-L.; Wang, P.; Huang, X.; Wang, Z.-Y.; Cao, D.; Liu, C.; Liu, Y.-Y.; Li, R.-L.; Chen, A.-J. Rapamycin Protects Skin Fibroblasts from UVA-Induced Photoaging by Inhibition of p53 and Phosphorylated HSP27. *Front. Cell Dev. Biol.* **2021**, *9*, 134. [CrossRef] [PubMed]
76. Frey, K.; Hafner, A.; Pucker, B. The Reuse of Public Datasets in the Life Sciences: Potential Risks and Rewards. *Peer J.* **2020**, *22*, e9954. [CrossRef]
77. Okonechnikov, K.; Golosova, O.; Fursov, M.; Team, U. Unipro UGENE: A Unified Bioinformatics Toolkit. *Bioinformatics* **2012**, *28*, 1166–1167. [CrossRef]
78. Grabherr, M.G.; Haas, B.J.; Yassour, M.; Levin, J.Z.; Thompson, D.A.; Amit, I.; Adiconis, X.; Fan, L.; Raychowdhury, R.; Zeng, Q. Full-Length Transcriptome Assembly from RNA-Seq Data without a Reference Genome. *Nat. Biotechnol.* **2011**, *29*, 644. [CrossRef]
79. Afgan, E.; Baker, D.; Batut, B.; Van Den Beek, M.; Bouvier, D.; Čech, M.; Chilton, J.; Clements, D.; Coraor, N.; Grüning, B.A. The Galaxy Platform for Accessible, Reproducible and Collaborative Biomedical Analyses: 2018 Update. *Nucleic Acids Res.* **2018**, *46*, W537–W544. [CrossRef] [PubMed]
80. Dereeper, A.; Guignon, V.; Blanc, G.; Audic, S.; Buffet, S.; Chevenet, F.; Dufayard, J.-F.; Guindon, S.; Lefort, V.; Lescot, M. Phylogeny. Fr: Robust Phylogenetic Analysis for the Non-Specialist. *Nucleic Acids Res.* **2008**, *36*, W465–W469. [CrossRef]
81. Dereeper, A.; Audic, S.; Claverie, J.-M.; Blanc, G. BLAST-EXPLORER Helps You Building Datasets for Phylogenetic Analysis. *BMC Evol. Biol.* **2010**, *10*, 8. [CrossRef]
82. Castresana, J. Selection of Conserved Blocks from Multiple Alignments for Their Use in Phylogenetic Analysis. *Mol. Biol. Evol.* **2000**, *17*, 540–552. [CrossRef] [PubMed]
83. Anisimova, M.; Gascuel, O. Approximate Likelihood-Ratio Test for Branches: A Fast, Accurate, and Powerful Alternative. *Syst. Biol.* **2006**, *55*, 539–552. [CrossRef]
84. Guindon, S.; Gascuel, O. A Simple, Fast, and Accurate Algorithm to Estimate Large Phylogenies by Maximum Likelihood. *Syst. Biol.* **2003**, *52*, 696–704. [CrossRef] [PubMed]
85. Chevenet, F.; Brun, C.; Bañuls, A.-L.; Jacq, B.; Christen, R. TreeDyn: Towards Dynamic Graphics and Annotations for Analyses of Trees. *BMC Bioinform.* **2006**, *7*, 439. [CrossRef] [PubMed]
86. Letunic, I.; Bork, P. Interactive Tree Of Life (ITOL) v4: Recent Updates and New Developments. *Nucleic Acids Res.* **2019**, *47*, W256–W259. [CrossRef] [PubMed]
87. Choi, Y.; Sims, G.E.; Murphy, S.; Miller, J.R.; Chan, A.P. Predicting the Functional Effect of Amino Acid Substitutions and Indels. *PLoS ONE* **2012**, *7*, e46688. [CrossRef]
88. Waterhouse, A.; Bertoni, M.; Bienert, S.; Studer, G.; Tauriello, G.; Gumienny, R.; Heer, F.T.; de Beer, T.A.P.; Rempfer, C.; Bordoli, L.; et al. SWISS-MODEL: Homology Modelling of Protein Structures and Complexes. *Nucleic Acids Res.* **2018**, *46*, W296–W303. [CrossRef]
89. Emamzadah, S.; Tropia, L.; Vincenti, I.; Falquet, B.; Halazonetis, T.D. Reversal of the DNA-Binding-Induced Loop L1 Conformational Switch in an Engineered Human p53 Protein. *J. Mol. Biol.* **2014**, *426*, 936–944. [CrossRef]
90. Pettersen, E.F.; Goddard, T.D.; Huang, C.C.; Couch, G.S.; Greenblatt, D.M.; Meng, E.C.; Ferrin, T.E. UCSF Chimera—a Visualization System for Exploratory Research and Analysis. *J. Comput. Chem.* **2004**, *25*, 1605–1612. [CrossRef]
91. Källberg, M.; Wang, H.; Wang, S.; Peng, J.; Wang, Z.; Lu, H.; Xu, J. Template-Based Protein Structure Modeling Using the RaptorX Web Server. *Nat. Protoc.* **2012**, *7*, 1511–1522. [CrossRef] [PubMed]
92. Simonetti, F.L.; Teppa, E.; Chernomoretz, A.; Nielsen, M.; Marino Buslje, C. MISTIC: Mutual Information Server to Infer Coevolution. *Nucleic Acids Res.* **2013**, *41*, W8–W14. [CrossRef] [PubMed]
93. Herrero, J.; Muffato, M.; Beal, K.; Fitzgerald, S.; Gordon, L.; Pignatelli, M.; Vilella, A.J.; Searle, S.M.; Amode, R.; Brent, S. Ensembl Comparative Genomics Resources. *Database* **2016**, *2016*, 96. [CrossRef] [PubMed]

MDPI
St. Alban-Anlage 66
4052 Basel
Switzerland
Tel. +41 61 683 77 34
Fax +41 61 302 89 18
www.mdpi.com

International Journal of Molecular Sciences Editorial Office
E-mail: ijms@mdpi.com
www.mdpi.com/journal/ijms

www.ingramcontent.com/pod-product-compliance
Lightning Source LLC
Chambersburg PA
CBHW041946260926
51850CB00018B/1650
* 9 7 8 3 0 3 6 5 7 4 4 5 5 *